Management of Water, Energy and Bio-resources in the Era of Climate Change: Emerging Issues and Challenges

N. Janardhana Raju • Wolfgang Gossel
AL. Ramanathan • M. Sudhakar
Editors

Management of Water, Energy and Bio-resources in the Era of Climate Change: Emerging Issues and Challenges

Springer

Editors
N. Janardhana Raju
School of Environmental Sciences
Jawaharlal Nehru University
New Delhi, India

Wolfgang Gossel
Hydrogeology & Environmental Geology
Martin Luther University
Halle, Germany

AL. Ramanathan
School of Environmental Sciences
Jawaharlal Nehru University
New Delhi, India

M. Sudhakar
Ministry of Earth Sciences
New Delhi, India

Co-published by Springer International Publishing, Cham, Switzerland, with Capital Publishing Company, New Delhi, India.

Sold and distributed in North, Central and South America by Springer, 233 Spring Street, New York 10013, USA.

In all other countries, except SAARC countries—Afghanistan, Bangladesh, Bhutan, India, Maldives, Nepal, Pakistan and Sri Lanka—sold and distributed by Springer, Haberstrasse 7, D-69126 Heidelberg, Germany.

In SAARC countries—Afghanistan, Bangladesh, Bhutan, India, Maldives, Nepal, Pakistan and Sri Lanka—printed book sold and distributed by Capital Publishing Company, 7/28, Mahaveer Street, Ansari Road, Daryaganj, New Delhi, 110 002, India.

ISBN 978-3-319-05968-6 ISBN 978-3-319-05969-3 (eBook)
DOI 10.1007/978-3-319-05969-3
Springer Cham Heidelberg New York Dordrecht London

Library of Congress Control Number: 2014942847

Printed on acid-free paper

Springer is part of Springer Science+Business Media (www.springer.com)

Message from Alexander von Humboldt Foundation

Alexander von Humboldt
Stiftung/Foundation

Maintaining a dynamic exchange of ideas and gaining new insights – this deep interest makes us human beings. Fostering and supporting people's scientific curiosity has been the Alexander von Humboldt Foundation's mission for 60 years now. Since its establishment in 1953, the Alexander von Humboldt Foundation sponsors top-level scientists and scholars from abroad who come to Germany within the scope of our fellowships and awards to work here in close cooperation with German colleagues. The fellowships and awards of the Alexander von Humboldt Foundation have earned a considerable reputation worldwide. We aim to support excellence and to create an expanding global network of cultural and scientific dialogue on highest levels. Until today, the Alexander von Humboldt Foundation has sponsored more than 26,000 scientists and scholars from all over the world embracing over 130 countries and including 49 Nobel Prize winners. We never set any quota for countries of origin nor fields of research in the selection of future Humboldt fellows. Our only criterion is scientific excellence. So far, the Alexander von Humboldt Foundation has granted well above 5100 research fellowships and awards to excellent scientists and scholars from Asia, amongst them 1781 from India. Today, roundabout 1200 Humboldt Alumni live in India. They form one of the largest regional Alumni networks in the world having established 16 active, self-organized Humboldt Alumni Associations in the country. The Humboldt fellows on the Indian sub-continent are vividly and enthusiastically participating in national and international Alumni activities.

"Once an Humboldtian, always an Humboldtian" – from the very beginning this was the hallmark of the Alexander von Humboldt Foundation. Humboldt sponsorship is enduring: the foundation is a lifetime partner, maintaining connections on a long-term basis through its alumni sponsorship programmes. Moreover, the foundation encourages its Alumni to undertake their own initiatives and collaborations across disciplinary and national borders. As a result, many Humboldtians make use of the foundation's extensive Alumni sponsorship offers. It was in this context that the Humboldt Kolleg "Management of Water, Energy and Bio-resources in Changing Climate Regime" took place in Delhi in February 2013. The Humboldt Kolleg was hosted by Humboldt Alumnus Professor Dr. Nandimandalam Janardhana Raju from the School of Environmental Sciences at the Jawaharlal Nehru University choosing a topic of major importance to the development in Asia for the conference. The Humboldt Kolleg served as a forum for scientific exchange and networking between Humboldtians and other young and experienced researchers from various disciplines. In total, 231 researchers participated in the conference, amongst them 35 Humboldt Alumni, 135 young academics, seven scholars from Germany and 54 other experienced researchers. A total of 135 presentations were given; another 63 scholars introduced the audience to their fields of research interest during a scientific poster session.

Dealing with the changes of our earth climate and its impacts on natural resources and the environment is one of the biggest challenges for mankind in this century. Worldwide, experts call for action against climate change and its negative environmental, fiscal, social, and cultural effects. As the organizers and presenters during the Humboldt Kolleg pointed out correctly, fragile and conflict-ridden societies will be especially prone to climate change and its impacts, as diminishing resources like groundwater and increasingly unequal distribution will tighten competition and will potentially evolve violent consequences.

On behalf of the Alexander von Humboldt Foundation I would like to thank Professor Dr. Janardhana Raju and the organizing committee at Jawaharlal Nehru University for their dedication and the initiative to conduct the Humboldt Kolleg whose outcome is published in the proceedings of this conference. The Alexander von Humboldt Foundation is most grateful to its Humboldtians, who support our aims and goals of fostering academic cooperation across borders and bringing forward the next generations of top-class international researchers. I wish all participants in the Humboldt Kolleg and the authors of this conference volume success and the best of luck for their future plans.

Dr. Thomas Hesse

Dpt. Secretary General
Alexander von Humboldt Foundation
Bonn, Germany
August 2013

Foreword

This volume entitled *"Management of Water, Energy and Bio-resources in the Era of Climate Change: Emerging Issues and Challenges"* contains many papers presented during the International Alexander von Humboldt Kolleg that was held at Jawaharlal Nehru University, New Delhi (India), on February 8–9, 2013. The meeting, convened by Dr. N. Janardhana Raju, School of Environmental Sciences, brought together about 200 scientists from different parts of India and overseas including Germany, USA, Brazil, Croatia, Taiwan, Tajikistan, Bangladesh, Iran, Ethiopia, Nepal and Sri Lanka. This book brings out different aspects of natural resources management addressed during the meeting and is divided into three broad sections: (i) Water Resources Management, (ii) Energy and Bio-resources Management and (iii) Climate and Natural Resources Management. The themes and topics covered amply show the broad spectrum of multidisciplinary scientific activities. Most of the papers are written by eminent scholars in their fields, but also by some young scientists, and are very informative with lots of data and methods with suggestions for improvement and conservation.

Environmental sciences require a broad knowledge that goes beyond the boundary of any single discipline and covers multiple objectives of researchers from various subjects. Thus, combining knowledge of different aspects of geosciences can greatly assist in coping with mechanisms for sustainable development and management of natural resources. Scientific evidence is needed to support decision making at national and international levels in order to protect and manage natural resources, to conserve ecosystem services and environmental challenges. Water shortages are caused mainly by increasing population, waste and pollution resulting in negative impacts on the environmental, socio-cultural, political and economic spheres of society. Water contamination can be due to geogenic and anthropogenic sources and continues to be one of the critical challenges adversely affecting natural ecosystems, agriculture and human health. The environmental problems caused by the increase of pollutant loads discharged into the natural environment require adequate legislation, too. An additional factor for future water shortages in certain areas that is currently much discussed may be climate change and its resultant impact on water resources.

Overall this book addresses water, biomass and energy needs, which must be the core objectives of all governmental policies and strategies in their future course of action. Future water shortages, their reuse and remediation methods, which challenge human health and the environment, are also discussed. The book holds interest for all those who are keen to know about the management of natural resources such as water, energy and bio-resources and should make an important contribution to a better understanding of natural resources management. I hope that this book will serve those concerned to acquire additional scientific information and experience required for ensuring natural quality and quantity aspects to protect natural resources from indiscriminate exploitation and consequent environmental degradation and to stimulate future work for sustainable development and management of natural resources.

I complement all the authors of the articles. I also congratulate the editorial team for their tremendous effort in bringing out this edited book. I trust the volume will serve for many years as a scientific information base for future planning of the management of water, energy and bio-resources and energize synergy among academicians, researchers, stakeholders and policy makers for documentation and dissemination of knowledge in natural resources management.

Martin-Luther-Universität Halle-Wittenberg
Germany

Gerhard H. Bachmann

Preface

Earth is a bounty for the sustenance of life and the very existence of human beings depend on natural resources and congenial environment on the planet. Water, biomass and renewable and non-renewable energy are the most fundamental resources for any civilization and demand for these resources is ever increasing at an alarming rate and is moving towards unsustainable levels. Degradation and erosion of natural resources, that are used to produce food and other valued goods and services essential for our survival and prosperity, are also the root causes of the agrarian crisis in the world. Diminishing water resources and their unequal distribution in the changing scenarios will increase competition for water which may turn potentially to violent events/wars in future. The water shortage that may be caused by a changing climate regime results in negative impacts on environmental, socio-cultural, political and economic spheres of the societies. No present or intended use of natural resources should condemn our children to endless toil or deprivation.

The natural resource management incorporates the understanding of the scientific and technical aspects of these resources distribution and ecological systems which help in supporting the healthy survival of life on the planet 'Earth'. The rising demand of water with increasing population density along with energy and biological resource management in the changing climatic scenario become prime concern for any country's economic growth and healthy environment for humans, fauna and flora to survive and flourish. The majority of the population are looking forward for energy efficient system to enhance the judicious conservation of water and bio-resources of our environment but the human pressure and their anthropogenic activities are slowly but steadily deteriorating these resource management capacities. In spite of the rapid development in technology the situation to conserve these resources are not improving; rather it is declining. It also involves the management of whole environment including social aspects and impact of climate change which are closely interrelated to the resource depletion and thus urges the need for an effective management plan. These aspects were kept in mind while bringing out this volume.

The book contains papers of multi-disciplinary views, discussing the management of water, energy and bio-resources for better management of such resources in the changing climate scenarios. It is thus aimed to interest all those who are keen to know about the management of natural resources and hence contribute to a better understanding of the Earth's resources. The papers are contributed by distinguished scientists and academicians from various important universities and institutions from all over the world including India who are contemporary workers in this field.

This edited book is the outcome of the International Humboldt Kolleg held from 8-9 February, 2013 at the Jawaharlal Nehru University (JNU), New Delhi, India. It contains twenty eight chapters which are grouped under three sections viz., (a) Water Resources Management, (b) Energy and Bio-resources Management and (c) Climate and Natural Resources Management. The volume presents recent case studies and examples from various parts of the world in the context of climate change scenarios and their management. Each chapter demonstrates the need for managing the demanding resources due to change in climate, land use, industrialization and the need by each country to take initiatives and commit themselves to manage these resources in a sustainable way.

We would like to thank all the contributors for expressing their individual views and also acknowledge our colleagues for their untiring efforts to review the manuscripts. One of the editors (N.J. Raju) would particularly like to thank his collaborators and research scholars for supporting his research activities over two decades which helped him in the process of bringing out this contribution. The generous support extended by the Alexander von Humboldt Foundation, Germany and the added support of other agencies in organizing the International Humboldt Kolleg (IHK2013) at JNU is gratefully acknowledged. We hope this book will be very useful for managers, environmentalists, hydrologists, water resource and energy managers, and for governmental and other regulatory bodies dealing with water, energy and bio-resources. Finally, we thank the publishers for taking efforts in bringing out this volume.

New Delhi, India N.J. Raju
Halle, Germany W. Gossel
New Delhi, India AL. Ramanathan
New Delhi, India M. Sudhakar

Contents

About the Editors

N. Janardhana Raju Dr. Nandimandalam Janardhana Raju did his PhD from Sri Venkateswara University, Tirupati, India and is currently Associate Professor at the School of Environmental Sciences, Jawaharlal Nehru University, New Delhi, India. He has vast experience in Hydrogeology and Environmental Geosciences and is engaged with research on rainwater harvesting systems, hydrogeochemistry and groundwater quality (arsenic and fluoride) in collaboration with a number of universities in India and overseas. He has been a visiting scientist at Ruprecht Karl University, Heidelberg, Germany; Martin Luther University, Halle, Germany; Asmara University, Eritrea, NE Africa; and Federal University of Fluminense, Rio de Janeiro, Brazil and is the recipient of the Alexander von Humboldt Fellowship, Germany. He has guided PhD scholars in hydrogeology and environmental geology and published more than fifty research papers in refereed journals. He has also contributed a chapter in the book 'Geochemical Processes: Conceptual models for reactive transport in soil and groundwater'. He has travelled widely participating in several national and international seminars/conferences/workshops. He is a member of the International Association for Mathematical Geosciences (IAMG); Geological Society of India (GSI); International Association of Hydrogeologists (IAH) and International Association of Hydrological Sciences (IAHS).

Wolfgang Gossel Dr. Wolfgang Gossel holds a PhD (1999) degree from the Freie University, Berlin, Germany and completed Habilitation (2008) from the Martin Luther University, Halle and is presently working as Senior Scientist, Hydrogeology and Environmental Geology, in the same university. He has contributed more than twenty research papers in refereed journals and has published a book 'Interfaces in coupling of hydrogeological modeling systems'. He has travelled widely participating in national and international conferences/workshops and also conducted training programmes in Egypt and India, and International Training Courses in Germany on GIS and Hydrogeological Flow and Transport Modeling. Dr Gossel has completed major research projects pertinent to groundwater flow modeling and salt water intrusion in Egypt, Mexico and Germany. He is a member of the International Association for Mathematical Geosciences (IAMG); and International Association of Hydrogeologists (IAH). He has served as referee and associate editor for many national and international journals.

AL. Ramanathan Dr. AL. Ramanathan did his PhD in sedimentary geochemistry from Centre of Advanced Studies in Geology, Panjab University. He is currently Professor in the School of Environmental Sciences, Jawaharlal Nehru University, New Delhi, India. Prof. Ramanathan has specialized in coastal biogeochemistry, hydrogeochemistry and glaciology aspects and worked extensively on the mangroves, estuaries, coastal groundwaters and glaciers of India for the past two decades. He is engaged in environmental research projects, with a number of universities and research organisations in India, Norway, France, Australia, Russia, USA etc. He has guided seventeen PhDs in the above subject and published more than seventy papers in reputed refereed journals. He has also published six books and several chapters in many books by reputed international publishers. He has received the IFS Sweden (Project) Award thrice on his work on mangrove biogeochemistry. He is a member of the editorial board of *Indian Journal of Marine Sciences* and served as referee for many national and international reputed journals as well.

M. Sudhakar Dr. Maruthadu Sudhakar did his PhD in Applied Geology from Indian School of Mines, Dhanbad, and M.Sc. in Law of the Sea and Marine Policy (1990) from London School of Economics and Political Science, UK. He worked in two premier institutions of the country i.e., National Institute of Oceanography (NIO) and National Centre for Antarctic and Ocean Research (NCAOR), Goa for 27 years in research and development, survey, planning, teaching and administration in the field of oceanography/offshore surveys/polar science/marine technology. Currently he is an Advisor to the Government of India, Ministry of Earth Sciences (MoES), New Delhi and heads the Outreach and Awareness, Research Vessels Programmes of the MoES. He has been the project leader of various major programmes and has led several oceanographic expeditions to the Indian and Southern Oceans and Antarctica. Dr. Sudhakar is an elected member of the International Seabed Authority (ISA) in Legal and Technical Commission from 2007 till 2016 and represents India in ISA. He is associated with many professional bodies and has published 50 research papers in refereed international/national journals and conference proceeding volumes and in seminar/symposia. He has been serving in the National Standing and Technical Committees of Govt. of India.

Section I
Water Resources Management

3D Geological and Hydrogeological Modelling – Integrated Approaches in Urban Groundwater Management

Peter Wycisk

Introduction

Urbanisation has a major impact on groundwater recharge in both quality and quantity as well as groundwater flow beneath cities. The impact is due to the import of large quantities of water as well as the extensive use of the ground for effluent discharge, waste disposal and groundwater extraction. Hence, effective management of urban aquifers has to incorporate the negative effects on groundwater resources in the underlying groundwater systems. The effect on recharge arises both from modifications to the natural infiltration system and changes in natural drainage. These changes are induced by leakage from water mains and by wastewater seepage. The resultant effect on the quality of recharge is generally adverse with urbanisation processes being the main causes of severe, but essentially diffuses pollution of groundwater and rising levels of salinity. Widespread groundwater contamination results from chlorinated hydrocarbons and other organic compounds. Additional adverse effects on a more localised basis are due to pathogenic agents in upper aquifer systems with insufficient sewage and waste-disposal infrastructure. Changing groundwater related issues could affect urban buildings and infrastructure resulting from lowering of groundwater levels by high extraction rates for water supply as well as by rising water tables. The general change in water quality can create significant problems especially in the latter situation.

Morris et al. (1997) point out that, for a sustainable urban groundwater management, there must be a coherent understanding of the local groundwater/aquifer system, which is unique at each location in respect of its particular groundwater setting and aquifer architecture. It implies an adequate knowledge of the subsurface

P. Wycisk (✉)
Martin Luther University, Halle-Wittenberg, Von-Seckendorff-Platz 3,
06120 Halle/Saale, Germany
e-mail: peter.wycisk@geo.uni-halle.de

N.J. Raju et al. (eds.), *Management of Water, Energy and Bio-resources in the Era of Climate Change: Emerging Issues and Challenges*,
DOI 10.1007/978-3-319-05969-3_1

geology that establishes the basis of the groundwater interaction influencing engineering, infrastructure and related water resources problems. Aquifer vulnerability mapping, a specific need in most of the developing countries, is based on land use as well as on the upper groundwater conditions defined by the aquifer geology. Therefore, land-use planning has to follow robust matrices that indicate what activities are acceptable and what has to be done to increase awareness of the role and susceptibility of the subsurface to contamination.

In order to provide subsurface geological data for urban groundwater management, over the last 10 years, digital 3D modelling of geological structures has markedly increased, moving from a 2D mapping to a 3D modelling culture related to advanced software and IT capabilities. There is natural progression in data handling and in subsequently coupled modelling strategies and decision support in regional management.

Recent examples from Brussels and Vienna show that data preparation of subsurface information is, to some extent, done using Geo-Information Systems, which allow only 2.5 D processing. Nevertheless, the increasing examples of 3D mapping and geological modelling of different cities in Europe, such as Glasgow by the BGS (UK) (Culshaw, 2005), Vienna (Pfleiderer and Hofmann, 2004), Magdeburg by the State Geological Survey of Saxony-Anhalt (Germany) (Neumann et al., 2006), as well as the city of Halle (Saale) (Wycisk and Schlesier, 2006) show the increasing applicability of this digital 3D modelling concept in practice.

Due to future needs involving consistent data management, developing countries such as Bangladesh are also implementing 3D geological mapping and modelling for the Greater Dhaka City area. The major objectives of this project are to provide subsurface information for hydrogeology and geotechnical applications as well as environmentally related planning issues using a flexible and dynamic model-based system. Objective and project defined results on demand indicate the progress of this approach rather than the static 2D analogue map printed solutions. This progress will be of future importance for most of the developing countries.

Depending on the available information and quality of data for urban water management, the coupling of urban drainage models with numerical groundwater simulations for integrated water management is becoming increasingly important (Schrage et al., 2005; Morris et al., 2007). This integration provides the needed basis for sustainable strategies using the 3D-generated subsurface digital information for integrated modelling and management. This paper describes examples of different and, to some extent, innovative applications from urban areas of Germany.

3D Geomodelling in Urban Management

Model Concepts and Requirements

With advances in 3D modelling software and visualisation tools over the last few years, as well as the increasing 2.5D and 3D GIS applications, digital subsurface information of local to regional scales has become more applicable and available (Culshaw, 2005; Wycisk et al., 2002). This aspect is in line with the increasing needs of conceptual and structural geological models in the broad fields of urban groundwater flow and transport simulation, of geotechnical investigations, as well as for predictive results of geological subsurface information for infrastructure planning.

Due to the complexity of 3D modelling software tools and the specific situations of regional geology, users might be not aware of the differences in their models or resulting limitations and disadvantages. The availability of subsurface data in quality and quantity could limit the modelling approach. Apart from this general question, the regional setting and coverage of subsurface drilling information has to often bridge large information gaps. 3D modelling software and visualisation tools are available using geostatistical algorithms (e.g. GoCAD, EVS/MVS, Rockworks) that are working completely differently than the TIN-based (triangulated irregular net) interpretation using intersecting cross-sections (GSI3D). The shortcomings of automatic contouring by statistical and geostatistical algorithms are compensated by the use of constructive intersected cross-sections and mapping information.

The major obstacles in regional 3D modelling have been summarized by Berg and Keefer (2004) and can be confirmed by our experience from different regional models. Geological data verification is time-consuming, and many projects that require groundwater modelling investigations have not had sufficient time or money to obtain the detailed geologic interpretations needed. The interpretation of complex geological sequences, especially in the Quaternary, requires experience in regional stratigraphy and sedimentology. Without this specific regional knowledge, any 3D mapping and modelling will be either incomplete or will fail.

The expected results of visualisation and analysis of geological 3D models offer more advantages when compared to classic two dimensional analogue methods of urban groundwater management. The main aspects are:

- Visualisation and analysis of the subsurface to deliver a decision support system for urban planners, managers and decision makers;
- Integration of various methods of investigation (hydrogeology, hydrochemistry, geophysics, geotechnics, infrastructural planning);
- Attributing and coupling to groundwater and surface water models;
- Analysis prediction and visualisation of subsurface information; and
- Consistent GIS-based IT-infrastructure, consistency of the data base, adapted to monitoring and management systems.

Constructive Cross-section Net-based Interpolation

The use of constructive cross-section net-based interpolation approaches are of advantage when one has unbalanced regional coverage of drilling information in complex and heterogeneous lithological subsurface units. The modelling process with GSI3D is based on the creation of a series of intersecting user-defined cross-sections. The entire stacking order of all deposits in the study area has to be defined stratigraphically and sedimentologically, and a generalized vertical section has to be created. The lithostratigraphic classification of the sedimentary succession within a consistent regional stratigraphic framework is more helpful than a pure grain-size of lithology-based approach for groundwater modelling. Most of the software tools allow the input of 2D mapping or surface information, especially from areas with a less dense borehole record. GSI3D allows modelling of the distribution and geometry of sedimentary layers by knowledge-based control of the modeller, which is especially needed for heterogeneous aquifer systems and/or artificially formed lithological units within urban settlements. GSI3D is currently intended for use in the near-surface modelling of surficial and Quaternary sediments (Neber et al., 2006). The software is used by the British Geological Survey on different mapping scales and, in combination with GoCAD for deep subsurface investigations and geological modelling (Culshaw, 2005), also for regional investigations of urban groundwater contaminated mega-sites (Wycisk et al., 2002; Wycisk et al., 2005).

Statistically Based 3D Interpolation

Uneven and spotty distribution of geological drilling information is one of the major obstacles in regional subsurface modelling with automatically contoured distributions and thickness of layers in urban areas. In the case of known geology or sufficient coverage of drilling information, the statistically based interpolation tools provide a less time-consuming modelling approach. As an example, EVS (Environmental Visualization System; C Tech Development Corp., Santa Barbara, Ca.) provides true 3D volumetric modelling together with 2D and 3D kriging algorithms with best fits of variograms to analyse and visualise geoscientific and environmental data. EVS/MVS allows the seamless integration with ArcView GIS, as well as with Modflow and MT3D. The 3D modelling of the subsurface geology with EVS is based exclusively on selected drilling information and the geostatistical interpolation of the individual layers. This procedure can lead to different results in cut-and-fill structures of Quaternary sedimentary channel fills, as well as any artificial structures e.g. subsurface dumps.

Interactive 3D-modelling of the Subsurface Geology

In conjunction with the City of Halle (Saale), a digital 3D spatial model was constructed for the total city area of 135 km^2. Consequently, Halle is one of the first cities in Germany to have available an extremely sharp and detailed 3D volume model of its geological subsurface, composed of 24 layers (Fig. 1A). The 3D spatial

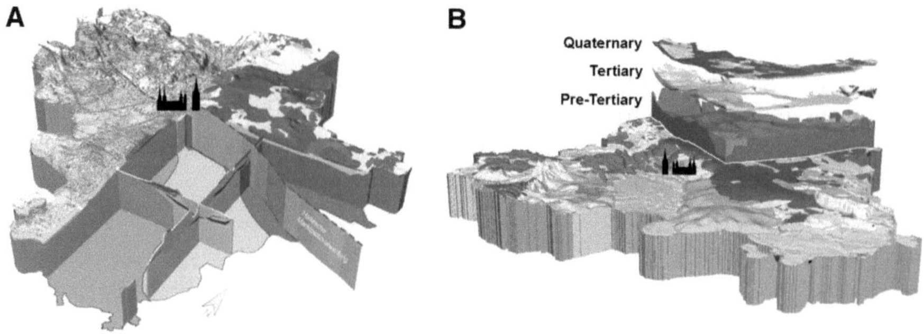

Fig. 1 Geological model of Halle/Saale. Fig. 1A shows virtual cross-sections and the main faults striking through the city. In Fig. 1B the distribution of 24 individual layers, e.g. aquifers and aquitards, is visualised to give insight to potential aquifer connections and layers with high adsorption capacities

model provides the basis for a future GIS supported information and prognosis system. Such a system also provides the basic framework for future extensions to information and decision-making systems. The possible applications lie in the areas of geological and geotechnical recognition of particular locations of interest, of groundwater and environmental protection, as well as for environmental and city planning.

Geodata and Model Construction

Innovative use of the 3D spatial geological model of Halle (Saale) City is based on the extremely precise modelling of the geological subsurface. At the same time the ordering and juxtaposition of the geological layers in the domain provides information on the most important groundwater aquifers and aquitards. The heterogeneity of aquifers and aquitards is very high and is due to the structural geological setting. The urban hydrogeology is characterized by different porous aquifers with frequent local aquifer/aquifer contact areas as well as by fractured aquifers of different lithologies (Fig. 1B).

The basic data for the model are obtained from a network of 32 cross-sections based on drilling information. The precision of the horizontal resolution of the spatial model is 40 m $\times$ 40 m, while the vertical resolution is of the order of cm depending on available core data.

Completely new interpretation and evaluation possibilities arise from the superposition of information at different depths in the subsurface in the form of spatial views of individual layers, “virtual” cross-sections, as well as hydrogeological analyses with “real” views from aerial photographs or maps.

In the future, this “augmented reality” will allow one to obtain a new level of quality in the geosciences and environmental disciplines. This can be presented in the forms of maps or interactive digital 3D visualisations, and can be used for specific technical questions in real time. The possible uses of an interactive device, such as a subsurface viewer, enable the direct investigation of geological and hydrogeological information. The results can be presented as virtual boreholes, as vertical and horizontal sections, as thematic maps, as well as locating 3D bodies contained in the modelled area with sharper spatial precision.

Interactive Modelling in Real Time

The advantage of the modelled system allows interactive modelling in real time using an independent viewer system. This subsurface viewer is based on the independent model construction by modellers and also on the application with the viewer system by users, who do not need specific modelling knowledge. The separation of the modelling system into an active part, which includes modelling and subsequent improvement by implementing new drilling information, and a user-oriented part, allowing only the analytical and predictive features of interpretation, ensures the integrity and quality of the primary 3D modelling database. The software-tool is linked to each respective model, enabling users to investigate the 3D setting of the geological subsurface immediately in an easy to understand format. The colour coding of each modelled geological unit is defined according to analogue standards of geological maps. In addition, it is possible to visualize the model colour scheme according to specific applied geological parameters such as hydraulic conductivity, geotechnical values, etc.

Integrative Modelling Concepts of Contaminated Urban Aquifers

Environmental Setting of the Former Industrial Mega-site

Large scale groundwater contamination sites such as the urban area of Bitterfeld-Wolfen in Germany are characterized by different environmental impacts caused by the former chemical industry and extensive landscape devastation by lignite mining over the last 100 years or more. Due to the multi-source regional contamination in the upper and lower aquifers, an integrated assessment is needed of the groundwater quality and of the local risk areas. Therefore, a regional 3D spatial model of distinct individual environmentally related core modules has been developed for the urban Bitterfeld-Wolfen area. Two investigation projects have assessed the long-term development for the urban and industrial areas in terms of a risk-based land management approach (Fig. 2).

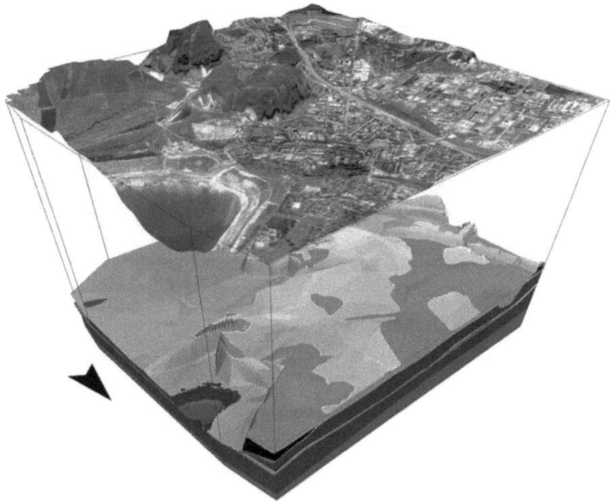

Fig. 2 Former and future land use are an important factor for contaminated upper urban aquifers. Geological structures defining aquifer/aquifer contacts, as shown here for the Bitterfeld area, are influenced by abandoned mining and infrastructure projects. Both aspects give a completely new view to management strategies

Due to the large affected urban area and rural vicinity of more than 60 km^2, an intensive groundwater monitoring programme has been operative for more than 15 years. More than 170 individual organic substances, as well as 30 inorganic substances, have been monitored. The importance of the former industrial area is characterised by more than 40,000 chemical plant workers during GDR time, by 15 industrial toxic land-fills, and by more than 5000 individual chemical substances during the nearly 80 years of industrial production in an urban environment.

The most frequent substances are Tetrachloroethene (PCE), Tri-chloroethane (TCE), cis-1,2 Dichloroethane (cis-1,2 DCE), Vinylchloride (VC), 1,4-Dichlorobenzene (1,4 DCB), 1,2-Dichlorobenzene (1,2 DCB), Monochlorobenzene (MCB), and Benzene. The regional distribution of contaminants reflects their different and multiple sources and pathway relations. The cities of Bitterfeld and Wolfen, as well as several villages and the rural landscape of the alluvial plain of the high flood affected River Mulde, describe the environmental sensitivity to humanity and natural resources (Heidrich et al., 2004; Wycisk et al., 2003).

GIS-based Multi-source Data Management

To assess the complex environmental situation of the Bitterfeld-Wolfen area, a GIS-based spatial model was required that includes the heterogeneous aquifer setting in 3D and in as much detail as possible (Fig. 3). The subsurface information had to be available for a GIS-based assessment and predictive calculations correlated to surface information of potential receptors.

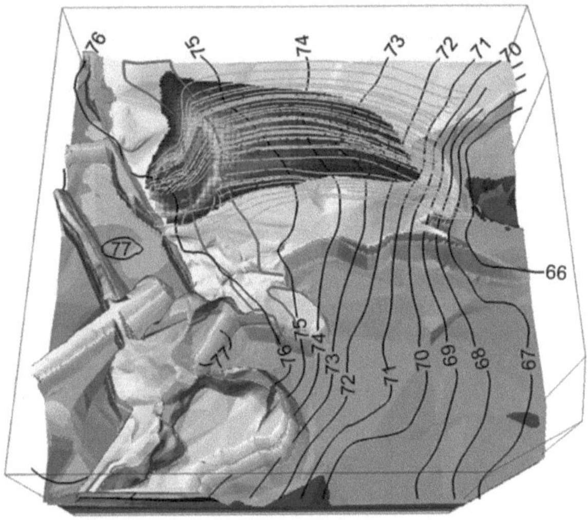

Fig. 3 Groundwater flow system in a complex geological structure in the case of the Bitterfeld area. The groundwater contours do not help in construction of the pathways. Only groundwater flow and transport models (based on detailed geological models) yield numerically appropriate results that can be used for the assessment of contaminant spreading

Therefore, the following major information modules have to be integrated into the spatial model on a local scale, including the specific objectives and used modelling tools:

- Land-use classification;
- Groundwater contaminants;
- Hydrogeological data; and
- 3D model of the subsurface geology.

The GIS data management for all hydrogeological and hydrochemical data was done with ArcView 3.x and ArcView 8.x (ESRI). The spatial model includes point data such as borehole data (lithology/stratigraphy), hydrochemistry, contaminants monitoring data, etc. (Wycisk et al., 2007). The geological cross-sections, with their vertical 2D structure, were held in a special device for geological 3D models. The geological structures had to be held in a GIS database to obtain an interface to numerical groundwater modelling tools such as Feflow or Modflow. These data are stored in GRID of point formats in ArcView.

Integrated Modelling of Urban Aquifers

To support the integrated modelling concept the high-resolution 3D structural model of the entire area was enlarged to about 65 km^2 (Wycisk et al., 2005). This model allows volumetric calculations of partial or distinct sedimentary units, such as lignite-bearing strata, that are relevant for assessing the natural absorption potential and retardation processes inside the aquifer. The digital data set of the true 3D structural geology was used with reference to the hydraulic characterization for subsequent flow and transport models. The numerical groundwater model was carried out with two objectives:

1. Description of the hydrodynamic system and the pathway prediction assessing the exposure route of local contaminants as well as for optimizing the well observation sites; and
2. Predictive calculations of the changed hydraulic situation after the once-in-a-century flooding of the Goitzsche mining lake and raised groundwater level in August 2002.

The numerical model consists of two parts: A groundwater flow model and a transport model based on the flow model. The modelling systems Modflow, ModPath and MT3D with the Visual Modflow 3.0 preprocessors and postprocessors were used for the studied areas. Figure 3 shows the integration of simulated results from high-resolution 3D geology as well as from flow and transport modelling. The resulting impacts of laterally changing hydraulic conductivity inside the aquifers by channel-fill structures or anthropogenic activities by humans are clearly shown and are underestimated in most hydrological models.

Conclusions

To generate an almost realistic scenario of urban areas, it is necessary not only to gather high-resolution land-use information, but also to have a model of the aquifer systems corresponding to the real world scenario of the geological subsurface setting. The "true" regionalization of hydraulic and contamination data is also needed. This statement is valid for most Quaternary sediments underlying urban areas representing heterogeneous aquifer conditions, especially from fluvial and deltaic environments. The 3D geological model also serves as a future consistent data base and for the prediction of modelling provides groundwater and engineering related deliverables for urban management. The new approach of digital 3D geo-data management also enables municipalities of developing countries to produce a capable upgrading system and forward looking management tool.

References

Berg, R.C. and Keefer, D.A. (2004). Three-dimensional Geologic Modeling: Challenging our Terminology and Understanding of Geological Maps. *In:* Berg, R.C., Russell, H. and Thorleifson, L.H. (Eds.). Three-dimensional Geological Mapping for Groundwater Applications – Open File Series 2004–8.

Culshaw, M.G. (2005). From concept towards reality; developing the attributed 3D geological model of the shallow subsurface. *Quart. J. of Engineering Geology and Hydrogeology*, **38:**231–284.

Heidrich, S., Schirmer, M., Weiss, H., Wycisk, P., Großmann, J. and Kaschl, A. (2004). Regionally contaminated aquifers – toxicological relevance and remediation options (Bitterfeld Case Study). *Toxicology*, **205(3):** 143–155.

Neumann, D., Schönberg, G. and Strobel, G. (2006). 3D-modeling of ground conditions for the engineering geology map of the city of Magdeburg. *In:* Culshaw, M., Reeves, H., Spink, T. and Jefferson, I. (Eds.). IAEG 2006 Engineering Geology for Tomorrow's Cities, The 10th IAEG International Congress, Nottingham, United Kingdom, 6–10 September 2006.

Pfleiderer, S. and Hofmann, T. (2004). 3D-Visualisation of Vienna's subsurface. *In:* Schrenk, M. (Ed.) *9. internationales Symposion zur Rolle der Informationstechnologie in der Stadt- und Regionalplanung sowie zu den Wechselwirkungen zwischen realem und virtuellem Raum*, Proceedings of the CORP2004 and GeoMultimedia04, Vienna.

Schrage, C., Wolf, L., Klinger, J. and Hötzl, H. (2005). Coupling Urban Drainage Models with Numerical Groundwater Simulations for Integrated Water Management. *In:* Sánchez-Vila, X., Cabrera, M.C. and Valverde, M. (Eds.). Proceedings of the IAH Workshop "From data gathering and groundwater modeling to integrated management", 4–8.10.2005, Alicante, Spain.

Wycisk, P., Stollberg, R., Neumann, Chr., Gossel, W., Weiss, H. and Weber, R. (2012). Integrated methodology for assessing the HCH groundwater pollution at the multi-source contaminated mega-site Bitterfeld/Wolfen-ESPR (online first).

Wycisk, P., Hubert, T., Gossel, W. and Neumann, Ch. (2009). High-resolution 3D spatial modelling of complex geological structures for an environmental risk assessment of abundant mining and industrial mega sites. *In:* Computers & Geosciences. Elsevier, Bd. 35.2009, 1.

Wycisk, P., Weiss, H., Kaschl, A., Heidrich, S. and Sommerwerk, K. (2003). Groundwater pollution and remediation options for multi-source contaminated aquifers (Bitterfeld/Wolfen, Germany). *Toxicology Letters*, **140–141:** 343–351.

Wycisk, P., Fabritius, H., Ruske, R. and Weiß, H. (2002). Das digitale geologische Strukturmodell Bitterfeld als neuer Baustein in der Sanierungsforschung. *Grundwasser*, **7(3):** 165–171.

Long-term Saltwater Intrusion Modelling – Case Studies from North Africa, Mexico and Halle

Wolfgang Gossel, Reiner Stollberg, Eduardo C. Graniel, and Thomas R. Ruede

Introduction

Coastal saltwater intrusions threaten drinking and irrigation water resources along coastlines. On the other hand submarine groundwater discharges in nearly the same areas. The interface between saltwater and freshwater shifts over time and is mainly influenced by the seawater level and aquifer characteristics. During the last 140,000 years, seawater level fluctuation was very rapid. After a fast rising of about 120 m in the last interglacial, a slow drawdown of the same magnitude followed during the last glacial within 100,000 years. In the last 10,000 years, the water level again rose very fast and then it remained stable. Different aquifer types are influenced by these fluctuations in different ways. The two examples from North Africa and Mexico show how not only shelf platforms are flooded but also the interface in the ground water shifts over time. Both investigation areas are transboundary aquifers and the water supply of the population is affected directly by the hydrogeological development.

W. Gossel (✉)
Dept. Hydrogeology and Environmental Geology, Martin Luther University Halle, Von-Seckendorff Platz 3, D-06120 Halle, Germany
e-mail: wolfgang.gossel@geo.uni-halle.de

R. Stollberg
Dept. Groundwater Remediation, Helmholtz Centre for Environmental Research - UFZ, Permoserstr 15, D-04318 Leipzig, Germany

E.C. Graniel
Universidad Autónoma de Yucatán, Facultad de ingenieria, Apdo. postal no 150, Cordemex 97111 Merida, Yucatan, Germany

T.R. Ruede
RWTH Aachen, LuFG Hydrogeology, Lochnerstr. 4-20, D-52064, Aachen, Germany

N.J. Raju et al. (eds.), *Management of Water, Energy and Bio-resources in the Era of Climate Change: Emerging Issues and Challenges*,
DOI 10.1007/978-3-319-05969-3_2

Nubian Aquifer System

The interface in the Nubian Aquifer System seems to be stable nowadays but the model results show the slow dynamic effect of seawater level changes over this time frame. The amount of intrusion water is limited by the hydraulic conductivities. The flow velocities are in a range of 0.002-0.003 m/day during the times of the fastest rising seawater levels at the end of the last glacial and beginning of the holocene as shown by Gossel et al. (2010). Figure 1 shows three steps of the

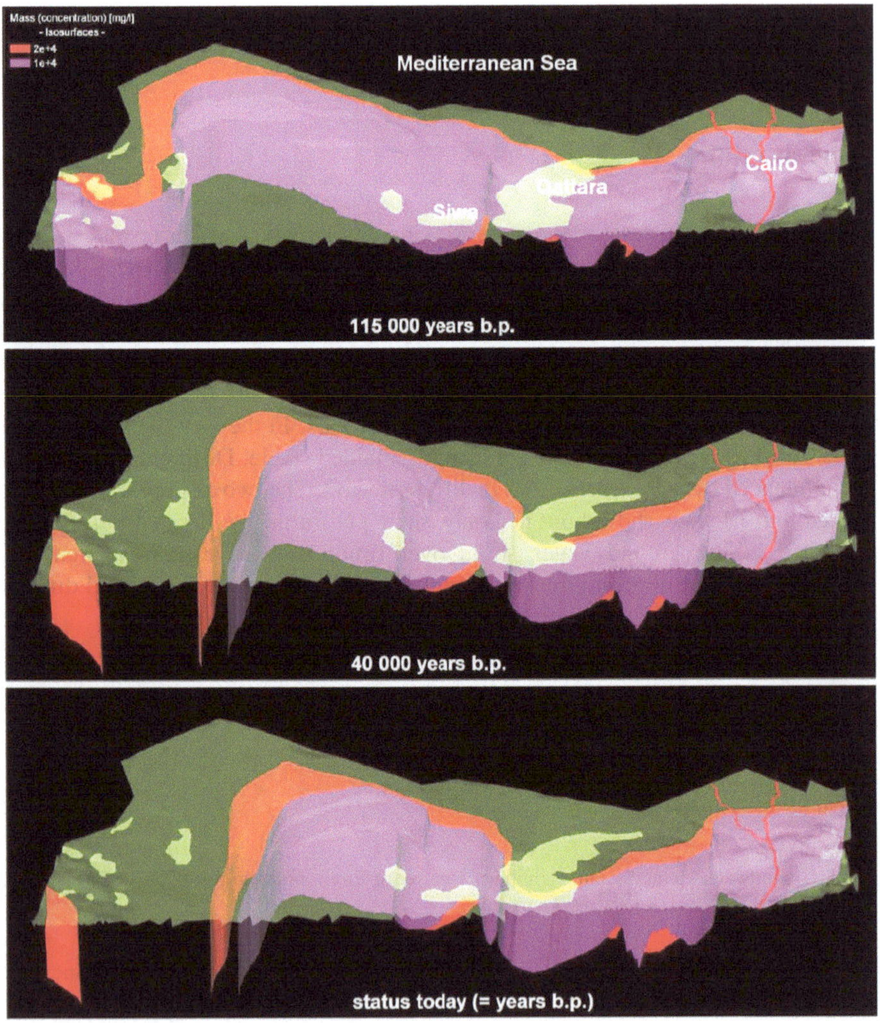

Fig. 1 Visualization of the interface at three steps during the last 140,000 years (isosurfaces of 20,000 and 30,000 mg salt/L). The first step (top) shows the situation about 120,000 years b.p. with high seawater levels at the end of the last interglacial or beginning of the last glacial. The second step shows the interface with lowest seawater levels (16,000 years b.p.) and the third outlines the interface about 10,000 years b.p. with a seawater level as today

interface movement: Infiltration of saltwater at the last interglacial about 115,000 years before present (b.p.), the deep seawater levels at about 40,000 years b.p. and after the rapid rising of the seawater levels during the last 5000 years.

Yucatan Peninsula

The karstified aquifers in Yucatan Peninsula (Mexico) react much faster on changes of climate and seawater levels. At the surface the carbonate platform of 350,000 km^2 is flooded half today so that the recent peninsula is only about 165,000 km^2 large. The rapid horizontal changes are connected to swapping of the interface in ground water, too. The fluxes of the numerical groundwater model show a high flux of fresh water out of the area (1.5-70 m/day) in times of decreasing seawater levels and in times of rising seawater levels a very high flux of saltwater into Yucatan Peninsula (1.75-75 m/day). The ingressions of several 100 km are shown in the three steps in Fig. 2.

Figure 3 shows the cross-sections of two situations: The first one during the fastest decrease of seawater level about 75,000 years b.p. and the second one about thousand years after the lowest water level (about 15,000 years b.p.). In both timesteps the depth dependent and (hydraulic) conductivity dependent distributions of saline water are clearly visible.

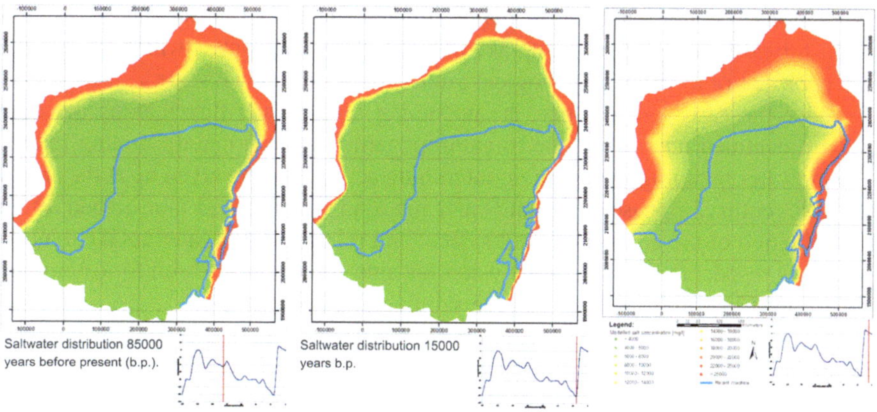

Fig. 2 Saltwater intrusion in Yucatan Peninsula over the last 140,000 years. The concentrations for the upper 100 m is responding after short times on the development at the surface

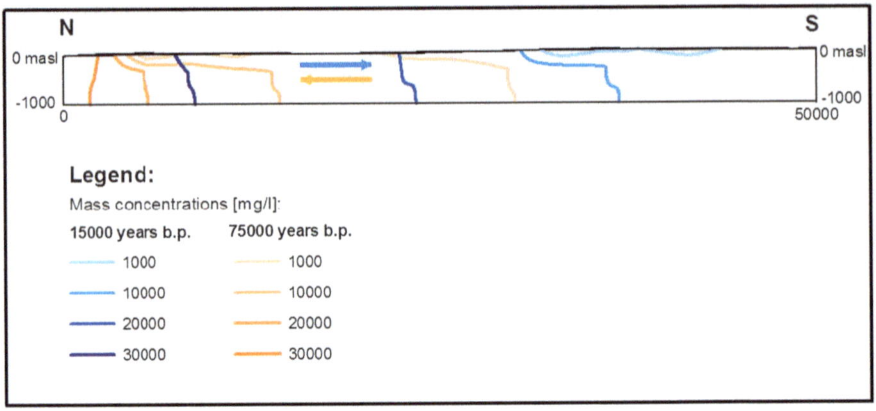

Fig. 3 Cross-section of saltwater intrusion in Northern Yucatan Peninsula. During rising seawater levels the flux into the aquifer leads to fast ingression of the saltwater with the higher density whereas the freshwater with lower density flows on top to the coast. During decreasing seawater levels the density effect is not so visible

Saltwater Upconing, Halle (Saale)

The numerical model conceptualization Halle (Saale) focuses on inland saltwater intrusion or saline deep waters along regional fault systems, respectively. This brine water uptake constitutes a danger for the shallow freshwater aquifers in terms of salinization and therefore harms potentially the local drinking water supply of the urban area. In historic ages, the occurrence of shallow accessible high saline waters has been an outstanding reason for tended population settlements and economical utilization. According to this, the major city Halle (Saale), Central Germany, owes its historical foundation on the uptake of high saline deep waters which is linked to the presence of a complex geological fault system of regional scale.

Since the early 19th century, the upcoming saline water fluxes along the fault system declined noticeably due to anthropogenic impacts, mainly mining related groundwater pumping. As the regional mining activities were given up at the end of the last century, a phase of groundwater rebound was initiated that is linked to a re-activation of the geological fault system and related saline water uptake dynamics. With respect to this, the saline water became subject of research for estimating their impact on probable near-surface salinization processes.

As the complex geological fault system defines and influences significantly the saline hydrodynamics, its spatial characterization and three-dimensional conceptualization constitutes the essential structural model input for numerical experimentations on related fault system hydrodynamics.

In recent years the geological structure of the fault system was revised multiply based on regional geophysical investigations which provided the conceptual model database of a previous 3days fault zone modelling. With respect to the three-

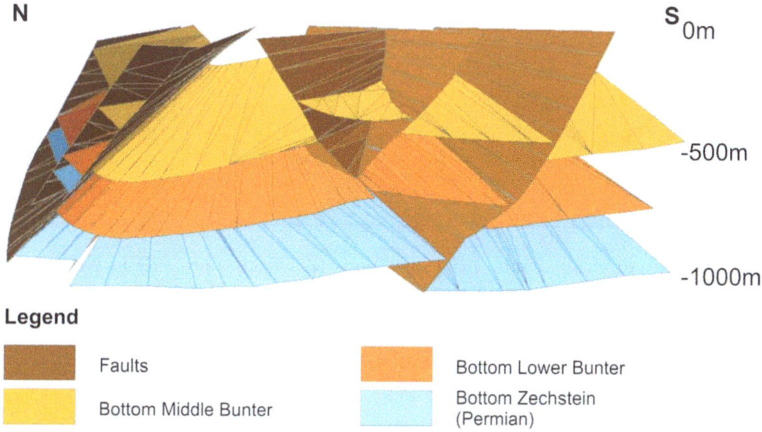

Fig. 4 View into the 3D structural model of Halle (Saale). Tectonic dislocations effect a displacement of the geological units along the steep fault planes. (Lähne et al., in prep.)

dimensional representation of the fault system (Fig. 4), information about the orientation of fault planes, structural intersections as well as volumetric and spatial information about involved geological units is available.

Based on the 3D fault model, a two-dimensional cross-section of about seven kilometres length was extracted that comprises geometries of respective geological units and regional faults. Subsequently, this section was transferred into a finite element model and was spatially discretized into 545,000 model elements.

Beside river level data and data of the regional groundwater monitoring, the numerical flow model was parameterized with characteristic hydraulic conductivity values which were derived from hydraulic laboratory tests, in-situ pumping tests and slug-and-bail tests.

The analysis of current and historical measurement data or literature indicates local highly confined aquifer conditions in the lower Zechstein formation which is interpreted simultaneously as the driving force of the regional saline water uptake.

The hydrodynamic modelling approaches under currently measured conditions (Fig. 5, Zechstein unit: +89 m NHN) shows almost no regional uptake dynamic along the fault system but confirms confined conditions ESE part of the cross-section.

The simulation results of the 2-D steady-state flow simulations were analysed and visualized regarding respective hydraulic head patterns and advective flow velocity fields. Moreover, conservative particle tracking was used to illustrate advective flow pathways of imaginary particles. Flow direction arrows describe the hydrodynamic situation along fault lines that were derived from previous simulated velocity fields.

For hydrodynamic experiments (Stollberg et al., 2012), the investigated and generic hydraulic head potentials were defined and assigned to the model, to study their hydrodynamic effects onto the groundwater flow regime. With respect to the high mineralisation of the deep waters a density dependent correction was required for the assignment of the hydraulic head scenarios.

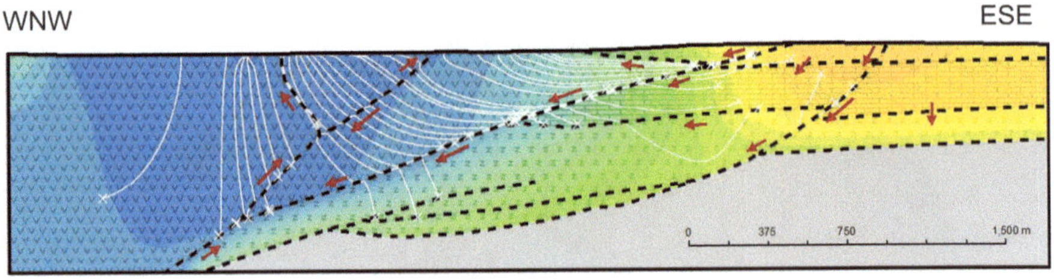

Fig. 5 Based on current hydraulic head conditions (Zechstein unit: +89 m NHN), simulation results do not indicate clear water uptake dynamics along geological unit boundaries, lineaments or tectonic fault lines. Rather, the flow velocity patterns signify a regional drainage effect of the studied fault system

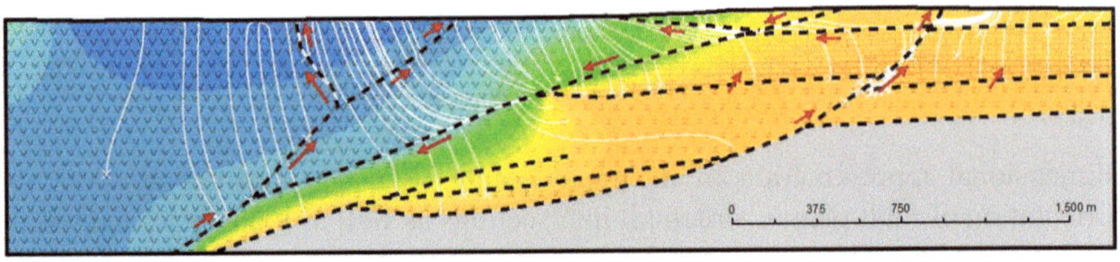

Fig. 6 A hydraulic head scenario of +101 m NHN indicates a clear water uptake along selected fault lines. Principally in the eastern part, high saline water uptake directly from the Zechstein unit is highly probable. Here, a local upward movement is additionally proven by generated pathlines

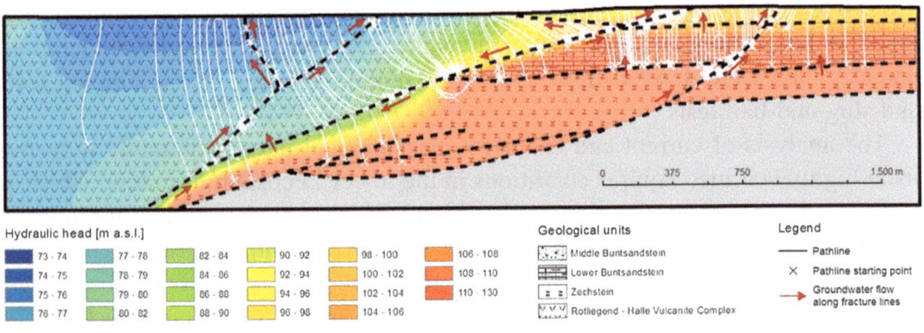

Fig. 7 Highly confined conditions of the Zechstein unit activate all fault lines for a general water uptake in the Permian and Triassic aquifers based on numerical experimentations. Here the sensitive hydraulic head range varies between +110 and +125 m NHN. In particular, the full hydraulic penetration of the dislocated Zechstein block is noticed

With respect to Fig. 6, a prescribed head of +101 m NHN was assigned to the top surface of the Zechstein unit which activates the most eastern fault line for water uptake. Moreover, fully upward oriented flow dynamics is implied by respective flow vectors and related pathlines along the western fault line.

With respect to the highly confined Zechstein scenario of Fig. 7, a general upward movement of potential deep saline waters of the Zechstein aquifer is observed. From a hydraulic point of view, all fault lines could be activated for deep water uptake, except a local fault segment which is characterised by a significant positive hydraulic conductivity contrast. This hydraulic fault line activation is observed within a hydraulic head range of the Zechstein unit from +110 to +125 m NHN.

Currently, local investigations are running to specify respective mixing effects along the deep saline water and freshwater interface based on a density coupled transport modelling. Here, the calibration of the density coupled transport model is quite difficult since available data about the mineralisation and distribution of deep waters is generally rare or challenging to estimate. The recent drawdown of uptake dynamics led to a regional freshwater intrusion which was observed over the last twenty years and could be verified by the modelling approach. Moreover, the re-activation of respective fault lines as the minor uptake pathways of saline deep waters in combination with realistic pressure conditions within the Permian and Triassic aquifers could be confirmed as well.

For giving an appropriate risk assessment concerning the salinization potential of the shallow freshwater resources, the on-going density-coupled transport study is inevitable.

References

Gossel, W., Sefelnasr, A. and Wycisk, P. (2010). Modelling of paleo-saltwater intrusion in the northern part of the Nubian Aquifer System, Northeast Africa. *Hydrogeology Journal*, **18:** 1447–1463.

Lähne, R., Stollberg, R., Gossel, W. and Wycisk, P. (in prep.). Geological fault zone modelling and its application to numerical experimentations on saline groundwater uptake – Case study, Central Germany.

Stollberg, R., Lähne, R., Rappsilber, I. and Gossel, W. (2012). Hydrogeologische Interpretationen zur Dynamik des historischen Salzwassseraufstieges der Stadt Halle (Saale). *In:* Liedl, R., Burghardt, D., Simon, E., Reimann, Th. and Kaufmann-Knoke, R. (Hrsg.): Grundwasserschutz und Grundwassernutzung – Modelle, Analysen und Anwendungen, Schriftenreihe der DGG, Heft 78, Dresden.

Hydrogeochemical Characterisation and Evaluation of Seasonal Variation in Groundwater Chemistry in Upper Panda River Basin, India

Sangita Dey, N. Janardhana Raju, Prahlad Ram, and Janmejoy Singh

Introduction

The shortage of water resources of good quality is becoming an important issue in hard rock and semi-arid zones and rapid declining of groundwater supplies are common (Raju and Reddy, 2007). Groundwater is the primary source of water for domestic, agricultural and industrial uses in many countries, and its contamination has been recognised as one of the most serious problems in India (Raju, 2007; Reddy et al., 2010; Raju et al 2009a). Major ion-chemistry of ground water provides the basis to investigate the weathering reactions in the basin (Das and Kaur, 2007; Raju et al., 2011). Each groundwater system, in the area, has a unique chemistry, acquired as a result of chemical alteration of meteoric water recharging the system (Back, 1966; Drever, 1997; Raju, 2012). The assessment of the suitability of groundwater for domestic water supply requires knowledge of the concentrations of inorganic constituents and their comparison with existing standards. Irrigation water quality concerns the amounts of salts present in ground water and their effects on crop growth and development. Since there is no adequate surface water supply, about 80-90 % of drinking and irrigation use is from available groundwater resources and the importance of groundwater utilization has increased at an alarming rate in parts of Panda River basin, Sonbhadra district of Uttar Pradesh (Dey, 2010).

S. Dey
Department of Geology, Banaras Hindu University, Varanasi 221005, India

N.J. Raju (✉)
School of Environmental Sciences, Jawaharlal Nehru University, New Delhi 110067, India
e-mail: rajunj7@gmail.com

P. Ram
Central Ground Water Board, Faridabad, Haryana, India

J. Singh
National Geophysical Research Institute, Uppal Road, Hyderabad, India

N.J. Raju et al. (eds.), *Management of Water, Energy and Bio-resources in the Era of Climate Change: Emerging Issues and Challenges*,
DOI 10.1007/978-3-319-05969-3_3

Some parts of the study area are facing severe groundwater problems i.e. fluoride contamination which makes the water unfit for human consumption (Raju et al., 2009b). The main objective of the study is to assess spatial distribution of hydrogeochemical parameters and evaluate seasonal variation in groundwater chemistry of the upper Panda River basin, Sonbhadra district of Uttar Pradesh, India.

The upper Panda River basin is located in the Chopan block (Fig. 1) in the Sonbhadra district, experiences semi-arid climatic condition and is bounded

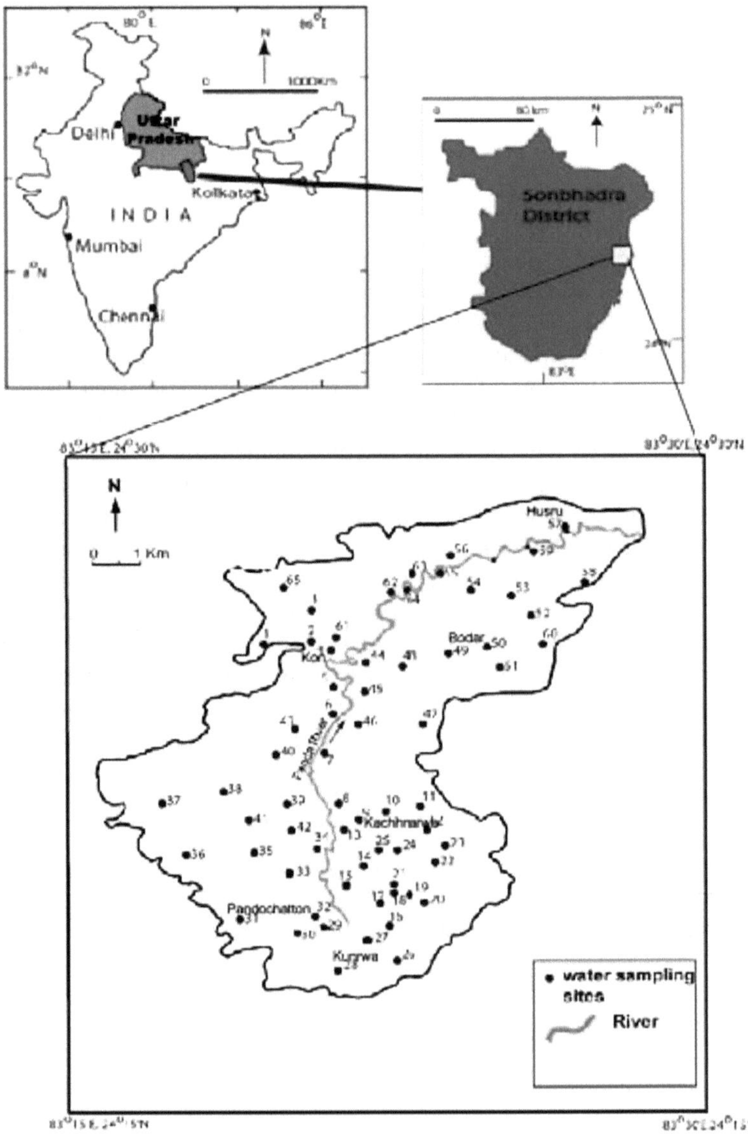

Fig. 1 Location map and groundwater samples of the upper Panda River basin

between north latitude 24°15' and 24°30' and east longitude 83°15' and 83°30'. The upper Panda River basin has a drainage area of 216 km^2 and the basin length of about 17 km. The average annual rainfall in the study area is 850 mm. The upper Panda River basin is dominated by metasedimentary rock sequences of Dudhi Granitoid Complex of Palaeo Proterozoic Age. The major litho units identified belong to Parsoi Formation, and mainly consist of phyllites. Dudhi Granitoid comprises of granite, granodiorite, gneisses and pegmatite etc. The ground water in these rocks is brought out for use by digging open dugwells and borewells. In general, the depth of water table ranges from 3 m to 34.3 m in pre-monsoon and 2.7 m to 30 m in post-monsoon season. The general flow directions are from N-S, SE-NW and SW-NE which is in accordance with the general slope of the area and also direction of stream flow. The maximum depth of the hand tubewell for drinking water use is varying 40-60 m below ground level.

Methodology

Groundwater samples were collected during April-May 2006 (pre-monsoon) and in the month of October 2006 (post-monsoon) from dug wells, bore wells and hand tubewells and were analysed for various physico-chemical characteristics of groundwater quality parameters (APHA, 1995). Electrical conductivity and pH were measured using EC and pH meters. Total dissolved solids were estimated by ionic calculation method. Total alkalinity (TA), CO_3 and HCO_3 were estimated by titrating with HCl. Total hardness (TH) and Ca were analysed titrimetrically using standard EDTA. Mg was computed, taking the difference between TH and Ca values. Na and K were measured by flame photometer, Cl was estimated by standard $AgNO_3$ titration and SO_4 was measured by the turbidimetric method. Fluoride was estimated by using an ion selective electrode (model 96- 09) with 720 pH/ISE meter (Orion, USA). Rock powder samples were analysed with the help of X-ray fluorescene.

Results and Discussion

Rock-chemical Analysis

XRF analysis was carried out for seven fresh rock samples and the degree of chemical weathering can be obtained by the measurement of chemical index of alteration (CIA). The weathering trend can be displayed on a ternary plot of $(CaO + Na_2O) - K_2O\text{-}Al_2O_3$ (Nesbit and Young, 1989).

$$\text{CIA} = [Al_2O_3/(Al_2O_3 + CaO + Na_2O + K_2O)] \times 100$$

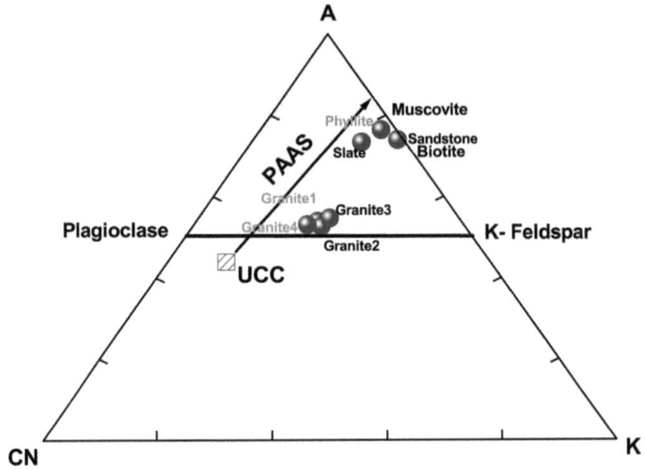

Fig. 2 Ternary plot of $(CaO + Na_2O)$-K_2O-Al_2O_3 for rock weathering trends

The resultant value is a measure of the proportion of Al_2O_3 versus the major oxides in the analysed samples. The ternary plot (Fig. 2) of $(CaO + Na_2O)$-K_2O – Al_2O_3 form a trend parallel to the $(CaO + Na_2O)$–Al_2O_3 side of the diagram in the initial stages of weathering whereas advanced weathering shows excess loss of K_2O and move towards the Al_2O_3 apex. Fresh granites have CIA values between 45 and 55 (Nesbit and Young, 1982). Dudhi granitoids of the study area show the values ranging between 53 and 55. It indicates that the granites of the study area have suffered low to medium extent of weathering. Intensely weathered rock yields mineral composition trending towards kaolinite or gibbsite and corresponding CIA approaches to 100. Phyllite, slate and sandstone show the high CIA value of 76, 73 and 74 respectively which indicate the high weathering conditions.

Major Ion Chemistry

Sixty five groundwater samples were collected randomly during pre-monsoon and post-monsoon season for physico-chemical analysis (Fig. 1) and the ranges of the quality parameters are shown in Table 1. Groundwater is generally alkaline in nature with an average pH of 7.13 and 7.45 in pre- and post-monsoon seasons, respectively. Based on the total dissolved solids classification (Davis and De Wiest, 1966), out of the 65 samples, 55 samples of pre- and 51 samples of post-monsoon are within desirable (i.e. upto 500 mg/l) and 10 samples of pre- and 14 samples of post-monsoon are within permissible for drinking (i.e. 500-1000 mg/l) and all the samples are useful for irrigation purpose (upto 3000 mg/l). The high value of TDS is generally indicative of the amount of contribution of the dissolved solids to these

Table 1 Ranges of hydro-chemical quality parameters in groundwater and their comparisons

	Pre-monsoon	Post-monsoon	WHO (1997)	BIS (1991)	Percentage of samples exceeding desirable limit	
Parameters	Range	Range	Desirable-Permissible		Pre-monsoon	Post-monsoon
Ca^{2+}	6-69	14-78	75-200	75-200	-	1.5
Mg^{2+}	10.8-56.8	4.02-59.2	30-150	30-100	45	28
Na^{+}	10.4-230	12.5-256	50-200	-	38.5	52.3
K^{+}	1.3-3.8	1.4-4.2	10-12	-	-	-
HCO_3^{-}	92-396	104-455	300-600	300-600	12.3	7.7
SO_4^{2-}	10-190	25-195	200-600	200-400	-	-
Cl^{-}	6-66	22-96	250-600	250-1000	-	-
F^{-}	0.37-5.6	0.08-6.7	0.9-1.5	1.0-1.5	30	45
TDS	206-724	140-800	500-1500	500-2000	15	22
pH	6.5-7.7	6.9-8.6	7.0-9.2	6.5-9.2	85	97
EC	342-1190	428-1280	-	-	-	-
TH	94-288	98-314	100-500	300-600	98	98

waters by the human activity such as sewage and irrigation practices. The total hardness of the analysed samples varies from 94 to 288 mg/l in pre-monsoon and 98 to 314 mg/l in post-monsoon season indicating moderately hard to hard type of waters (Sawyer and McCarty, 1967). Among the cationic concentrations in the pre- and post-monsoon seasons are: sodium is the dominating ion followed by calcium, magnesium and potassium. In general weathering, dissolution and base-exchange processes control the levels of cationic concentrations in ground water. High sodium waters can be explained by the combination of dilution factors, ion-exchange and sulphate reduction (Krothe and Oliver, 1982). Among the anionic concentrations in the pre- and post-monsoon seasons are: bicarbonate is the dominating ion followed by sulphate, chloride and fluoride. The high bicarbonates in the ground waters are derived mainly from the soil zone CO_2 and at the time of weathering of parent minerals.

Concept of hydrochemical facies offers a mass regional relation between chemical character, lithology and regional flow pattern. Based on relative dominance of major cations and anions in terms of their reacting values, six hydrochemical facies (Table 2) have been identified. Na-HCO_3 is dominated most part of the study area, which represents 60 % of the total number of water samples for both the seasons. The Mg-HCO_3, Ca-HCO_3, and Na-SO_4 facies each represent 18.5 %, 15.3 % and 6.2 % of total number of water samples analysed in pre-monsoon, while in post-monsoon Ca-HCO_3, Mg-SO_4, Ca-SO_4 and Na-SO_4 each represent 35.4 %, 1.5 %, 1.5 % and 1.5 % of total number of water samples analysed, respectively. Sulphate facies are increased in the post-monsoon season because sulphate dissolution is more during the rainy season due to mixing of infiltrated water. The plot of chemical data (Fig. 3) in the Piper diagram (1944) reveals that the majority of the

Table 2 Hydrochemical facies of the study area

Sl. No	Facies	Groups	Pre-monsoon		Post-monsoon	
			No. of samples	%	No. of samples	%
1	Na-HCO_3	Na-Ca-HCO_3-SO_4	12, 14, 17, 62, 63	**60**	6, 9, 11, 12, 18, 20, 24, 32, 33, 42, 45, 47, 48, 57, 58, 62, 63	**60**
		Na-Mg-Ca-HCO_3-SO_4	-		38, 50, 51	
		Na-Ca-Mg-HCO_3-SO_4	9, 26, 39, 41, 51		55	
		Na-Ca-HCO_3	11, 20, 34, 47		23, 31, 35, 36, 59	
		Na-HCO_3-SO_4	15, 16, 18, 19, 25, 29		14, 16, 17, 19, 25, 27, 28, 29, 37	
		Na-Ca-Mg-HCO_3	21		34	
		Na-Mg-HCO_3-SO_4	23, 24, 30, 31, 32, 59		10, 30	
		Na-Mg-Ca-HCO_3	55		53	
		Na-Mg- HCO_3	8, 10, 27, 28, 35, 37, 40, 53, 57, 58		-	
		Na- HCO_3	48		-	
2	Ca-HCO_3	Ca-Mg-HCO_3-SO_4	2, 4, 22	**18.5**	22, 49, 54, 56, 60	**35.4**
		Ca-Na-Mg-HCO_3-SO_4	7, 49		26, 39	
		Ca-Na-HCO_3-SO_4	13, 43, 56, 65		1, 3, 7, 8, 13, 21, 43, 44, 46, 61	
		Ca-Na-HCO_3	3, 5		5, 40, 41	
		Ca-Mg-Na-HCO_3-SO_4	-		65	
		Ca-Mg-Na-HCO_3	-		52	
		Ca-Mg-HCO_3	46		-	
		Ca-Na- HCO_3-Cl	-		4	

3	Mg-HCO_3	Mg-Ca-HCO_3-SO_4	1	**15.3**	-	-
		Mg-Na- Ca-HCO_3- SO_4	38		-	
		Mg-Na-HCO_3-SO_4	6, 44, 50, 64		-	
		Mg-HCO_3	52		-	
		Mg-Ca-HCO_3	54, 60		-	
		Mg-Na-HCO_3	61		-	
4	Na-SO_4	Na-SO_4-HCO_3	33	**6.2**	15	**1.5**
		Na-Ca-SO_4-HCO_3	36, 45		-	
		Na-Mg-SO4-HCO_3	42		-	
5	Mg-SO_4	Mg-Na-SO4-HCO_3	-	-	64	**1.5**
6	Ca-SO_4	Ca-SO_4-HCO_3	-		2	**1.5**

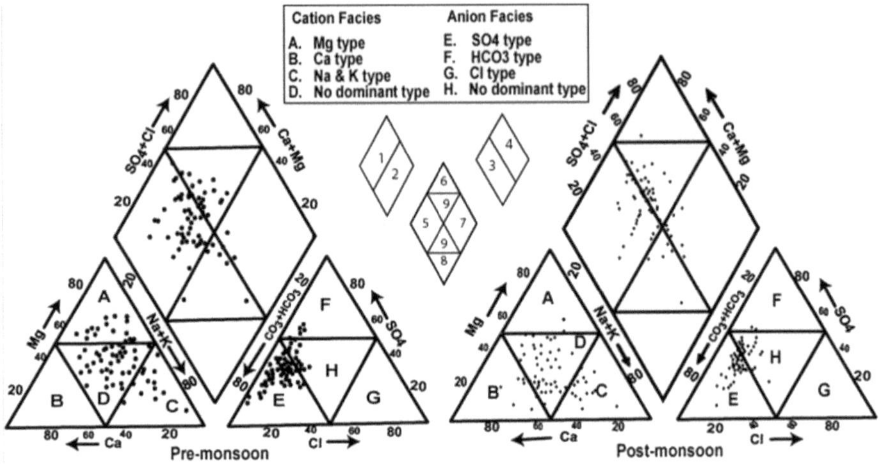

Fig. 3 Diagram showing different groundwater types (Piper, 1944)

samples of pre-monsoon fall in the fields of 1, 3 and 5 suggesting that alkaline earth exceeds alkalis; weak acids exceed strong acids and the ions representing carbonate hardness (secondary alkalinity) exceed 50 %, respectively. In pre-monsoon season evaporation and evapotranspiration bring the SO_4^{2-} and Cl^- bearing salts into the surface zone but during post-monsoon, these salts mixed/dilute with rainwater and infiltrates into the groundwater during rainy season. The majority of the samples in post-monsoon falls into the fields of 1, 4 and 9 suggesting that alkaline earth exceeds alkalis, strong acids exceed weak acids and none of cation and anion pairs exceeds 50 %, respectively. Most of the samples of pre-monsoon occur in Ca-Mg-HCO_3 facies while post-monsoon season shows mixed hydrochemical facies.

Inter-ionic Relationships and Saturation Indices

The Na^+/Cl^- molar ratio is >1 (Table 3) in almost all the samples of both the seasons indicating that silicate weathering was the primary process responsible for the excess release of Na^+ into the ground water (Stallard and Edmond, 1983; Pophare and Dewalkar, 2007). In the study area $(Na^++K^+)/TZ^+$ ratio is 0.36 in pre-monsoon and 0.38 in post-monsoon seasons. The low molar ratio of $(Ca^{2+}+Mg^{2+})/TZ^+$ and $(Na^++K^+)/TZ^+$ also supports the contribution of cation via silicate weathering. The $(Ca^{2+}+Mg^{2+})/(Na^++K^+)$ ratio indicates 2.4 and 2.1 in pre-monsoon and post-monsoon, respectively (Table 3). It suggests that the ground water is largely influenced by silicate weathering and little contribution of carbonate weathering. The proportion of HCO_3^- and SO_4^{2-} in the water reflects the relative dominance of the two major sources of protons during the chemical weathering. The relative

Table 3 Mean chemical indices of groundwater of the study area

Chemical Indices (meq/l)	Pre-	Post-
Na^+/Cl^-	4.06	2.59
Ca^{2+}/Mg^{2+}	0.70	1.62
$Ca^{2+}+Mg^{2+}/ HCO_3^-$	1.20	1.31
Na^++K^+/TZ^+	0.36	0.38
$Ca^{2+}+Mg^{2+}/TZ^+$	0.64	0.61
$Ca^{2+}+Mg^{2+}/ Na^++K^+$	2.4	2.1
Ca^{2+}/Na^+	0.94	1.16
Mg^{2+}/Na^+	1.58	1.03
Mg^{2+}/HCO_3^-	0.75	0.61
$HCO_3^-/ (HCO_3^- + SO_4^{2-})$	0.63	0.59
Ca: HCO_3	0.15	0.25
Ca: SO_4	0.41	0.47
Na^+/Ca^{2+}	2.3	1.5
CAI I	−3.15	−1.63
CAI II	−0.29	−0.33

importance of the two major proton producing reactions i.e. carbonation and sulphide oxidation can be signified on the basis of the $HCO_3^-/(HCO_3^- + SO_4^{2-})$ equivalent ratio, called C-ratio (Brown et al., 1996). C-ratio varies from 0.37 to 0.88 with an average of 0.63 in pre-monsoon and varies from 0.38 to 0.89 with an average of 0.59 in post-monsoon (Table 3). The C-ratio value signifies that coupled reactions involving the carbonic acid weathering are the major proton producer in the ground waters. The Ca^{2+} and Mg^{2+} together constitute 50.1 % of the total cations of pre- and 50.2 % of post-monsoon. A major proportion of HCO_3^-, Ca^{2+}, Mg^{2+} and Na^+ ions are derived from the weathering of Ca-Mg silicates, chiefly from the Ca-plagioclase, amphiboles, pyroxenes and biotite. Average Na^+/Ca^{2+} ratio is more than unity (Table 3) indicating a deficiency of Ca^{2+}. This may be caused by precipitation of $CaCO_3$ and/or ion-exchange process. Under the circumstance, the ratio of Ca:HCO_3 (mean 0.15 in pre- and 0.25 in post-monsoon) would not be simply controlled by the dissolution of carbonate minerals. Hence, the soil CO_2 is a preferred source rather than dolomite dissolution since ground waters are undersaturated with calcite and dolomite minerals.

The precipitation of mineral saturation can be assessed in terms of a saturation index (SI), using a simple thermodynamic approach (Garrels and Christ, 1965). The plot of SI of minerals against TDS (Fig. 4) shows groundwater samples less saturated with respect to carbonate minerals but few samples show oversaturation in the post-monsoon groundwaters. This may be due to surface water that percolates into the subsurface zone and enrich the carbonate concentration by the dissolution of carbonate minerals (i.e. lime kankar). All the samples except two in the pre-monsoon season show negative SI values whereas in post-monsoon 75.4 % of calcite and 88 % of dolomite show negative SI values. The majority of the ground waters are undersaturated with carbonate and sulphate minerals, but oversaturated with the silicate minerals (i.e. chalcedony and cristobalite) for both the seasons.

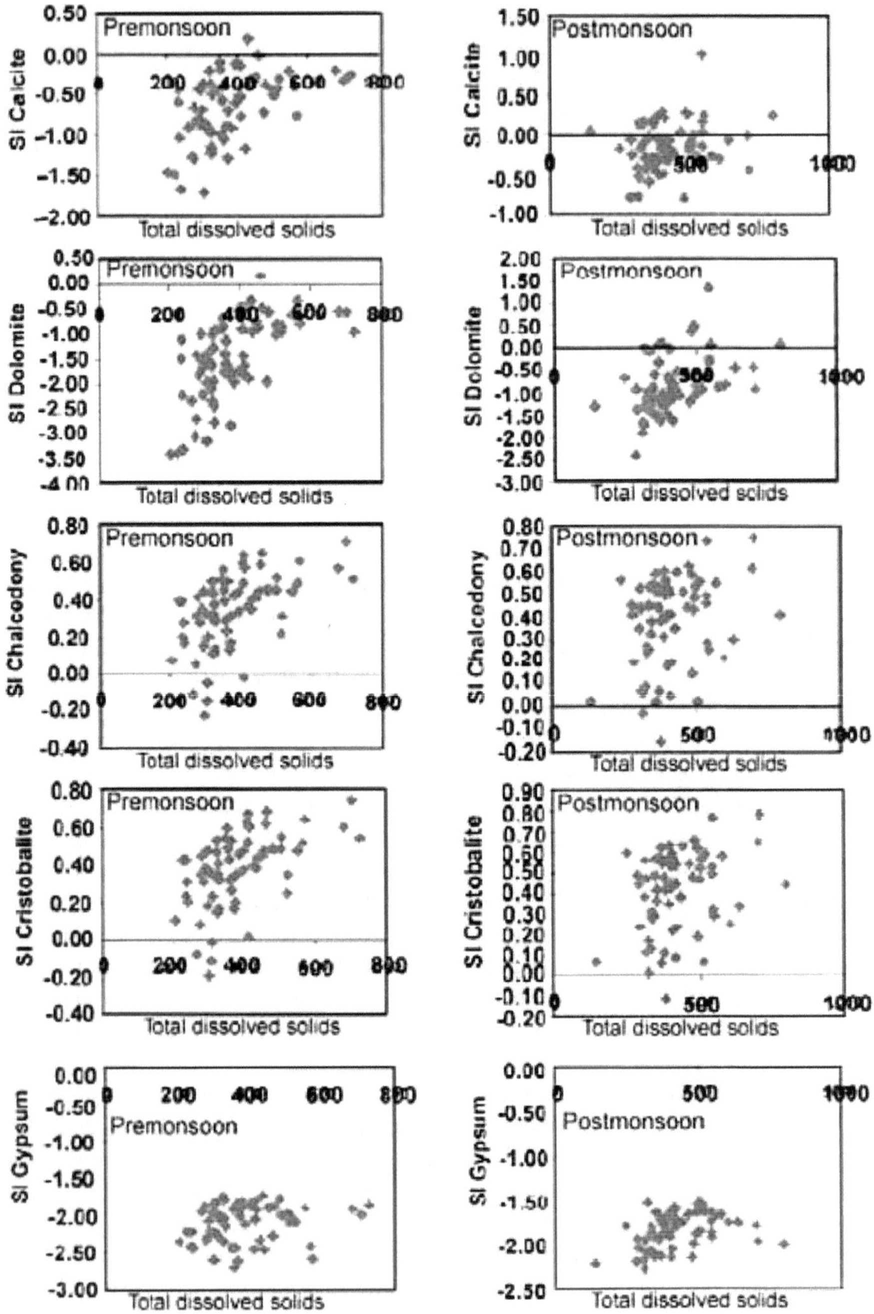

Fig. 4 Scatter plots of different mineral saturation indices against total dissolved solids

Solute Acquisition Process

The proportion of HCO_3 and SO_4 in the water reflects the relative dominance of the two major sources of protons (carbonation and sulphide oxidation) during chemical weathering. The relative high ratio of $HCO_3/(HCO_3 + SO_4)$ in most of the groundwater samples (>0.5) signified that carbonic acid weathering was the major proton producer. The dissolved bicarbonate (HCO_3) in the ground water originates mainly from the biologically active layers of the soil and the dissolution of carbonates and/or silicate minerals by the carbonic acid. Feldspars are more susceptible to weathering and converted to clay minerals such as kaolinite, smectite etc.

$$\underset{\text{Limestone}}{CaCO_3 + H_2CO_3} \rightarrow Ca^{2+} + 2HCO_3^-$$

$$\underset{\text{Albite}}{2NaAlSi_3O_8} + 2H_2CO_3 + 9H_2O \rightarrow \underset{\text{Kaolinite}}{Al_2Si_2O_5(OH)_4} + 2Na + 4H_4SiO_4 + 2HCO_3^-$$

Some samples contain nearly equal amounts of HCO_3 and $(Cl + SO_4)$ indicating inputs from the weathering of pyrites. The weathering reaction involved in oxidation of pyrites can be written as:

$$\underset{\text{Pyrite}}{2FeS_2} + 7O_2 + 2H_2O = 2Fe^{2+} + 4SO_4^{2-} + 4H^+$$

The contribution of chloride may result from pollution by domestic sewage wastes and leaching of saline residues in the soil (Appelo and Postma, 1993). Highly weathered porphyritic granitic gneiss, which contain good amount of fluor-apatite and biotite mica minerals occur most prominently in the southern parts of the study area, which forms the source of fluoride to percolating ground water in this recharge zone (Raju et al., 2009b). While reacting with granite gneissic rocks for a prolonged period, the fluoride concentrations of ground water are continuously enriched even after the ground water reaches an equilibrium state with respect to fluor-apatite due to removal of Ca by precipitation of calcite ($CaCO_3$). The weathering activity characterised by alternate wet and dry conditions of the semi-arid climate is responsible for leaching fluoride from the mineral present in the soils and rocks. High concentration of Ca and Mg is attributed to weathering of calcium carbonate nodules (lime kankar) and Ca-Mg silicates (amphiboles, pyroxenes, micas etc.). The major possible source of Na in groundwaters is from sodic plagioclase because it occurs in all types of rocks and also Na-Ca silicate minerals weather most rapidly (Berner and Berner, 1987). Potassium is not as abundant as sodium due to its fixation in the formation of clay minerals.

Groundwater Quality Assessment

The range of hydrochemical parameters and their comparison with the prescribed specification of Indian Standard (BIS, 1991) and WHO (1997) have been summarized in Table 1. Based on TDS content (Fetter, 1994), all groundwater samples are within permissible limits of WHO and BIS which fall under fresh water category. Hardness of groundwater shows (Table 4) slightly to moderately hard in nature. The desirable limits of calcium and magnesium are 75 mg/l and 30 mg/l, respectively. Almost all the calcium values in groundwater samples are within the desirable limit for both the seasons, whereas 45 % of pre-monsoon and 28 % in post-monsoon samples exceeded desirable limits of magnesium ions. In the case of sodium 38.5 % in pre-monsoon and 52.3 % in post-monsoon samples are exceeding the desirable limit (50 mg/l) and only one sample is exceeding the permissible limit (200 mg/l). The higher sodium intake may cause hypertension, congenial heart diseases and kidney problems. HCO_3 exceeded the desirable limit (300 mg/l) in 12.3 % of the pre-monsoon and 7.7 % of post-monsoon groundwater samples. Bicarbonate has no

Table 4 Irrigation classifications of groundwater samples using different methods

Percent Sodium Classification			
		Pre-	*Post-*
Excellent	0-20	10	6
Good	20-40	34	34
Permissible	40-60	17	19
Doubtful	60-80	3	5
Unsuitable	>80	1	1
Sodium Adsorption Ratio Classification			
		Pre-	*Post-*
Excellent	0-10	64	64
Good	10-18	1	1
Fair	18-26	–	–
Poor	>26	–	–
Electrical Conductivity Classification			
		Pre-	*Post-*
Excellent	<250	–	–
Good	250-750	49	45
Fair/Medium	750-2250	16	20
Poor/Bad	2250-5000	–	–
Hardness Classification			
		Pre-	*Post-*
Soft	<75	–	–
Slightly hard	75-150	12	9
Moderately hard	150-300	53	56
Very hard	>300	–	–

known adverse health effects on human beings. All groundwater samples of the study area are within the chloride and sulphate desirable and permissible limit. Fluoride has identified health effect and benefits for human beings (BIS, 1991). 30 % of pre-monsoon and 45 % of post-monsoon samples are exceeding the desirable limit (Table 1) and 11 % of pre-monsoon and 18 % of post-monsoon are exceeding the permissible limit of fluoride concentration in the study area.

Salinity hazard and sodium hazard are the important water quality issues especially in irrigated areas of arid and semi-arid regions (Jalali, 2007; Raju et al., 2012). Salinization of irrigated lands is a major cause of loss of production and it has adverse environmental impacts on irrigation processes. As per Richard's (1954) classification on the basis of EC values, 75 % in pre-monsoon and 69 % in post-monsoon samples are good category and 25 % in pre-monsoon and 31 % in post-monsoon groundwater samples are medium category (Table 4). Sodium reacts with the soils causing soil aggregates to disperse and ultimately reduces its permeability (Tijani, 1994). In all natural water percent sodium (%Na) is a parameter to evaluate its suitability for agricultural purposes (Wilcox, 1948). The sodium percentage (%Na) ranges from 8.16 % to 84.28 % in pre-monsoon and 8.71 % to 85.12 % in post-monsoon season. A high sodium percent causes deflocculation and impairment of tilth and permeability of soils (Karanth, 1989). The chemical quality of groundwater samples was studied from plots of percentage of sodium and electrical conductivity on the Wilcox (1948) diagram (Fig. 5). Out of total samples 73.8 % in pre-monsoon and 69.2 % in post-monsoon are excellent to good, 18.5 % in pre-monsoon and 24.6 % in post-monsoon are good to permissible, 6.15 % in pre-monsoon and 4.62 % in post-monsoon are permissible to doubtful and only one sample of both the season is doubtful to unsuitable (Table 4). The agricultural crop yields are generally low in lands irrigated with waters belonging to doubtful to unsuitable category. This is probably due to the presence of excess salts, which cause osmotic effects on soil-plant system. The relative activity of sodium ion in the exchange reaction with soil is expressed in terms of a ratio known as sodium adsorption ratio (SAR). As per the SAR values, almost all groundwater samples are in excellent category except one sample in both seasons falling in good category (Table 4).

Conclusions

Evaluation of hydrogeochemical parameters reflects the weathering of silicate and other rocks which are responsible for major ion chemistry of groundwater in the upper Panda River basin. Post-monsoon groundwaters are more alkaline than the pre-monsoon season. The groundwater samples collected from the study have high concentrations of Na constituting 48 % and 47.8 % among cations, and HCO_3 forms 63.4 % and 57.4 % among anions in pre- and post-monsoon, respectively. Na-HCO_3 facies (60 %) is dominated in both the seasons in most parts of the area. Piper diamond field classification revealed that most of the groundwaters have temporary hardness in pre-monsoon whereas in post-monsoon season none of the cations and

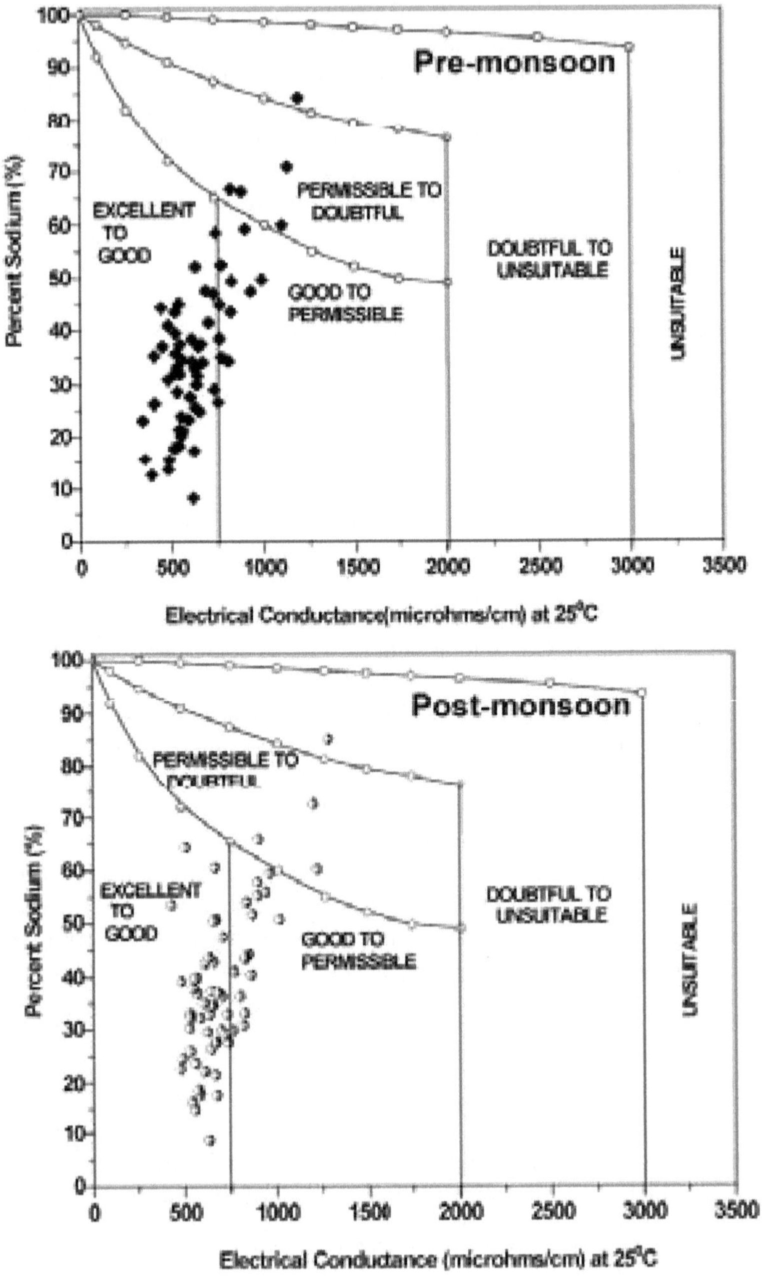

Fig. 5 Rating of groundwater samples on the basis of electrical conductivity and percent sodium (after Wilcox, 1948)

anions exceeds 50 %. Inter-ionic relationships suggests that the water is largely influenced by silicate weathering along with little contribution of carbonate weathering. Majority of the samples show positive saturation indices with respect to cristobalite and chalcedony and the negative saturation indices with respect to calcite, dolomite and gypsum. High fluoride content in the southern part could be attributed to the longer residence time and percolation of water through fluoride bearing rocks. Except the fluoride contaminated southern part of the area, all the groundwater samples for both the seasons are suitable for various purposes.

References

APHA (1995). Standard methods for the examination of water and wastewater. 19th edn. American Public Health Association, Washington DC.

Appelo, C.A.J. and Postma, D. (1993). Geochemistry, Groundwater and pollution. A.A.Balkema, Rotterdam.

Back, W. (1966). Hydrochemical facies and ground flow patterns in northern part of Atlantic coastal plain. USGS Prof. Paper, 498–A.

Berner, E.K. and Berner, R.A. (1987). The global water cycle. Prentice-Hall Inc., Englewood Cliffs, N.J.

BIS (1991). Bureau of Indian Standards – Indian standard specification for drinking water. IS:10500.

Brown, G.H., Sharp, M.J. and Tranter, M. (1996). Subglacial chemical erosion: Seasonal variations in solute provenance, Haunt Glaciard Arolla, Valais, Switzerland. *Ann. Glaciol.*, **22:** 25–31.

Das, B.K. and Kaur, P. (2007). Geochemistry of surface and subsurface waters of Rewalsar Lake, Mandi district, Himachal Pradesh: Constraints on weathering and erosion. *J. Geol. Soc. India*, **69(5):** 1020-1030.

Davis, S.N. and De Wiest, R.J.M. (1966). Hydrogeology. Wiley, New York.

Dey, S. (2010). Hydrogeological and hydrogeochemical studies with reference to fluoride contamination in the Panda River Basin, Sonbhadra district, Uttar Pradesh, India. Unpubl. Ph.D thesis, Banaras Hindu University, Varanasi.

Drever, J.I. (1997). The geochemistry of natural waters. 3rd edn. Prentice-Hall, New Jersey.

Fetter, C.W. (1994). Applied hydrogeology. 3rd edn. Macmillan College Publication, New York.

Garrels, R.M. and Christ, C.L. (1965). Solutions, minerals and equilibria. Freeman, Cooper, San Franciso.

Jalali, M. (2007). Salinization of groundwater in arid and semarid zones: An example from Tajarak, western Iran. *Env. Geol.*, **22:** 1133–1149.

Karanth, K.R. (1989). Hydrogeology. Tata McGraw Hill Publication Ltd., New Delhi.

Krothe, N.C. and Oliver, J.W. (1982). Sulfur isotopic composition and water chemistry from the high plains aquifers, Oklahoma panhandle and southwestern Kansas. *US Geol. Surv. Wat. Res. Investigation*, **82(12):** 1.

Nesbit, H.W. and Young, G.M. (1982). Early Proterozoic climates and plate motions inferred from major element chemistry of lutites. *Nature*, **299:** 715–717.

Nesbit, H.W. and Young, G.M. (1989). Formation and diagenesis of weathering profiles. *Jour. Geol.*, **97:** 129–147.

Piper, A.M. (1944). A graphical procedure in the geochemical interpretation of water analysis. *Am. Geophys. Union. Trans.*, **25:** 914– 928.

Pophare, M.A. and Dewalkar, S.M. (2007). Groundwater quality in eastern and south eastern parts of Rajura Tehsil, Chendrapur district, Maharashtra. *Gondwana Geological Magazine Special*, **11:** 119–129.

Raju, N.J. (2007). Hydrogeochemical parameters for assessment of groundwater quality in the upper Gunjanaeru River basin, Cuddapah District, Andhra Pradesh, South India. *Environ. Geol.*, **52:** 1067–1074.

Raju, N.J. and Reddy, T.V.K. (2007). Environmental and urbanization affect on groundwater resources in pilgrim town of Tirupati, Andhra Pradesh, South India. *Jour. Appl. Geochem.*, **9 (2):** 212–223.

Raju, N.J., Ram, P. and Dey, S. (2009a). Groundwater Quality in the lower Varuna River basin, Varanasi district, Uttar Predesh, India. *Jour. of Geol. Soc. of India*. **73:** 178–192.

Raju, N.J., Dey, S. and Das, K. (2009b). Fluoride contamination in groundwaters of Sonbhadra District, Uttar Pradesh. *India Current Science*, **96(7):** 979–985.

Raju, N.J., Shukla, U.K. and Ram, P. (2011). Hydrogeochemistry for the assessment of groundwater quality in Varanasi: A fast urbanizing center in Uttar Pradesh, India. *Envn. Monit. Asses.*, **173:** 279–300.

Raju, N.J. (2012). Evaluation of hydrogeochemical processes in the Pleistocene aquifers of middle Ganga Plain, Uttar Pradesh, India. *Environmental Earth Sciences*, **65:** 1291–1308.

Raju, N.J., Dey, S., Gossel, W. Wycisk, P. (2012). Fluoride hazard and assessment of groundwater quality in the semi-arid upper Panda River basin, Sonbhadra District, Uttar Pradesh, India. *Hydrological Sciences Journal*, **57(7):** 1433–1452.

Reddy, A.G.S., Reddy, D.V., Rao, P.N. and Maruthy Prasad, K. (2010). Hydrogeochemical characterization of fluoride rich groundwater of Wailpalli watershed, Nalgonda district, Andhra Pradesh, India. *Envir. Monit. Assess.*, **171:** 561–577.

Richard, L.A. (1954). Diagonosis and improvement of saline and alkali soils. US Department of Agriculture, Hand Book No. 60.

Sawyer, C.N. and McCarty, P.L. (1967). Chemistry for sanitary engineers. 2nd edn. McGraw-Hill, New York.

Stallard, R.F. and Edmond, J.M. (1983). Geochemistry of Amazon: 2. The influence of geology and weathering environment on the dissolved load. *Jour. Geophys. Res.*, **88:** 9671–9688.

Tijani, M.N. (1994). Hydrochemical assessment of groundwater in Moro area, Kwara State, Nigeria. *Environmental Geology*, **24:** 194–202.

WHO (1997). Guidelines for drinking water quality. Vol. 1, Recommendations. 2nd edn. Geneva.

Wilcox, L.V. (1948). Classification and use of irrigation waters. US Department of Agriculture, Washington DC. Circular 962.

Assessment of Groundwater Vulnerability in the Borazjan Aquifer of Bushehr, South of Iran, Using GIS Technique

Jaber Mozafarizadeh and Zahra Sajadi

Introduction

Ground water is treated as an important source of water due to its large volumes and its low vulnerability to pollution when compared to surface waters (USEPA, 2006). Bushehr is an arid region with low rainfall (100 mm/year) and very high evapotranspiration (3000 mm/year). Like other arid countries, ground water is the main source of water for different purposes. Pollution vulnerability assessment requires in-depth knowledge of the hydrogeological, hydrodynamic and hydrochemical characteristics of aquifers. Several methods and simulation models have been developed since the 1970s. In the last 15 years, the advent of GIS, permitting the inventory, archival, retrieval and display of spatial data and the link to numerical rating systems, has resulted in the widespread use of parametric methods, based on the hydrogeological setting: e.g. DRASTIC (Aller et al., 1987) commonly used in the USA, and SINTACS (Civita and De Maio, 2000), an Italian modification.

The vulnerability of an aquifer, as the sensitivity to surface pollution, is based on the occurrence of natural attenuation process pollution of the soil. The degree of attenuation is defined by hydrogeological characteristics and geological intrinsic of an aquifer system and the medium overlying (Doerfliger et al., 1999; Leitao et al., 2003; Sinan and Razack, 2009). For the danger of pollution, this is defined as a potential source of contamination as a result of anthropogenic activities on the soil surface (Mimi and Assi, 2009). Risk pollution can therefore be estimated by an interaction between the contaminant load and vulnerability to pollution. With the

J. Mozafarizadeh
Bushehr Regional Water Authority, Iran

Z. Sajadi (✉)
Shahid Chamran University, Iran
e-mail: zahrasajadi92@yahoo.com

N.J. Raju et al. (eds.), *Management of Water, Energy and Bio-resources in the Era of Climate Change: Emerging Issues and Challenges*,
DOI 10.1007/978-3-319-05969-3_4

knowledge of the behaviour of water quality and vulnerability factors associated with use, development scenarios can be generated that lead to the protection of the aquifer system.

This study addresses the influence of geological uncertainties in determining the SINTACS index (Civita and De Maio, 2000) and the method used to determine the intrinsic vulnerability of aquifers. It has as objective to apply this methodology to the area of the aquifer in the Borazjan, bounded by the River Basin Helleh, Dalaki and Shapour, Iran.

Hydrogeological Setting of the Study Area

The Borazjan Sampling Area

Borazjan study area is located in Helleh basin (Fig. 1). This area is restricted in longitudes 49° 29′ 84″ up to 53° 00′ 40″ East and latitudes 32° 22′ 10″ up to 32° 45′ 93″ and Genaveh study area in North, Ahram study area in the South and Shahpour and Dalaki watershed in the West. Climatically, this region is located in area with mean annual rainfall of 259.8 mm and mean temperature of 29.9°C and according to Emberger divisions is a desert district with middle warm climate.

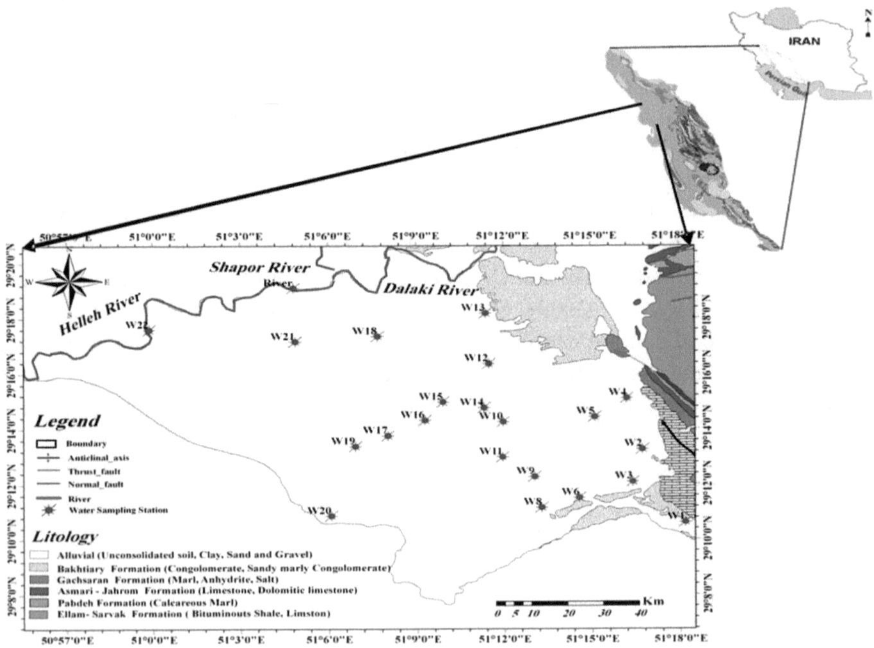

Fig. 1 Location map of the study area and sampling stations

The study area, with its almost flat morphology, covers approximately 418 km^2. The Borazjan Plain is crossed by the upper part of the Helleh, Dalaki and Shapour rivers. It is delimited by the Zagros mountains and by the Gisakan ridge to the east. The Borazjan Plain consists of alluvial deposits (thickness 0-80 m).

The aquifer in the plain can be considered a single continuous body. Water circulation takes place mainly in the permeable layers, with sand and gravel-sized grains. The aquifer is unconfined; locally, pyroclastic or clayey layers generate confined sectors. Borazjan area is located in one of the syncline of folded belt of Zagros mountain. The effect of tectonic forces on this region caused anticlines, tectonization formation and to form thrust faults and Qatar-Kazerun fault which with northern-southern direction passes through eastern part of plain. Qatar-Kazerun fault has divided study area into mountains and plain district; mountains district is located in eastern part of region. Surrounding heights of region in northern and southern areas include hills and high grounds of Bakhtiyari formation and in eastern area include Gisakan anticline with north western-south eastern direction and maximum height of 1785 metres from sea level with river water from Cretaceous to Pliocene. Stratigraphy of this study area has been shown in Fig. 1.

Methodology

The SINTACS procedure incorporates seven parameters relevant for the contaminant attenuation and vertical flow capacity (Table 1). In the vulnerability assessment procedure, the Borazjan plain was discretized with a regular mesh grid of 100×100 m. The grid square cell structure of the SINTACS input data has been designed in order to use several weight strings. The weight strings have been prepared in order to satisfactorily describe the effective hydrogeological and impacting situation as set up by the sum of data. The present release of SINTACS presents five weight strings for normal impact, relevant impact and drainage from the surface network. For each of the 7887 grid squares, element normalized SINTACS index was calculated and differently vulnerable areas were assessed using SINTACS R5 parametric methods (Civita and De Maio, 2000).

Results and Discussion

Soggiacenza as Depth to Water

Data from 22 water wells in the study area were used to prepare the depth to water table layer. The depth to water table was obtained by subtracting the water table level from the elevation of the piezometer wells and averaging over a three-year period (2009-2012). Generally, the depth to water in the centre part of study area is

Table 1 The SINTACS model parameters (Modified from Babiker et al., 2004)

Factor	*Description*	*Relative weight*
Soggiacenza as Depth to Water	Represents the depth from the ground surface to the water table, deeper water table levels imply lesser chance for contamination to occur.	5
Infiltrazione as Net Recharge	Represents the amount of water which penetrates the ground surface and reaches the water table, recharge water represents the vehicle for transporting pollutants.	4
Non Saturo as Impact of the Vadose Zone	Refers to the saturated zone material properties, which controls the pollutant attenuation processes.	3
TipologiaCopertura as Soil Media	Represents the uppermost weathered portion of the unsaturated zone and controls the amount of recharge that can infiltrate downward.	2
Caratteristiche Acquifero as Aquifer Media	Refers to the slope of the land surface, it dictates whether the runoff will remain on the surface to allow contaminant percolation to the saturated zone.	1
Conducibilita Idraulica as Hydraulic Conductivity	Is defined as the unsaturated zone material, it controls the passage and attenuation of the contaminated material to the saturated zone.	5
Acclivita Superficie Topografica as Slope	Indicates the ability of the aquifer to transmit water, hence determines the rate of flow of contaminant material within the groundwater system.	3

high (>30 m), and gradually decreases to boundaries. Even in the southwest of the study area depth to water table decreases to less than two metres. The depth to water layer was classified from 2 (least effect on vulnerability) to 10 (most effect on vulnerability) with regard to SINTACS classification (Fig. 2a).

Infiltrazione as Net Recharge

To prepare net recharge layer Piscopo (2001) method was used. In order to calculate the recharge value, a digital elevation model (DEM) of the study area was generated from the topographic map. The slopes in the study area were then derived from the DEM and classified according to the criteria given in Table 2a. The resulting slope map was converted into a grid coverage taking into consideration that the pixel values in this grid coverage are based on the slope ratings. The soil map was classified into two classes based on the criteria given in Table 2c and was then converted into grid coverage. This process was essential in order to perform arithmetic operations within the GIS. Finally, both grids were added together

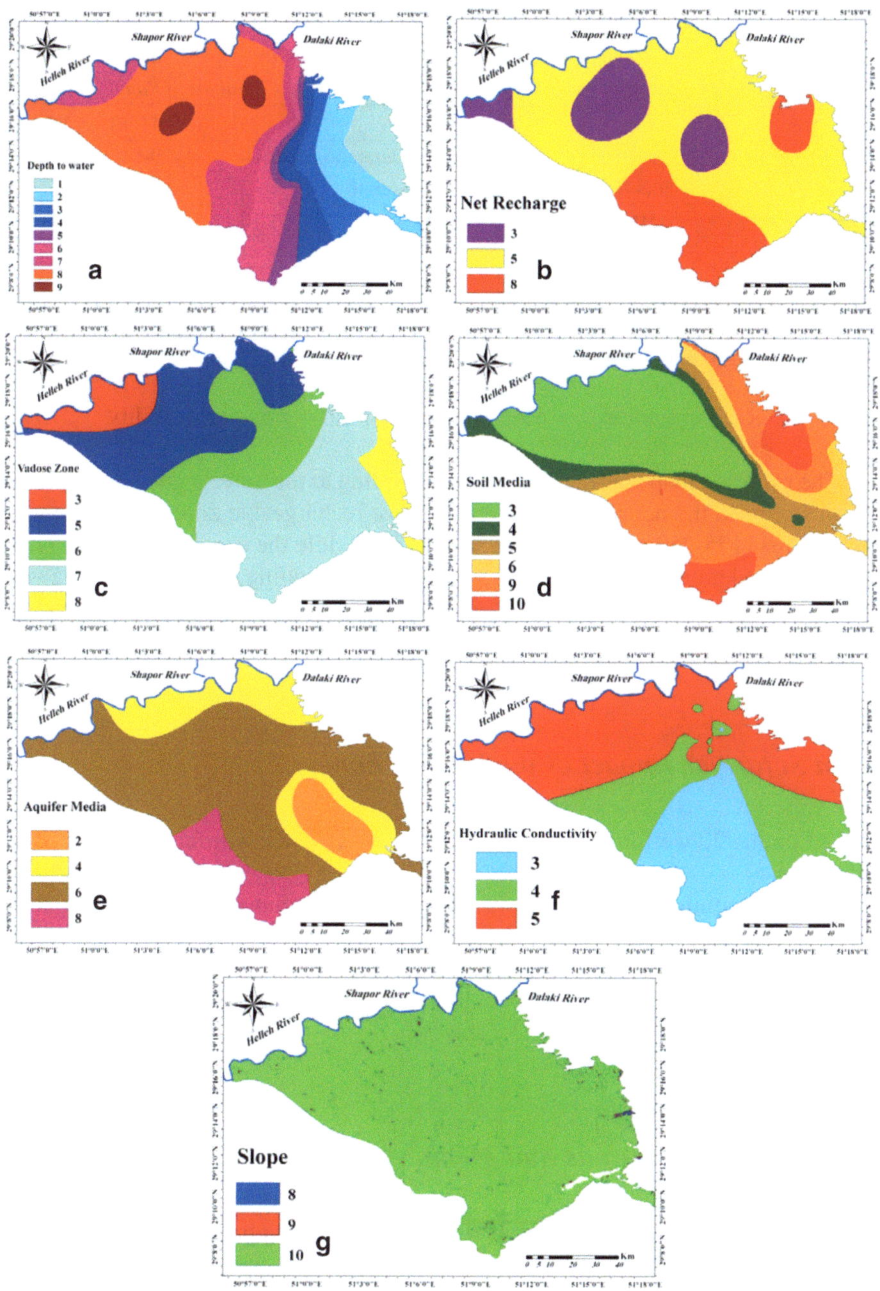

Fig. 2 Parametric maps of SINTACS groundwater vulnerability model for the Borazjan aquifer

Table 2 The recharge ratings and weightings for the study area

a) Slope		*b) Rainfall*		*c) Soil permeability*		*d) Recharge rate*	
Slope %	*Factor*	*Rainfall (mm)*	*Factor*	*Range*	*Factor*	*Range*	*Rating*
<2	4	>850	4	High	5	11-13	10
2-10	3	700-850	3	Moderate to high	4	9-11	8
10-33	2	500-700	2	Moderate	3	7-9	5
>33	1	<500	1	Low	2	5-7	3
				Very low	1	3-5	1

with the rating value of the rainfall, which is equal to 1 in the study area (Table 2b). Recharge index was then calculated as:

$$\text{Recharge Index} = \text{Slope } (\%) + \text{Rainfall} + \text{Soil permeability}$$

The resulting map was then classified according to the criteria given in Table 2d. With regard to the net recharge layer, the most rechargeable zone belongs to sand dunes area (175-250 mm/year, with rating of 8), while the central part of the study area has moderate recharge (100-175 mm/year, with rating of 5) and the area with low recharge (50-100 mm/year, with rating of 3) (Fig. 2b) surround the moderate zone.

Non Saturo as Impact of the Vadose Zone

Based on the 22 well logs available in the study area, the aquifer media layer was prepared.

First, the aquifer media rating was calculated for each well and then using these ratings and well locations, the aquifer media layer was prepared and finally converted to grid coverage. Aquifer media layer exhibits that most parts of the study area have the rating value equal 6 (clay and silt) and some small areas have 3 and 8 rating values (Fig. 2c and Table 3).

Tipologia Copertura as Soil Media

Soil map of the study area (1/50,000) that had been prepared by Environmental Source Office of Bushehr Province was used to prepare the soil media layer. First the hard copy of this map was scanned and was digitized using ENVI software and then the vector file of soil map was prepared in Arc View GIS software. The soil media types were then assigned ratings from 3 to 9 according to their permeability. Coarse soil media have high rates in comparison to fine soil media. The vector layer of soil was converted to a raster grid which has produced the map as presented in

Table 3 The SINTACS index for the study area

Parameter	*Rating*
Soggiacenza as Depth to Water (m)	
7-10	6
10-13	5
13-20	4
<56	1
Infiltrazione as Net Recharge (mm/year)	
175-250	8
100-175	5
50-100	3
Non Saturo as Impact of the Vadose Zone	
Sand and gravel	9
Sand with some clay/silt	7
Clay and silt with some sand/gravel	5
Clay and silt	2
Tipologia Copertura as Soil Media	
Silty Sand	7
Clayey sand	6
Clay with high sand	5
Clay with low sand	4
Clay and silt	3
Confined aquifer or compact clay	1
Caratteristiche Acquifero as Aquifer Media	
Sand dunes	9
Loamy sand to sandy loam	6
Silty-clayey loam to clay loam	3
Conducibilita Idraulica as Hydraulic Conductivity (m/day)	
28-40	6
12-28	4
4-12	2
<4	1
Acclivita Superficie Topografica as Slope	
0-2	10
2-6	9
6-12	5
12-18	3
>18	1

Fig. 2d. According to the soil media layer, sand dunes with high permeability are located in hydrologic group of A, the central part of the study area with texture of loamy sand to sandy loam is located in hydrologic group of B and other parts of the study area with texture of silty-clayey loam are located in hydrologic group of C.

Caratteristiche Acquifero as Aquifer Media

The preparation of the impact of vadose zone layer was exactly the same as the aquifer media layer. Moreover, this layer was prepared based on logs of 22 piezometers (Fig. 2e). According to this layer deposits of unsaturated zone in the north and northeast of the study area are highly discontinuous and have many interbeds of sand and clay. Existence of clay interbeds has high impact on decreasing of vulnerability. Conversely the unsaturated zone in most parts of south and southwest part of the study area have formed from continuous sandy deposits.

Conducibilita Idraulica as Hydraulic Conductivity

Pumping test data of 22 deep wells in the study area were used for preparation of this layer. Transmissivity values were calculated for all wells by using Mace method (Mace, 1996). Then with regard to the thickness of saturated zone, hydraulic conductivity was calculated for all wells and according to SINTACS model classification, rating values were obtained for the total of the study area. Figure 2f shows that most parts of the study area have hydraulic conductivity value between 4 and 12 (m/day) with rating value of 2. In the southern part of the study area there is a small part with hydraulic conductivity of less than 4 (m/day) and its rating value is 1.

Acclivita Superficie Topografica as Slope

In order to prepare the topography layer, a digital elevation model (DEM) of the study area was generated from the topographic map. The slopes in the study area were then derived from the DEM and classified according to the criteria of SINTACS model (Fig. 2g). Figure 2e shows that the slope value in most parts of the study area is less than 2 % (rating value = 10) and only a small part between 2 and 6 % (rating value = 9). Generally slope in the study area is low and therefore increases the groundwater vulnerability.

The SINTACS Vulnerability Index

The GIS coverage were all in raster format and values for each overlay were summed in ArcView GIS according to the pixel value of each area that resulted from multiplying the ratings with its appropriate SINTACS weight (Table 3).

Since the minimum possible SINTACS index for using these parameters is 63 and the maximum is 180, this range was divided into eight classes. These classes were (Civita and De Maio, 2000):

(a) Less than 63 (No risk)
(b) 64–77 (Very low)
(c) 78–99 (Low)
(d) 100-119 (Moderate)
(e) 120-139 (Moderate-High)
(f) 140-159 (High)
(g) 160-179 (Very high)
(h) More than 180 (Completely vulnerable)

The resulting SINTACS values in this application lay between 97 and 171. This range was classified on the basis of the above classification as: (c) 97-110, which was assigned a low vulnerability, (d) 111–126, which was assigned a moderate vulnerability, (e) 126–141, which was assigned a moderate to high vulnerability, (f) 141-156, which was assigned a high vulnerability, and (g) 156-171 which was assigned a complete vulnerability as shown in Table 4. Table 4 shows that 12.4 % of the study area has low vulnerability to contamination, 31 % of the study area has moderate vulnerability groundwater, 38.9 % of the study area has moderate to high vulnerability groundwater, 9.2 % of the study area has high vulnerability to contamination and the remainder of the study area (less than 8.5 %) has a very high vulnerability.

The SINTACS aquifer vulnerability map clearly shows the dominance of "Moderate-High" vulnerability classes in most parts of the study area, particularly in the centre, east and northeast of the study area. The western and southwestern parts are characterized by "moderate" vulnerability (Fig. 3). This pattern is mainly dictated by the variation in impact of vadose zone media from east to west. Also the soil media and depth to water table have high impact, but their importance are less than impact of vadose zone media. The parts at south of the study area display "High risk to pollution". This is due to the combination of deep water table, low permeability of vadose and aquifer media.

Table 4 The SINTACS index for the study area (Civita and De Maio, 2000)

SINTACS index	*SINTACS range*	*Area (km^2)*	*% of the total area*
Low vulnerability	97-110	52	12.4
Moderate vulnerability	111-126	129.9	31
Moderate to high vulnerability	126-141	162.6	38.9
High vulnerability	141-156	38.3	9.2
Very high vulnerability	156-171	35.6	8.5
	Total	418.4	100

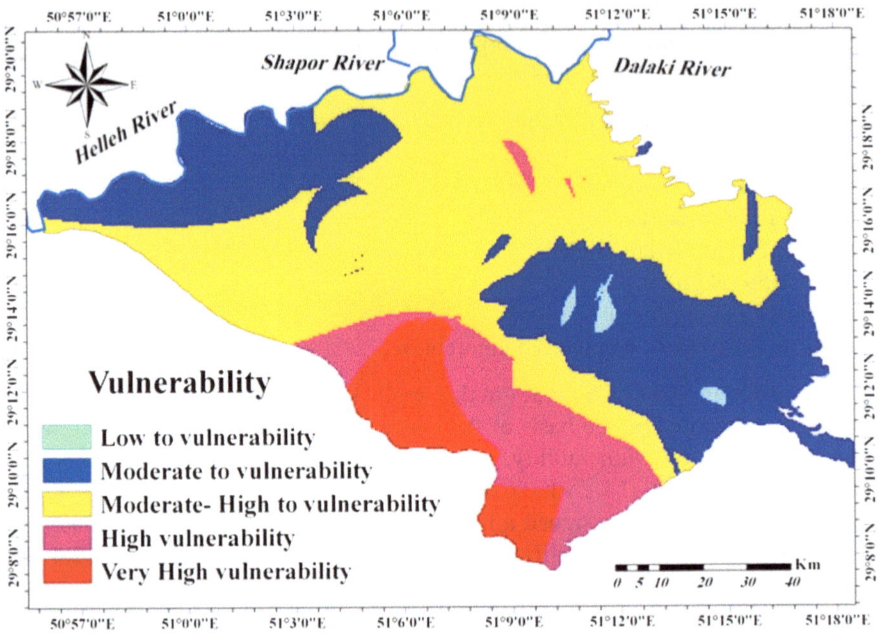

Fig. 3 Distribution of the groundwater vulnerability index in the Borazjan aquifer

SINTACS Model Validation

Twelve water samples were collected from agriculture wells and analyzed to validate the result of SINTACS model. With regard to the fact that the most-used fertilizers at the study area are animal or nitrate mucks, the nitrate anion was analyzed (Table 5). Chemical analysis was performed at laboratory of Zagros Abshenase Fars Company, Iran, using Metrohm 761 apparatus. It is a sophisticated and careful water analysis machine. Wells location and iso-nitrate map are shown in Fig. 4. According to Fig. 4 nitrate concentrations at the south parts of study area that have high vulnerability is more than nitrate concentration at the north, east and west parts that have moderate to high vulnerability; therefore the results of vulnerability assessment using SINTACS model are confirmed.

Correlation coefficient between SINTACS model parameters and nitrate layer was calculated using Arc GIS 9.3 software to identify the most effective parameter on groundwater vulnerability at the study area. The results of this calculation are presented in Table 6. These results show that the most correlation is between Non Saturo as Impact of the vadose zone and Caratteristiche Acquifero as Aquifer media parameters and nitrate layer. This means that impact of vadose zone and Aquifer media parameters have the most impact on groundwater vulnerability to pollution in comparison to other SINTACS parameters in the study area.

Table 5 Concentration of nitrate in groundwater

Well	*NO_3 (mg/l)*	*Well*	*NO_3 (mg/l)*
W_1	1400	W_7	51.7
W_2	5.17	W_8	1.64
W_3	3.29	W_9	2.88
W_4	94.8	W_{10}	4.15
W_5	17.31	W_{11}	49.9
W_6	162.4	W_{12}	22.34

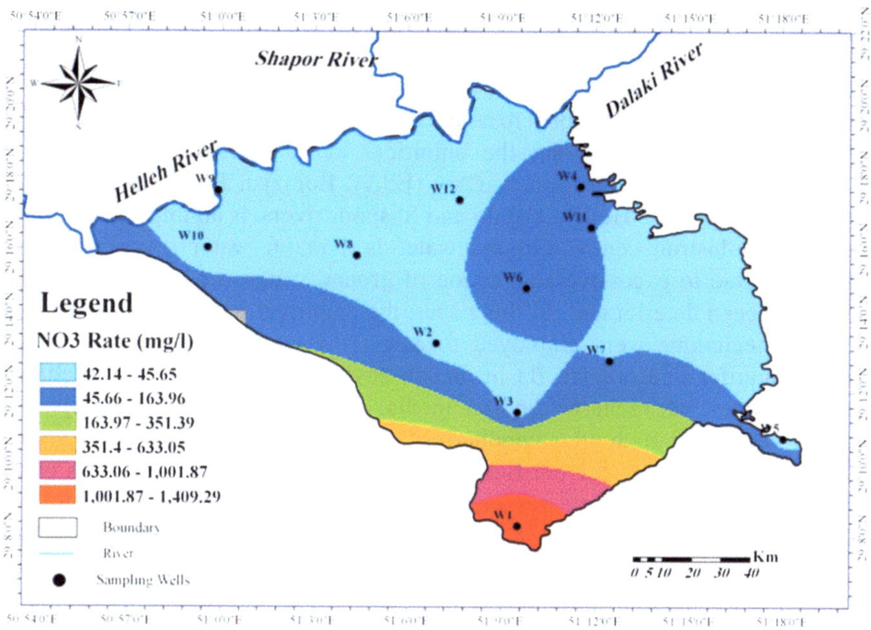

Fig. 4 Wells location and nitrate rate in groundwater at the study area

Table 6 Correlation coefficient between nitrate layer and SINTACS model parameters

Layer	*S*	*I*	*N*	*T*	*A*	*C*	*S*
Nitrate	-0/05259	0/53770	0/30323	0/46755	0/54989	-0/53895	0/022170

The reasons for this high correlation is that the most sediment at south part of study area are sandy and therefore surficial pollutants; especially nitrate ion can easily pass from sand filter and reach groundwater. While there are fine grain sediments silt and clay at the other parts of study area and therefore pollutants cannot easily pass from these fine grain filters and generally are removed from infiltrating water before reaching groundwater.

Aquifer media parameter also has high correlation with nitrate layer that can be explained by the fact that at low water table areas, pollutants have less time to be removed by physical (filtering), chemical (reaction with existing material at vadose zone) and biological (biodegradation) processes and hence reach groundwater rapidly. But at high water table areas, the mentioned processes have more time for removing the pollutants and therefore aquifer is subjected to less pollution risk. Superposition of nitrate layer on depth to water table layer at the study area confirms the results of SINTACS model clearly.

Conclusions

In this paper, we have attempted to assess the aquifer vulnerability of the Borazjan groundwater basin employing the empirical index SINTACS model of the U.S. Environmental Protection Agency (EPA). Borazjan Basin, located in the middle reaches of the Helleh, Dalaki and Shapour rivers, is an important agricultural and industrial centre. Ground water is a major water source for these activities. Due to excessive abstraction of ground water, the quality of ground water has been deteriorated. In this paper, the empirical index SINTACS model and GIS technique were employed to assess the aquifer vulnerability of the shallow aquifer of Borazjan Basin. Seven environmental parameters including depth to water, net recharge, aquifer media, soil media, topography, impact of vadose zone, and hydraulic conductivity were used to represent the natural hydrogeological setting of the Borazjan Basin. According to the results of the groundwater vulnerability assessment, the study area can be divided into five zones. The results show that 8.5 % of Borazjan aquifer area has very high contamination potential, 9.2 % high, 38.9 % moderate to high, 31 % moderate and 12.4 % has very low potential to be polluted. Southern areas of this plain has the higher contamination potential than other parts of it located in the groundwater recharge zones. Correlation analysis between SINTACS parameters and nitrate ion layers showed that nitrate layer has the best correlation with impact of vadose zone parameter followed by Acquifero as Aquifer Media parameter. This shows that most effective parameters on groundwater vulnerability to pollution at the study area are impact of vadose zone and depth to water table. According to the results, if current trends of contamination due to leaching of inorganic fertilizers which applied to agricultural land, domestic sewage and livestock and poultry wastewater are continued, the plain will reach the worst condition of pollution. The results of this study indicate that groundwater vulnerability and risk mapping is an efficient tool to assist with planning and decision-making processes.

References

Aller, L., Bennet, T., Leher, J.H., Petty, R.J. and Hackett. G. (1987). DRASTIC: A standardized system for evaluating ground water pollution potential using hydro-geological settings. EPA 600/2-87-035.

Babiker, I.S., Mohamed, A.A.M., Tetsuya, H. and Kikuo, K. (2004). A GIS-based DRASTIC model for assessing aquifer vulnerability in Kakamigahara Heights, Gifu Prefecture, central Japan. *Science of the Total Environment*, **345(1–3):** 127–140.

Civita, M. (1994). Le carte della vulnerabilit'a degli acquiferi all'inquinamiento: teoria e pratica [Contamination vulnerability mapping of the aquifer: theory and practice]. Quaderni di Tecniche di Protezione Ambientale, Pitagora Editrice.

Civita, M. and Maio, M. de (2000). SINTACS R5-Valutazione e cartografia automatica della vulnerabilità degli acquiferi all'inquinamento con il sistema parametrico. Pitagora Bologna.

Daly, D., Dassargues, A., Drew, D., Dunne, S., Goldscheider, N., Neale, S., Popescu, C. and Zwhalen, F. (2002). Main concepts of the "European Approach" for (karst) groundwater vulnerability assessment and mapping. *Hydrogeol. J.*, **10(2):** 340–345.

Doerfliger, N. and Zwahlen, F. (1998). Groundwater Vulnerability Mapping in Karstic Regions (EPIK) – Application to Groundwater Protection Zones. Swiss Agency for the Environment, Forests and Landscape (SAEFL), Bern.

Leitao, T.E., Ferreira, J.P.L., Oliveira, M.M. and Moinante, M.J. (2003). Groundwater Pollution deáguas: Key Issues, process of prevention and rehabilitation. 6th Symposium on Hydraulics and Water Resources of the Portuguese Speaking Countries, Cape Verde, 10-13 November 2003.

Mace, E.R. (1996). Determination of transmissivity from specific capacity tests in a karst aquifer. *Ground Water*, **35(5):** 733–742.

Mimi, Z.A. and Amjas Assi, A. (2009). Intrinsic vulnerability, hazard and risk mapping for karst aquifers: A case study. *Journal of Hydrology*, **364:** 298–310.

Piscopo, G. (2001). Groundwater vulnerability map, explanatory notes, Castlereagh Catchment, NSW. Department of Land and Water Conservation, Australia.

Sinan, M. and Razack, M. (2009). An extension to the DRASTIC model to assess groundwater vulnerability to pollution: Application to the Haouz aquifer of Marrakech (Morocco). *Environmental Geology*, **57:** 349–363.

U.S. EPA (2006). Hydrogeological assessment (Groundwater quality) guideline. EPA Victoria, Publication 888, ISBN 0730676587, Australia.

Van Stempvoort, D., Ewert, L. and Wassenaar, L. (1992). AVI: A Method for Groundwater Protection Mapping in the Prairie Provinces of Canada. PPWD pilot project, Sept. 1991 - March 1992. Groundwater and Contaminants Project, Environmental Sciences Division, National Hydrology Research Institute.

Geochemical Variations of Groundwater Quality in Coastal and Karstic Aquifers in Jaffna Peninsula, Sri Lanka

K. Gunaalan, H.B. Asanthi, T.P.D. Gamage, M. Thushyanthy, and S. Saravanan

Introduction

Chemical constituents of ground water evolve due to the interaction with aquifer minerals and mixing of different groundwater reservoirs along the flow path in the subsurface. Consequently, geochemical properties of ground water depends on chemistry of water in the recharge area as well as on the different geochemical processes that take place in the subsurface of the locality, including the presence of possible contamination sources. The increasing contamination of ground water by nitrate is primarily from the widespread use of commercial fertilizers which is an evolving public health concern factor in many agricultural regions of the world (Gulis et al., 2002; Raju et al., 2009).

A five-year study on the geographical pathology of malignant tumours in Sri Lanka from 1973–1977 had confirmed that the incidence of cancer is relatively higher in Jaffna district and one of the reasons for the esophagus cancer incidences could be due to NO_3^--N in groundwater (Panabokke, 1984). Later on, Gunaalan et al. (2011) have observed a relative risk factor for esophagus and stomach cancers associated with the consumption of NO_3^--N concentration higher than the recommended level (which is >10 mg/L for WHO standards) at the three areas in Jaffna peninsula. In many countries nitrate levels of ground water have increased significantly due to extreme use of nitrogenous fertilizers (Kross et al., 1992;

K. Gunaalan • H.B. Asanthi (✉) • T.P.D. Gamage
Department of Limnology, Sciences & Technology, University of Ruhuna, Sri Lanka
e-mail: asanthi@fish.ruh.ac.lk

M. Thushyanthy
Department of Agricultural Engineering, University of Jaffna, Jaffna, Sri Lanka

S. Saravanan
Regional office, National Water Supply & Drainage Board, Jaffna, Sri Lanka

N.J. Raju et al. (eds.), *Management of Water, Energy and Bio-resources in the Era of Climate Change: Emerging Issues and Challenges*,
DOI 10.1007/978-3-319-05969-3_5

Raju et al., 2011) and it is becoming major factor for polluting ground water. Therefore, consumption of well water rich in nitrogenous compounds has been considered as a possible risk factor for esophageal cancer (Zhang et al. 1996). Nitrates which could be converted into carcinogenic substances such as nitrosamines within the body are of importance in the carcinogenesis of esophageal and stomach cancers (Dissanayake, 1988).

Inhabitants in Jaffna peninsula mainly depend on groundwater sources because other freshwater sources are not available (Rajasooriyar et al., 1999). This is especially true in the regions of arid to semiarid climate where average annual rainfall is less than 1000 mm and the recharge of aquifer is limited (Udayalaxmi et al., 2010). It has been reported that the Jaffna peninsula has the highest levels of total hardness in ground water in Sri Lanka and there are cumulative effects of the carbonate factors, such as Cl^-, NO_3^--N, and SO_4^{2-} (Rajasooriyar et al., 1999).

The whole Jaffna peninsula is underlined by Miocene limestone formations, which are generally 100 m to 150 m thick and this layer is distinctly bedded and well jointed while highly karstified. The shallow aquifer of the peninsula occurs in the channels and cavities (karsts) of this Miocene limestone. All the shallow groundwater found within the karstic cavities originates from the infiltration of rainfall, and this shallow groundwater forms mounds or lenses floating over the saline water. These water mounds or lenses reach their peak during the monsoon rains which falls in between November and December. However, monitoring studies have confirmed a significant imbalance between the draw-off and recharge rates of groundwater (Balendran, 1969). The variations in the amount of precipitation, the timing of precipitation events, and the form of precipitation are the key factors in determining the amount and timing of recharge to aquifers and ultimately change the groundwater quality.

Approximately 80 percent of ground water is being used for high-value agriculture and the remaining 20 percent is for domestic uses including flushing of toilets in urban areas of Jaffna. Water quality studies have shown that the enhanced levels of nitrate pollution in domestic wells have been situated in more densely settled municipal areas of Jaffna peninsula (Nagarajah et al., 1988). Jaffna Feasibility Report (2006) disclosed that the thin cover (0 to 3 m) over the groundwater table consisting mainly of sandy soils with an infiltration capacity of 50,000 $Lm^{-2}.day^{-1}$ can provide no protection against pollution of the groundwater infiltrate from the surface.

Although the karstic aquifers are providing more favourable conditions for groundwater recharge compared to other aquifer types, karstic systems are more vulnerable to land surface-originated contamination due to fast transport ($>$100 m/day) and limited attenuation of contaminants (Alper and Rahime, 2011). For these reasons, it is important to assess the groundwater quality particularly in karstic aquifer functions as a primary water source in Jaffna peninsula. Therefore the prime objective of this study is to provide a statistical evaluation of groundwater quality in the selected areas in Jaffna peninsula using principle component matrix to select what are the most suitable parameters to explain selected aquifer types and to evaluate the water quality using the Sri Lankan standards for drinking water. This study is important as a step of continuous monitoring programme conducted by National Water Supply and Drainage Board (NWSDB) of Sri Lanka to draw attention towards the most vulnerable wells in Jaffna Peninsula.

Materials and Methods

Study Area

Three divisional secretaries in Jaffna peninsula, Vadamaradchi, Chunnakam and Jaffna, were selected for this study. Chunnakam area can be considered as an intensive agricultural area, Jaffna is a densely populated urban area and Vadamaradchi is considered as no relevant survey was found in literature on groundwater quality variations for the last few years. Vadamaradchi and Jaffna areas are located in close proximity to the coastline other than Chnunnakam area. Data obtained from Meteorological Department of Sri Lanka indicated that Jaffna peninsula has received mean annual precipitation of 1350 mm during the period of 2000 to 2011. The highest monthly average precipitations were observed in December, January and April owing to 532, 112 and 108 mm respectively in Jaffna peninsula. Typically, February and March were dry months during the period of study and the mean monthly rainfalls were lower than 3 mm. The entire drinking water is supplied from nearly 80,000 shallow wells in the Jaffna peninsula, in close proximity to sanitation facilities. Sivarajah (2003) has reported that the underground water is being continuously polluted as a result of human activities in the form of biological and chemical sources of pollutants in Jaffna peninsula. Also, with more and more refugees returning to their homes after thirty years long turbulent period the population is likely to increase rapidly, putting additional pressure on the already vulnerable groundwater resources.

Water Sampling and Analysis

Twenty wells were selected randomly from Chunnakam area (karstic aquifer), Vadamaradchi (coastal aquifer) and Jaffna area (coastal aquifer) and mapped their respective locations using Global Position System (GPS), as given in Fig. 1. The distance between the wells were not less than 100 m and each water sample was collected at 20 cm below the water surface using a depth sampler on monthly basis during the period of December 2010 to April 2011. Characteristics of the wells such as well depth, well diameter and height of the protection wall from the ground level were measured. Chemical parameters such as pH (Jenway CE 370 pH/mV meter, with an accuracy of 0.01 pH/$\pm$0.02 Jenway CE470) and conductivity (Meter with an accuracy of 0.01–0.1 mS/cm) were measured in the field using portable meters.

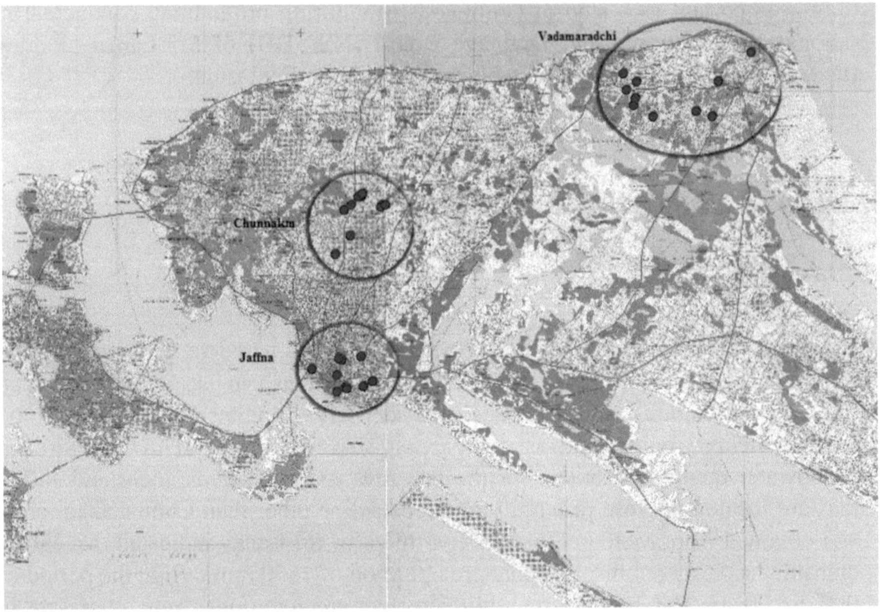

Fig. 1 Locations of wells at the three selected areas in Jaffna peninsula

Groundwater samples were collected into cleaned polyethylene bottles and were transported to the laboratory and filtered using 0.45 μm Millipore filter papers for further analysis. The chemical and biological parameters of NO_3^--N (photometry method using DR/2000 spectrophotometer (at an λ accuracy ±2 nm for 400–700 nm and ±3 nm for 700–900 nm), NO_2^--N (by diazotization method), ammonia (by Nessler's Reagent), phosphate (by PhosVer3 (ascorbic acid) method), sulphate (Turbidity meter 2100P), total hardness (EDTA titration), total alkalinity (acid-base titration) and chloride (Mohr's titration method) were analyzed excluding cations within 48 hrs after collection based on Sri Lankan standard 614 (1983). Since, the rainfall is the major factor of groundwater recharge in the areas the rainfall data were collected from meteorological department, Jaffna.

One-way ANOVA test and multiple comparison tests were used to study the variations of physico-chemical characteristics between the areas and correlations were studied between the parameters and between parameters and rainfall in Jaffna peninsula. Meanwhile, Principal Component Analysis (PCA) was conducted for explaining the large variance of the analytical data into single matrix. The matrices were used to find out the sets of parameters most related with the type of aquifers. Two components were separated based on the presenting variance from the total variance of the data. All statistical analysis was performed in SPSS (version 16.0).

Results

The minimum and maximum values of general characteristics of the selected wells are displayed in Table 1. Mean values of the well depth from ground level to bottom at Vadamaradchi area was significantly different from Chunnakam and Jaffna areas ($p < 0.05$). Relatively lower well depth and higher well depths were observed in Jaffna and Vadamaradchi areas respectively (Table 1).

The minimum and maximum values of physical characteristics of ground water in the three areas are given in Table 2. The pH concentration of well water samples in the three areas was generally alkaline in nature especially at Jaffna area. EC of the water samples has reached the highest value of 9700 μS/cm at Vadamaradchi area due to the salt water intrusion in January just after the high rainfall in December. Also, according to Tukey HSD the EC value in the wells at Chunnakam area of karstic aquifer was significantly different from other two areas ($p < 0.05$). Turbidity of the ground water was significantly different in Jaffna area from other two areas and the highest turbidity was observed at Jaffna area of mostly coastal aquifer. However, turbidity of all the wells did not exceed the maximum permissible level (SLS 614, 1983).

The minimum and maximum concentrations of chemical parameters in ground water at the three areas are given in Table 3. The maximum value of alkalinity,

Table 1 The minimum and maximum values of some important properties of wells located in the three study sites

Important properties of wells	Chunnakam	Vadamarachchi	Jaffna
Height of the protection wall from the ground level (m)	0.25–0.75	0.1–0.8	0.1–0.8
Well depth from ground level to bottom (m)	2.4–12.5	2.8–15	2.1–5.7
Depth from ground level to water level (m)	1.8–10.7	1.0–13	1.3–4.2
Water level (m)	0.6–2.2	0.3–3.5	0.6–2.8

Table 2 Minimum and maximum values of physical characteristics in ground water at the three areas (min-max)

Area	pH	EC (μS/cm)	Turbidity (NTU)
Chunnakam	6.84–7.74[a]	589–2880[a]	0.24–2.55[a]
Vadamaradchi	6.62–7.88[a]	260–9700[b]	0.12–3.12[a]
Jaffna	7.04–8.20[b]	669–6930[b]	0.19–3.86[b]
Drinking water quality standards (SLS 614, 1983)			
Highest desirable level	7.0–8.5	750	2
Maximum permissible level	6.5–9.0	3500	8

Note: a and b indicate the significant difference along rows

Table 3 The minimum and maximum values of chemical parameters recorded in ground water at the three areas

Area	Total alkalinity (mg/L)	Total hardness (mg/L)	Chloride (mg/L)	NO_3^{-}-N (mg/L)	NO_2^{-}-N (mg/L)	PO_4^{3-} (mg/L)	SO_4^{2-} (mg/L)
Chunnakam	136–880[b]	80–414[b]	46–946[b]	0.0–19.0[b]	ND–0.114[b]	0.00–2.56[ab]	42–347[b]
Vadamaradchi	80–504[b]	72–862[a]	51–1725[a]	0.0–10.1[a]	ND–0.019[a]	0.00–1.72[a]	4–500[a]
Jaffna	320–776[a]	96–691[c]	70–1527[a]	0.0–11.3[a]	ND–0.330[b]	0.00–2.38[b]	48–532[c]
Drinking water quality standards (SLS 614, 1983)							
Highest desirable level	200	250	200	–	–		200
Maximum permissible level	400	600	1200	10	0.01	2	400

Note: a and b indicate significant difference of the detected parameter at the three locations while ND is for not detected values

NO_3^--N and PO_4^{3-} were observed at Chunnakam area and the maximum value of total hardness and chloride were observed at Vadamaradchi area. Similarly, the maximum value of NO_2^--N and SO_4^{2-} were observed at Jaffna area during the study period. The range of minimum and maximum concentrations of chloride and SO_4^{2-} have been recorded as 0–355 mg/l and 1–405 mg/l respectively in ground water of Nalgonda district, Andhra Pradesh, India and these concentrations were comparatively lower than the observed values in present study (Rajesh et al., 2011).

The both ions of chloride and SO_4^{2-} were relatively higher at Vadamaradchi and Jaffna areas than at Chunnakam area due to salt water intrusion of those coastal aquifers. Total alkalinity, NO_3^--N and NO_2^--N in some wells at the three areas exceeded the maximum permissible levels of SLS 614 (1983). Also, total hardness, chloride and SO_4^{2-} in some wells exceeded the maximum permissible level of SLS 614 (1983) only at Vadamaradchi and Jaffna areas of coastal aquifers. Table 3 explains the significant difference of each chemical parameter in the three studied areas according to Tukeys HSD ($p < 0.05$). Total hardness and SO_4^{2-} concentrations were the only parameters varied significantly among the three areas. However, total hardness, total alkalinity and SO_4^{2-} concentrations varied significantly in the coastal aquifer of Jaffna area from other two areas. The intensive agricultural area of Chunnakam area is significant due to chloride and NO_3^--N concentrations in ground water during the study period.

Three components were extracted from the total variance of all data and the component 1 explains 44.34 %, 38.49 %, and 50.54 % of the total variance in Jaffna, Chunnakam and Vadamaradchi areas respectively. The component 2 explains 20.15 %, 24.42 % and 18.24 % of the total variance in Jaffna, Chunnakam and Vadamaradchi areas respectively. Therefore, Table 4 gives the component matrix for all parameters in the three studied areas under the above two components. According to the component matrix the contribution were higher than 0.8 for EC, SO_4^{2-}, total alkalinity and chloride concentrations and those can be considered as major parameters for explaining the water quality in karstic and coastal aquifers in Jaffna peninsula. Meantime, the component matrix shows the contribution of 0.963 for total hardness and it is also an important parameter for the water quality in coastal aquifer of Vadamaradchi area. Similarly, NO_3^--N concentration is an important parameter especially at Chunnakam and Vadamaradchi areas presenting the contribution higher than 0.8 in the matrix due to the effects of agricultural runoff on ground water in the areas.

Some of the authors have explained the relationships between the groundwater level and physico-chemical characteristics in the water (Rajesh et al., 2011). However, there were no significant correlations between the general characteristics of the wells (Table 1) and physico-chemical characteristics in water (Tables 2 and 3) at the studied areas.

Table 4 Component matrix for the water quality parameters in the studied areas

Parameter	Chunnakam area		Vadamardchi area		Jaffna area	
	Component 1	Component 2	Component 1	Component 2	Component 1	Component 2
pH	0.321	−0.482	−0.518	0.346	−0.016	0.935
EC	0.922	0.277	0.926	−0.180	0.965	−0.121
Turbidity	−0.016	0.477	−0.616	−0.179	0.535	0.475
Sulphate	0.912	0.233	0.961	0.040	0.889	−0.019
Alkalinity	0.904	−0.312	0.861	−0.075	0.822	0.472
Hardness	0.434	0.674	0.963	−0.044	0.585	−0.708
Chloride	0.934	0.157	0.890	−0.212	0.936	−0.251
NO_3-N	−0.370	0.825	0.221	0.843	0.315	0.188
NO_2-N	−0.231	0.581	0.148	0.627	0.658	0.141
PO_4^{-3}	0.042	−0.507	0.308	0.695	0.025	0.256

Discussion

The depth range of the dug wells has been recorded from 1.45 m to 20 m below the ground level in Andhra Pradesh, India (Rajesh et al., 2011). Since the maximum depth did not exceed 15 m in the present study the groundwater level fluctuation was varied within a very narrow range and the maximum groundwater level was 3.5 m at Vadamaradchi area. Narrow range of variation of groundwater level during the study period explains no significant correlation between important physical properties of the wells and physico-chemical characteristics in water.

Rajesh et al. (2011) has also recorded the range of EC value of ground water from 375 to 2500 μS/cm in Andhra Pradesh, India and another study in Madras aquifer, India has reported EC value of groundwater from 400 to 4200 μS/cm (Elango and Manickam, 1987). However, EC value in some wells of Vadamaradchi and Jaffna areas of coastal aquifers exceeded the maximum permissible level of 3500 μS/cm (SLS 614, 1983). Ramkumar et al. (2010) have noticed that the overexploitation of coastal aquifers for agricultural and drinking purposes, along with structural interventions and climatic conditions increases the possibility of seawater intrusion into the groundwater reservoirs. In fact, the maximum values of total hardness, chloride and SO_4^{2-} at Vadamaradchi and Jaffna areas of coastal aquifers were comparatively higher than the maximum values that have been observed in ground water of a karstic aquifer system near Izmir, Turkey (Alper and Rahime, 2011). In areas of limestone geology (Karst topography), sinkholes provide a direct link between the surface-applied nitrogen fertilizers and subsurface water. Thus, areas with a shallow water table or sinkholes are more vulnerable to nitrate contamination. Since, Chunnakam aquifer is a shallow limestone aquifer; it could be one of the reasons for high nitrate-N in the ground water.

Rainfall is one of the important climatic variables responsible for qualitative and quantitative changes of ground water. After the rainfall, 10–15 % of rain water runs off and about 40–48 % is lost by evaporation, only 30–32 % of rainfall is left over for groundwater recharge in Jaffna area (Navaratnarajah, 1994). The greater variability in rainfall could mean more frequent and prolonged periods of high or low groundwater levels and saline intrusion in coastal aquifers due to sea level rise and resource reduction in Jaffna peninsula that have been observed by Navaratnarajah (1994) and Rajasooriyar et al. (2002). However, no significant correlations were observed between physico-chemical characteristics and rainfall in the studied areas except for PO_4^{3-} concentration ($p < 0.05$) and those physico-chemical characteristics did not change significantly with the water level in the wells.

Conclusion

Since, there were no significant temporal variations of physico-chemical characteristics during the study period a long-term study on dependence of climatic variables on groundwater quality is recommended especially at water scarcity areas as Jaffna peninsula. Also, further studies should be focused on chemical nature of ground water specially with respect to EC, SO_4^{2-}, total alkalinity and chloride concentrations because those parameters were identified in the principal component matrix as the most specific water quality parameters in the studied areas of karstic and coastal aquifers in Jaffna peninsula.

Acknowledgement The authors wish to thank Research and Development section of National Water Supply & Drainage Board (NWS&DB) for providing financial assistance to conduct this research and for providing laboratory facilities in the regional office of NWS&DB in Jaffna.

References

Alper, E. and Rahime, P. (2010). Assessment of the statistical significance of seasonal groundwater quality change in a karstic aquifer system near Izmir-Turkey. *Environment Monitoring Assessment*, **172:** 445-462.

Balendran, V.S., Sirimanne, C.H.I. and Arumugam, S. (1968). Groundwater resources of Jaffna Peninsula. Water Resource Board, Colombo.

Balendran, V.S. (1969). Salt water fresh water interface studies in the Jaffna Peninsula. Report III. Geological survey department, Colombo, Sri Lanka.

Dissanayake, C.B. (1988). Nitrate in the groundwater in Sri Lanka – Implication for community health. *Journal of the Geological Society of Sri Lanka*, **1:** 80-84.

Elango, L. and Manickam, S. (1987). Hydrogeochemistry of the Madras aquifer, India: Spatial and temporal variation in chemical quality of groundwater, The role of Geology in Urban development. *Geological Society of Hong Kong Bulletin*, **3:** 525-534.

Feasibility report (2006). A feasibility study for water supply and sanitation system for the Jaffna Peninsula. ADP project. Project No: 5068020. National Water Supply and Drainage Board, Jaffna, Sri Lanka.

Gulis, G., Czompolyova, M. and Cerhan, J.R. (2002). An ecologic study of nitrate in municipal drinking water and cancer incidence in Trnava District, Slovakia. *Environ Res*, **88:** 182-187.

Gunaalan, K., Asanthi, H.B., Gamage, T.P.D., Thyshyanthy, M., Saravanan S. and Jayakumaran, N. (2011). Incidence of oesophageal and stomach cancer related to Nitrate-N concentration in groundwater in Jaffna peninsula. *In:* The water resources research in Sri Lanka. Dayawansa N.D.K. and Ranjith Premalal De Silva (eds). Symposium proceedings of the Water Professionals' Day, 95-103.

Kross, B.C., Hallberg, G.R., Bruner, D.R., Cherryholmes, K. and Johnson, J.K. (1993). The nitrate contamination of private well water in Iowa. *Am J Public Health*, **83:** 270-272.

Nagarajah, S., Emerson, B.N., Abeykoon, V. and Yogalingam, S. (1988). Water quality of some wells in Jaffna and Kilinochchi with special reference to nitrate pollution. *Tropical Agriculture*, **44:** 61-73.

Navaratnarajah, V. (1994). Water problems in the Jaffna peninsula. 20th WEDC International Conference in Affordable water supply and sanitation. 1994, Colombo, Sri Lanka.

Panabokke, R.G. (1984). The geographical pathology of malignant tumors in Sri Lanka—A 5-year study. *Ceylon Medical Journal*, **29**.

Rajasooriyar, L., Mathavan, V., Dharmagunawardhane, H.A. and Nandakumar, V. (2002). Groundwater quality in the Valigamam region of the Jaffna Peninsula, Sri Lanka. *In:* Sustainable groundwater management. Hiscock, K.M., Rivett, M.O. and Davison, R.M. (eds). Geological Society, London, **193:** 181-197.

Rajesh, R., Brindha, K., Murugan, R. and Elango, L. (2011). Influence of hydrogeochemical processes on temporal changes in groundwater quality in a part of Nalgonda district, Andhra Pradesh, India. *Environment Earth Science*, DOI 10.1007/s12665-011-1368-2.

Raju, N.J., Ram, P. and Dey, S. (2009). Groundwater Quality in the Lower Varuna River Basin, Varanasi District, Uttar Pradesh. *Jour. Geol. Soc. India*, **73:** 178-192.

Raju, N.J., Shukla, U.K. and Ram, P. (2011). Hydrochemistry for the assessment of groundwater quality in Varanasi: A fast urbanizing center in Uttar Pradesh, India. *Environ. Mont. Assess.*, **173:** 279-300.

Ramkumar, T., Venkatramanan, S., Anitha Mary, I., Tamilselvi, M. and Ramesh, G. (2010). Hydrogeochemical Quality of Groundwater in Vedaraniyam Town, Tamil Nadu, India. *Research Journal of Environmental and Earth Sciences*, **2(1):** 44-48.

Sivarajah, N. (2003). Health related problems of water pollution in Jaffna. International workshop on environmental management in north-east Sri Lanka. 1st- 4th December, 2003. Jaffna, Sri Lanka.

Sri Lanka standard 614: Part 1, 1983. Specification for potable water physical and chemical requirements.

Udayalaxmi, G., Himabindu, D. and Ramadass, G. (2010). Geochemical evaluation of groundwater quality in selected areas of Hyderabad, A.P., India. *Indian journal of Science and Technology*, **3(5):** 546-553.

Zhang, W.L., Tian, Z.X., Zhang, N. and Li, X.Q. (1996). Nitrate pollution of groundwater in northern China. *Agric. Ecosyst. Environ.*, **59:** 223-231.

Water in Ancient Indian Perspective and Ponds of Varanasi as Water Harvesting Structures

K.N. Prudhvi Raju and Diva Bhatt

Introduction

Rivers and river-valleys had been centres of development of civilisation. The importance of water as a precious resource had been realised by all civilisations to sustain life on earth. In ancient texts of India, it has been said, "*where there is water, there is food; where there is food, there is life*". What was once a ubiquitous resource has become scarce and even in places where water is everywhere, it had become scarcer because of quality deterioration. The causes for the present state of affairs are: population increase, industrialisation, urbanisation, poor governance and management and peoples' apathy and ignorance.

The importance and value of water had been realised very well in ancient India. The ancient seers expressed on the importance and value of water thus: *achamanam*—sipping of water thrice, *aapo marjanam*—sprinkling of water over the body, a symbolic expression of immersing oneself to be one with the universe, *punar achamanam*—sipping of water once again, *pratah prasanam*—sipping of water in the morning, *punar marjanam*—repetition of '*marjanam*', *arghya pradanam*—offering of water to the Sun God, *deva tarpanam*–offering of water to gods etc., are different steps in daily prayer—*sandhyavandanam*—morning, afternoon and evening prayer—which is intimately linked with water. At every step in this day to day prayer, water is used as the most precious gift that one can give to gods. This is the type of relationship that ancient Indians developed with

K.N.P. Raju (✉)
Department of Geography, Banaras Hindu University, Varanasi 221005, India
e-mail: knpraju1954@gmail.com

D. Bhatt
Institute of Environment and Sustainable Development, Banaras Hindu University, Varanasi 221005, India

N.J. Raju et al. (eds.), *Management of Water, Energy and Bio-resources in the Era of Climate Change: Emerging Issues and Challenges*,
DOI 10.1007/978-3-319-05969-3_6

water and this relationship and veneration for water are responsible for keeping water sources full and pure. It is because of the importance of water as life giving and life-saving and life sustaining element, it is venerated by Indians as the most precious liquid and hence have evolved a code of conduct near water sources especially near the river Ganga. So the problems the present population is facing with water are connected with ignorance of people about their own *sastras*—texts, about the roots of their own gods and goddesses, value of water and callous attitude towards long established traditions and culture.

The present study is an attempt to map the ponds of Varanasi and the sites of garbage collection and disposal sites and to give suggestions as to what is to be done to harvest the water and how to meet the future water needs of the people of Varanasi city in a scenario of climatic and environmental change.

Results and Discussion

The Case of Varanasi City

Varanasi City, geographically located between 82° 57′ to 83° 01′ E Longitude and 25° 15′ to 25° 22′ N Latitude, has an aerial extent of about 1550 sq km (Fig. 1). It was known as a city of ponds (*kunds*) and streams (*nalas*). The ponds, streams and a number of wells served the needs of the people of Varanasi till protected water supply system came into existence in 1892 to cater to about two lakhs of population. The water supply system was expanded to cater to 4.60 lakhs of population, latest in the year 1954. Now, the population of the city is about 14.34 lakhs. With people readily going for tap water, the ponds and streams went into disuse and abuse. The waters of the Ganga too turned bad, unfit for direct consumption. With supply systems deficient and failing, most of the people (about 80 %) depend on ground water to meet their needs. So, naturally the ground water is a stressed resource and in every summer water table goes down. With water supply not very much dependable, situation can turn towards the worst with the expected climate and environmental changes.

During the years between 1954 till date, the water supply sources are augmented by punching 80 minor and 141 major (as per Varanasi Nagar Nigam records) tube-wells (Fig. 2). Currently the water is supplied at a rate 165 LD^{-1} per capita whereas the WHO norm is 270 LD^{-1} per capita (Mohan et al., 2011). This figure of 165 LD^{-1} per capita is on paper; actually, in reality the water is supplied at a much lower rate. These tube-wells run on power and with power supply playing truant for most of the time, water supply through these tube-wells is not dependable. With insufficient public water supply, people are forced to go for their own shallow to deep tube wells. There are innumerable private hand-pumps and shallow to deep tube-wells. Most of these groundwater sources go dry in summer because of lowering of water table. So, with tremendous quantities of withdrawal of ground water (about

Fig. 1 Location of Varanasi in India

180 MLD by public tube wells) and with concrete covered urban surface being impervious and with ponds shrunk in size and dwindled in numbers the recharge rate is very poor and hence most of the bores go dry in summer because of falling water table. With climate and environmental changes looming large, the problem of water supply gets accentuated.

Geomorphology of Varanasi vis-a-vis Ponds and Streams

Varanasi is situated on the left bank of the river Ganga over a prominent natural levee built up along the concave bank of a meander loop. The natural levee though is a result of deposition by flood waters, it remains high and dry above the flood

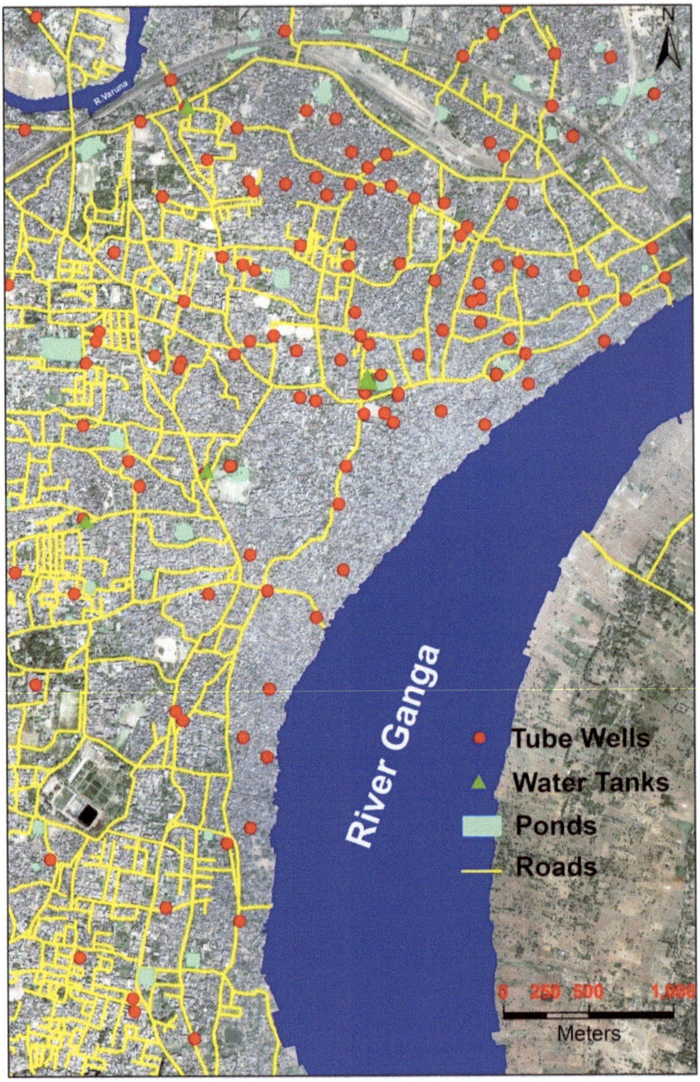

Fig. 2 Ponds, tube wells and garbage collection points in a part of Varanasi city as in 2011 with remote sensing (Google Earth) data in the background

basin and offers a favourable site (higher ground) for the establishment of human habitations. Moreover, a natural levee built-up by coarse deposits is a place of good groundwater source. So with two important conditions for the development of human habitations fulfilled, Varanasi has developed on this natural levee. When there is a prominent natural levee sloping gently away from the river, there is always a flood basin beyond with many streams and flow lines of water moving down from the natural levee into the flood basin. In a similar way, there can be many small streams flowing down the levee into the river. Quite often, the streams flowing into the river from the steeper slope of the natural levee can capture streams flowing into the flood basin resulting in long streams cutting across the natural levee. It is for these natural causes there are many natural ponds, streams and flow-lines (*nalas*) in Varanasi. This is exactly the reason why Varanasi is called the "Blessed Land" for it offered a higher ground over the normal flood waters and many natural ponds of water. This natural levee offering good aquifer deposits, a number of wells were dug up as water sources at a later stage. Many of the ponds and wells in Varanasi are connected with the religious places. They are connected with the religious places with a clear intention of keeping them clean and pure. Most of the ponds and *kunds* in Varanasi used to have natural inlets and outlets to receive waters from the surrounding catchment. These ponds together with wells served wonderfully well the needs of the people and pilgrims of and to Varanasi.

With the starting of water supply system in 1892, gradually, these ponds and wells went into disuse and misuse and were spoiled and filled in due course of time. The number of ponds has dwindled, their surroundings are occupied and some of the ponds are completely filled. According to Ajai and Mohan (2011) there are 118 ponds and tanks now in the city, out of which 39 ponds and tanks are not in the revenue records of local government and 54 ponds are in the final stage of disappearing because of garbage filling and encroachment. It is unfortunate that Varanasi Development Authority (VDA) and Varanasi Municipal Corporation (VMC/Nagar Nigam) have filled some of the ponds to either bring up residential buildings or to use them as garbage dump-yards (Aagrah, 2011). Figure 3 shows the ponds, *kunds* and wells of Varanasi as recorded and mapped by James Prinsep in 1822 (Kejariwal, 2009) and Fig. 2 shows the ponds of Varanasi mapped on high resolution remote sensing data of 2011. A comparison would reveal that the number of ponds has come down from 94 ponds in 1882 to just 49 ponds in 2011 (here comparison between Varanasi of 1882 and Varanasi of 2011 is made for the same extent of area as in 1822 map though Varanasi is spread far and wide towards north, south and west of 1882 extent) and particularly the catchment areas around ponds have been built-up into a dense urban landscape (Fig. 4). Of course, ninety nine per cent of old wells are filled and covered up and the land are occupied for some purpose or the other. In place of 191 open dug wells present in 1882, there are now 125 shallow to deep public tube wells (for the part of Varanasi shown in Figs. 2 and 3) to supply water. Depth of general water level in this area of the Ganga Basin varies between 5 and 29 m bgl (Raju, 2012). But unfortunately the ground water up to these levels has become unfit for consumption and it is now generally extracted from depths ranging from 60 m to 120 m and beyond. Most of the old city of Varanasi is situated over a natural levee composed in depth of coarse sandy soils (Prudhvi Raju and Pandey, 2013) where percolation is somewhat quicker and the leeward side of the natural levee slopes gradually into the flood basin beyond, over which a greater extent of modern Varanasi city has developed, is made up of alternating sand and clay layers creating a multi-layered aquifer system (Shukla and Raju, 2008).

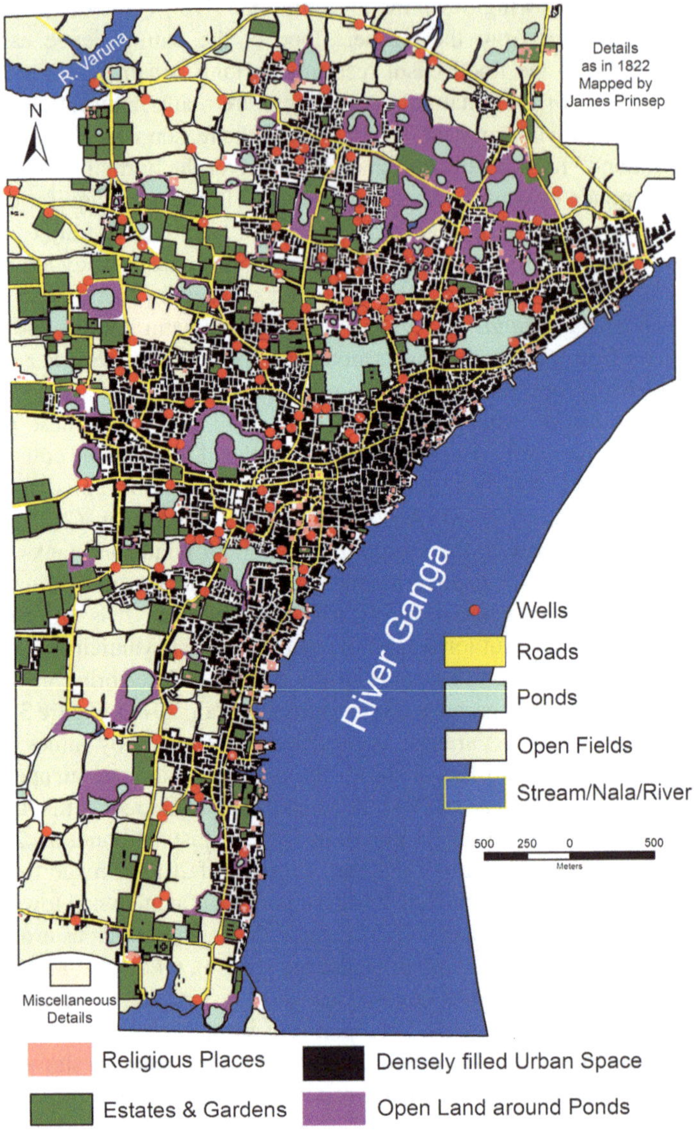

Fig. 3 A map showing (same extent of area as in Fig. 2) various details in Varanasi city as in year 1822 (mapped by James Prinsep)

Conclusion and Suggestion

The ponds of Varanasi offer an excellent opportunity to harvest rainwater for general use and to recharge ground water. But, as for using the existing ponds for water harvesting, there is a big impediment in the form of poor sewage and garbage

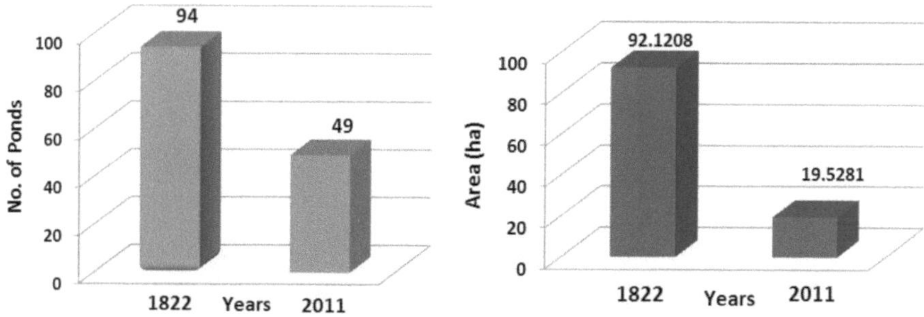

Fig. 4 Number and cumulative area of ponds in Varanasi city in years 1822 and 2011

management. So, unless the sewage and garbage is taken care of and unless the ponds are renovated after de-silting and with protection walls, water harvesting will not yield good results.

Since a large chunk of population of Varanasi is dependent on ground water, one solution to water problem is to allow for good recharge of ground water. The ponds of Varanasi offer an excellent opportunity to harvest the rainwater to recharge ground water. Many of these ponds are being rampantly occupied by encroachers and a few ponds linked to temples are not able to receive any discharge into them because of occupation of catchment. If the local government has a will, it can bring back the areas around most of the ponds by clearing the encroachments. But, there is a problem of garbage and sewage to harvest rain (storm) water into these ponds. Roof-top harvesting can be implemented later on. Figure 2 gives the original extent of ponds as in 1822. There is no proper scientific way of collection and disposal of garbage and sewage in the Varanasi city.

Garbage is disposed into many of these ponds (Figs. 5 and 6) and the stipulation of locating garbage disposal sites beyond 2 km of distance from the existing water sources—river and ponds—is given a go by and garbage is being dumped very near to surface water sources (Fig. 3). Garbage and sewage is grossly mismanaged inside the city. Varanasi produces 600 MT of solid waste per day and out of which about 150 MT remains uncollected every day (City Development Plan for Varanasi, JNNURM, 2006). The uncollected 150 MT (in total) per day is found dumped near, around and within the ponds. It is for these reasons of sewage leakage and decomposition of garbage, nitrate concentration in ground waters is found to be between 66-199 mg/l (Raju et al., 2009). Unless the stipulations and guidelines of locating garbage disposal are strictly followed, water harvesting can't be successful. But such programmes and projects cannot be implemented with success sans the co-operation and participation of the people.

Another major impediment for rain (storm) water harvesting is city's sewage. Varanasi has a very poor sewage disposal system and sewage is directly and quite often indirectly released into the various existing surface water systems of the city.

The city of Varanasi was well equipped with many water sources (streams, ponds and wells) like any other settlements in the past. These ponds and wells made

Fig. 5 Part of pond in Varanasi being filled with garbage and sewage

Fig. 6 Close-up of the same pond showing some details of the dumped garbage

it sure that the city never felt shortage of water. But with protected water supply system coming in place, ponds and wells went into disuse and misuse. With the result groundwater levels have also gone down because of high withdrawal of ground water and for want of recharge (from the ponds). As water supply system cannot be augmented at short intervals, now a time has come to look towards supplementing the water supply and recharging groundwater by revival of ponds as water harvesting structures. It can be done successfully if the ponds are reclaimed and revived after taking care of proper garbage and sewage disposal. It is time we should own the disowned infrastructures of the past. Revival of the ponds is the only refuge to stave off the looming water crisis for not only the holy city of Varanasi but also for other such settlements in India where ponds went into disuse.

Acknowledgements The authors wish to thank Ms. Shruti and Ms. Chanchal for helping in the preparation of maps. This work is undertaken as a part of UGC SAP DRS Programme of the Department of Geography.

References

Abhilash (2010). The meaning of the word Bhagawan, http://www.hindu-blog.com/2010/06/bhagavan-meaning-of-term-bhagawan.html. Accessed 2 April 2013.

City Development Plan for Varanasi, Jawaharlal Nehru National Urban Renewal Mission (JNNURM) Municipal Corporation, Varanasi Final Report (2006). Varanasi, India http://gangapedia.iitk.ac.in/sites/default/files/CDP_Varanasi.pdf. Accessed on 4 December 2012.

Aagrah (2011). Some important sacred ponds and places in Kashi. *Editorial*, **1(8):** 12.

Kejariwal, O.P. (2009). Benaras Illustrated, James Prinsep with James Prinsep and Benares. Pilgrims Publishing, Varanasi.

Mohan, K., Srivastav, A. and Rai, P. (2011). Ground Water in the City Varanasi, India: Present Status and Prospects. *Quaestiones Geographcae*, **30(3):** 47-60.

Prudhvi Raju, K.N. and Pandey, M.K. (2013). Varanasi: Origin and Growth from a Geomorphic Perspective. *In:* Jayaswal, V. (ed.), Varanasi: Myths and Scientific Studies. Aryan Books International, New Delhi.

Raju, N.J., Ram, P. and Dey, S. (2009). Groundwater Quality in the Lower Varuna River Basin, Varanasi District, Uttar Pradesh. *Journal Geological Society of India*, **73:** 178-192.

Raju, N.J. (2012). Evaluation of Hydrogeochemical Processes in the Pleistocene Aquifers of Middle Ganga Plain, Uttar Pradesh, India. *Jour. Envi. Earth Sci.*, **65:** 1291-1308.

Shukla, U.K. and Raju, N.J. (2008). Migration of the Ganga river and its Implication on Hydro-geological Potential of Varanasi area, U.P., India. *Jour. Earth System Sci.*, **117:** 489-449.

Singh, R. (2012). www.manmaujisansthan.info Issues 5A and 5B. Accessed 2 April 2013.

Srivastava, A. and Mohan, K. (2011). Ponds and Water Bodies in the City of Varanasi: State of the Art. *In:* Kumra, V.K. et al. (eds), Strategic Management of Environment for Sustainable Development. Shree Publishers and Distributors, New Delhi.

Glacier Mass Balance and Its Significance on the Water Resource Management in the Western Himalayas

Shruti Dutta, AL. Ramanathan, Anurag Linda, Jose George Pottakkal, Virendra Bahadur Singh, and Thupstan Angchuk

Introduction

The global climate system is a consequence of and a link between the atmosphere, oceans, ice sheets (cryosphere), living organisms (biosphere), and soils, sediments and rocks (geosphere). The role of glaciers as sensitive climatic indicators and natural buffers of hydrological cycle is well established. In this era of much talked about climate change and global warming, glaciers have been correctly recognized as a thermometer and a crucial freshwater resource. Glaciers release meltwater during summers and early autumn, and, therefore, act as a water resource downstream by recharging river fed aquifer and influencing runoff (Bolch et al., 2012).

The Himalaya and associated mountains of high Asia, together with the immensely significant Tibetan Plateau constitute a major hydrographic part of

S. Dutta
Glacier Research Group, School of Environmental Sciences, Jawaharlal Nehru University, New Delhi 110067, India

Department of Environmental Sciences, Amity School of Applied Sciences, Amity University, Haryana, India

AL. Ramanathan (✉) • J.G. Pottakkal • V.B. Singh • T. Angchuk
Glacier Research Group, School of Environmental Sciences, Jawaharlal Nehru University, New Delhi 110067, India
e-mail: alrjnu@gmail.com

A. Linda
Glacier Research Group, School of Environmental Sciences, Jawaharlal Nehru University, New Delhi 110067, India

School of Earth and Environmental Sciences, Central University of Himachal Pradesh, Dharamshala, Himachal Pradesh, India

N.J. Raju et al. (eds.), *Management of Water, Energy and Bio-resources in the Era of Climate Change: Emerging Issues and Challenges*,
DOI 10.1007/978-3-319-05969-3_7

planetary climatic control, as well as a huge catchment of snow and ice, and watershed for the surrounding. The runoff generation in these glaciated regions is a major source of water for population settled in the mountains and downstream. A study suggests that the Himalayan glaciers release about 0.5 km^3 of water on an average during summer and contribute a major share of water supply in the north Indian rivers (Kaul, 1990).

According to the studies conducted by Ageta and Pokhral (1999), approximately 80 per cent of the mass balance input is contributed by the monsoonal precipitation in Eastern Himalaya. But in Central Himalaya, it has been observed that the monsoonal precipitation contributes about 15 per cent of the mass balance influx, whereas the rest can be attributed to the westerlies disturbances. Further it has been observed that the mass balance characteristics of Western Himalaya are largely controlled by winter accumulation. In this region, 85 per cent of the influx is through winter precipitation. Henceforth, we see a range of climatic variability throughout Western Himalaya, Central Himalaya and Eastern Himalaya. Based on this, the Himalayan region can be classified in three zones depending on the amount of monsoon precipitation and the snowfall they receive. The Eastern/Central Himalaya is dominated by monsoon precipitation. The western Himalayan region witnesses equal or sub-equal monsoon and winter precipitation (areas of Ganga basin and parts of Himachal Pradesh) and dominant winter precipitation (Ladakh, Spiti and Tibet).

The snow and glaciated terrains in Western Himalaya are particularly important because rivers such as Indus, Brahmaputra and to a modest level, the river Ganga (Immerzeel et al., 2010) are dependent on the melt water catering to the needs of about 700 million population (Prasad and Singh, 2007). The variations in glacier dimensions give valuable inputs about the micro-climate, and observations become more significant where meteorological measurements are scarce and sparse. Western Himalaya is influenced by the mid latitude westerlies and the south Asian summer monsoon. However, the meteorological database remains rare because of logistics and unavailability of monitoring stations and observatories at higher elevations.

Various studies have shown the trend of temperature fluctuation over Western Himalaya. Observations and reconstruction from tree rings during pre-monsoon period have shown that rate of decrease of minimum temperature is thrice the rate of decrease of maximum temperature (Yadav et al., 2004). This indicates the greater contribution and role of minimum temperature to the cooling trend. The rate of decrease of minimum temperature during pre-monsoon (March–May) temperature using observations and reconstruction from tree rings decreased during the late 20th century. They also found that the rate of decrease of minimum temperature is three times that of the rate of decrease of maximum temperature, indicating that the minimum temperature is the larger contributor to the cooling trend in the pre-monsoon mean temperature (Bhutiyani et al., 2007). A regional climate model used by simulation of winter-time temperature and precipitation over Western Himalaya (Dimri and Ganju, 2007) concluded an over estimated temperature and under estimated precipitation over the western Himalayas. Another study reports a rise of 0.9 °C of annual maximum temperature over Western Himalaya (Dash et al., 2007). A recent study by Shekhar et al. (2010) reveals that the seasonal mean temperature for the period 1984–1985 to 2007–2008 in Western Himalaya has increased by ~2 °C.

The western Himalayan region has witnessed fluctuations as reported by various studies. The major reason being assigned as a controlling factor is the air temperature (Berthier et al., 2007; Kulkarni et al., 2007, 2011; Wagnon et al., 2007; Bhambri et al., 2011, Dutta et al., 2012). Snow cover monitoring for the period 1997–2001 has been carried out in the Beas basin (Kulkarni et al., 2011). Bhambri et al. (2011) used Corona and ASTER images of 1968 and 2006 to map the glaciers in Bhagirathi and Saraswati/Alaknanda basins of Garhwal Himalaya. The study revealed that the glacier vacated an area of about 4.6 ± 2.8 % with the recession rates increasing more recently (1990–2006). The basin showed fragmentation which led to an increase in the number of glaciers from 82 (1968) to 88 (2006). Some landmark studies using remote sensing data have been done to estimate the glacier retreat for 466 glaciers in Chenab, Parbati and Baspa basins of the western Himalayas (Kulkarni et al., 2007) showing an overall deglaciation of 21 % during 1962–2001. Glacier elevation changes and mass balance in the Spiti/Lahaul region was monitored by comparing a 2004 digital elevation model (DEM) to the 2000 SRTM of the region. A distinct thinning has been observed at low elevations, even on debris-covered tongues. This rate of ice loss is twice higher than the long-term (1977 to 1999) mass balance record for Himalaya indicating an increase in the pace of glacier wastage (Berthier et al., 2007). A few previous studies have also reported glacier mass balance in Indian Himalaya (Table 1).

Table 1 Various studies on glacier mass balance in Himalaya

Sl. No.	*Name of the Glacier*	*Location*	*Period of study*	*Sp.Bn. (m w.e.a^{-1})*	*Source*
1	Gara	Himachal	1974–1983	–0.324	Raina et al., 1977
2	Gor–Garang	Himachal	1976–1985	–0.572	Shankar, 2001
3	Shaune Garang	Himachal	1984–1989	–0.407	Singh and Sangewar, 1989
4	Neh Nar	Kashmir	1976–1984	–0.535	Srivastava et al., 1999
5	Changme Khangpu	Sikkim	1979–1986	–0.298	Sharma et al., 1999
6	Rulung	Ladakh	1980–1981	–0.105	Srivastava et al., 2001
7	Tipra Bamak	Uttrakhand	1981–1988	–0.241	Gautam and Mukherjee, 1989
8	Dunagiri	Uttrakhand	1984–1990	–1.038	Srivastava and Swaroop, 1989
9	Chhota Shigri	Himachal	1987–1988	–0.154	Dobhal, 1993
10	Dokriani	Uttrakhand	1993–2000	–0.320	Dobhal et al., 2008
11	Chhota Shigri	Himachal	2002–2010	–0.671	Wagnon et al., 2007; Ramanathan, 2011

Geochemical Studies in the Western Himalayas

The glacierised areas experience high chemical weathering rates and minimal anthropogenic impacts and thus act as an ideal environment to study the water-rock interactions. Various researches on meltwater quality variations have suggested the importance of these environments in estimates of terrestrial chemical erosion and global biogeochemical cycles (Brown, 2002). About 75 per cent of the runoff in three major river systems, the Brahmaputra, Ganga and Indus, occurs between June and September, in response to the snow and glacier ice-melt (Collins and Hasnain 1995). The water supply to these rivers from glaciers during the melting and lean season is more stable and predictable than that from rainfall.

An investigation involving discharge autocorrelation and correlations between discharge and meteorological factors was done to assess the role of meteorological parameters for discharge forecast. Changes in correlations between discharge and meteorological variables, lagged by 0-3 days, were determined. The summer months and the melt season showed a high discharge correlation suggesting substantial melt water storage in the glacier from the highly glacierized basin being dependent on the discharge of the previous day. A comparison between discharge-temperature correlation and discharge-precipitation correlation shows interesting results. Temperature shows a better correlation with discharge during June and September whereas discharge-precipitation gives better correlation in July and August. A number of linear regression equations were developed for individual months and also for the melt season. A better forecast was obtained using the seasonal regression equation (Singh et al., 2000).

The hydrological characteristics of the Gangotri Glacier through collection and analysis of streamflow records near the snout during four consecutive ablation seasons (May–October) (2000–2003) were monitored. The discharge showed an increase from May, reaches its highest value in July and then starts decreasing. Thus, the runoff distribution shows a maximum contribution from July (30.2 %) followed by August (26.2 %) with May and September reflecting maximum diurnal variability in runoff. The diurnal variation in runoff reflects the storage characteristics being maximum in the evening and minimum in the morning alongwith almost similar pattern from year to year. The beginning of the melt season and the close of ablation season shows almost negligible discharge whereas the peak melt period (July-August) shows noticeable discharge. Since the region witnesses very less rainfall, it is evident that most of the discharge has been derived from the melting of glacier (Singh et al., 2006).

Another study using the water balance approach with a water budget period of 10 years (October 1986-September 1996) was considered for the analysis. The isohyets of rainfall data of 10 stations located at different elevations in the basin was used for computing the rainfall input over the water budget period. The total volume of flow was computed using observed flow data of the Satluj river at Bhakra Dam for the above mentioned period. The temperature-evaporation relationship was used to estimate the evapotranspiration losses and at the same time the rainfall input to the study basin over the water budget period was computed from isohyets using rainfall data of ten stations located at different elevations in the basin. The total volume of flow for the same period was computed using observed flow data of the Satluj river at Bhakra Dam. A relationship between temperature-evaporation was developed and used to estimate the evapotranspiration losses. Satellite-based data were used to deduce the snow-covered area and its depletion with time. The study concludes that the snow and glacier runoff account for about 59 per cent of the annual flow of the Satluj river at Bhakra Dam, the remaining 41 per cent being from rain (Singh and Jain, 2002).

A study of major ions (Mg^{2+}, Ca^{2+}, Na^{+}, K^{+}, HCO^{3-}, SO_4^{2-} and Cl^{-}) of melt-water (area of study Chhota Shigri glacier) reveals the weathering and geochemical processes controlling the major ion chemistry of the glacier melt-water (Singh et al., 2013). A high ratio of $Ca + Mg/TZ^{+}$ (0.77) and low ratio of $Na + K/TZ^{+}$ elucidates the role of carbonate weathering followed by silicate weathering as the dominant mechanism controlling major ion chemistry. There are a number of processes responsible affecting the chemistry. A comparison of the datasets of the hydrochemistry of August 2008 with previous study carried out in 1987 attributed to increased weathering rates due to climate warming, atmospheric deposition or a combination of both.

Remote Sensing and Field-based Studies in the Western Himalayas

There has been a global consensus about the role of glaciers as the best reflectors in the regional climatic changes. However, the harsh environmental conditions, terrain characteristics and the inaccessibility pose logistic constraints in the consistent monitoring and collection of data on a regular basis. Field verifications cannot be carried out throughout the year. Keeping these limitations in mind, we take the help of Geographic Information System (GIS) and Remote Sensing techniques as a tool to monitor and observe the dynamics related with the glaciers. The Remote Sensing techniques are indeed useful to catalogue the changes in glacier because they provide a vast range of time scale and at the same time provide synoptic, multi-spectral and repetitive coverage. The implementation of satellite-based data can be useful in providing a complete and representative picture of any vast region which becomes a little difficult keeping in consideration the logistics.

An extensive mapping has been done in the Beas Basin, Himachal Pradesh (Dutta et al., 2012) to estimate the glacial fluctuations during the period 1972-2001. The investigations were carried out using Landsat (Multi Spectral Scanner MSS, Thematic Mapper TM and Enhanced Thematic Mapper ETM+) images of 1972, 1989, 2001 and Indian Remote Sensing Satellite (IRS) images of 2001 and 2006. The IRS images were rectified using ground control points of known geographical identifiers through the DGPS survey (Spectra Precision EPOCH 10). Field verification was carried out in the glaciers of adjacent basins, viz., Chhota Shigri glacier (Chandra basin), Panchnala glacier (Bhaga basin) and Patsio glacier (Bhaga basin). A mapping of 224 glaciers during this period concludes a loss of the glacier cover from 419 to 371 km^2, which is about 11.6 % of the basin area. A higher rate of retreat of the glaciers was observed during 1989–2006 as compared to the retreat during 1972–1989. The average elevation has also been observed to shift upward from 3898 m (1972) to 4171 m (2006) which may be a consequence of a shift in Equilibrium Line Altitude (ELA) reflecting imbalance.

Another study has also been conducted in the Ravi basin taking into account 60 glaciers. It has been observed that the glaciers have receded by 16.65 km^2 during the period 1972-2006. Therefore, about 16.37 % of the area has been lost in the last four decades. The investigation also elucidates a higher rate of retreat in the period 1972-89 (9.77 %) as compared to that during 1989-2006 (3.08 %). The study also reveals an increase in the number of the glaciers, although not very significantly (Dutta et al., 2013).

Therefore, it is evident that satellite imageries can be an effective tool to address the key issues associated with glacier changes and comprehensive and frequent sampling of glacier evolution (e.g. Bishop et al., 2000; Kulkarni et al., 2007). However, the calibration and validation of the data are extremely essential and that can be verified only from the field-based measurements and monitoring. Keeping this in view, a long-term monitoring programme was initiated on the benchmark glacier in the western Himalaya. The Chhota Shigri glacier (Fig. 1) was chosen as the benchmark because it has already been monitored during 1987–89 (Dobhal et al., 1995). Moreover, it proved to be convenient for observations of mass, energy and hydrological balances. Chhota Shigri Glacier (32.28 N, 77.58 E), Lahaul and Spiti district, Himachal Pradesh, India is easily accessible (around 100 km from Manali by road), of medium size (15.7 km^2 over a 34.7 km^2 catchment), mostly free of debris cover and lies in the crucial area alternately (Table 2).

The annual mass balance measurements have been done at the Chhota Shigri glacier using the traditional Glaciological method (Paterson, 1994) at the end of September or beginning of October. Stakes were inserted in the ice at 10 m depth in the ablation zone (ice density is taken as constant at 900 kg m^{-3}) to estimate the annual mass balance. In the accumulation area, cores are drilled to measure the annual net accumulation from snow layering (stratigraphy) and for density measurements. With this methodology, Wagnon et al. (2007) presented four years of the annual specific mass balance data of the Chhota Shigri glacier which was found to be mostly negative, with –1.4, –1.2, +0.1 and –1.4 m we in 2002-03, 2003-04, 2004-05 and 2005-06, respectively. The observation years were too short to derive any conclusion from this trend of mass balance data.

In the next phase, the mass balance of the glacier between 1999 and 2010 was estimated to be -4.8 ± 1.8 m we corresponding to -0.44 ± 0.16 m we yr^{-1}. The volume change between 1988 and 2010 has been estimated using the geodetic method. A slight loss of mass has been observed between 1988 and 2010 (-3.8 ± 2.0 m we corresponding to –0.17 pm 0.09 m we yr^{-1}).

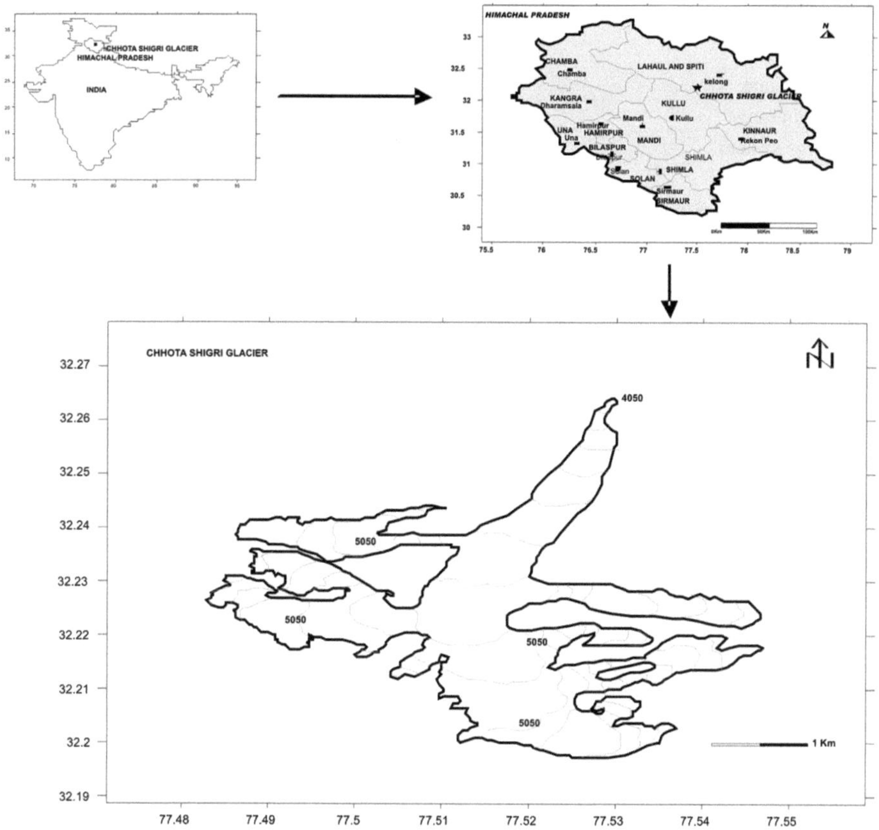

Fig. 1 Location map of Chhota Shigri glacier

The differencing of satellite DEM yields a slightly less negative MB of the Chhota Shigri Glacier (–0.39 pm 0.15 m we yr^{-1}) as compared to the MB of a 2110 km^2 glaciarized area in the Lahaul and Spiti region (–0.44 ± 0.09 m we yr^{-1}) during 1999–2011. The results in the above mentioned study indicate moderate ice wastage over the last 22 years with near equilibrium conditions during the nineties, and an ice mass loss after. The late nineties (latest in 1999) is probably the crucial junction from balanced to negative mass budget (Vincent et al., 2013).

Role of Western Himalayas in Water Resource Management

Trend analyses of the maximum, minimum and mean temperatures in the north-western Himalayan region during 1901–2002 (Bhutiyani et al., 2007) reveal that there has been a rise by 1.6 °C in the air temperature over the region and the warming rate has been highest during 1991–2002. Dash et al. (2007) have

Table 2 Geographical and topographical characteristics of Chhota Shigri glacier

General Features	
Country, state	India, Himachal Pradesh
Mountain range	Western Himalaya, Pir Panjal range
Drainage system	Chandra River-Indus River (Chenab branch)
Climate	Monsoon-arid transition zone
Glacier Characteristics	
Latitude	32.19°-32.288° N
Longitude	77.49°-77.558° E
Maximum elevation	6263 m
Snout position	4050 m
Basin area	34.7 km^2 (47 % glacierized) at
	3900 ma.s.l. on the proglacial stream
Total glacierized area	16.3 km^2
Chhota Shigri Glacier area	15.7 km^2
Glacier length	9 km
Mean orientation	North
Mean annual temperature at ELA	Between -5.5 and -6.58 °C

Source: Wagnon et al., 2007

suggested that there has been a rise of 0.9°C of annual maximum temperature over the western Himalaya. A recent study by Shekhar et al. (2010) reveals that the seasonal mean temperature for the period 1984–1985 to 2007–2008 in the western Himalaya has increased by ~2 °C. A rise in air temperature serves as an indicator to change in climate and this change is manifested in the regional climate. A rise in temperature and thus subsequent increase of solar radiation and albedo is likely to play an important role in the melting of glacier and ultimately to negative mass balance. Negative glacier mass balance is also the result of glacier vacating part of the area occupied by it, thus enhancing the weathering processes and the ionic concentrations in the meltwater stream. Warming also gives rise to an increase in meltwater temperatures, which would accelerate chemical weathering, releasing more ions into the portal meltwater.

The rivers originating from the western Himalayas are the cradles of human civilizations. In the peak months of summers, the perennial rivers receive substantial contribution from the glacier melt. The above mentioned observations and data clearly indicate that if there is a fluctuation in the glacier melt, it is bound to manifest in the hydro-electric power potential. The power generation from several projects depends heavily on the melt runoff generated from snow and glacier.

Therefore, a proper research on the glacier hydrometry along with forecasting of the expected runoff has become inevitable. A consistent monitoring of fluctuation of the glacier melt and its relationship with the mass balance is of utmost significance. A negative mass budget yields an increased runoff from the glacial melt whereas a positive mass budget would yield a deficit of runoff. Keeping in view the role of glaciers in the hydrological cycle, the knowledge gaps can be filled by

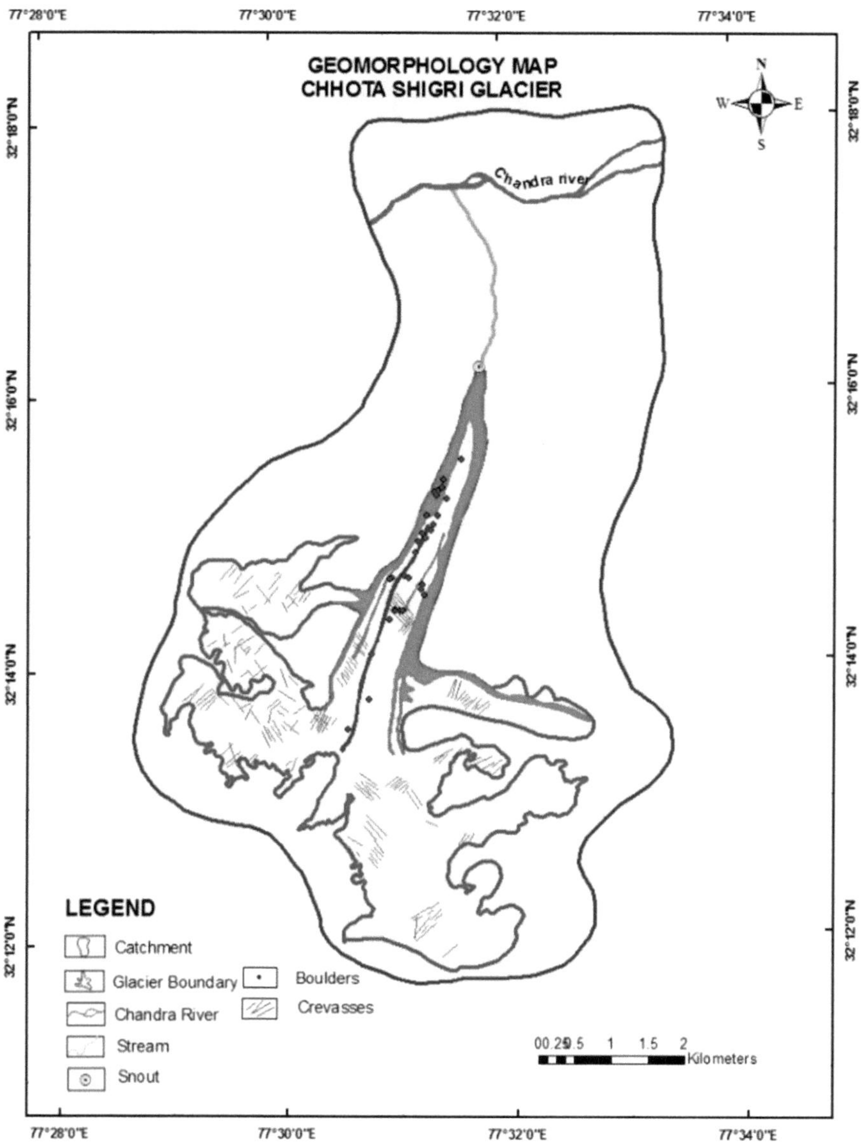

Fig. 2 Geomorphological map of Chhota Shigri glacier

setting of a regionally complete, updated and accurate glacier inventory integrating it with all the vital topographic parameters. A proper management of the water resources in designing the project should further be carried with the help of modelling and isotopic studies to have a holistic view of the glacier health.

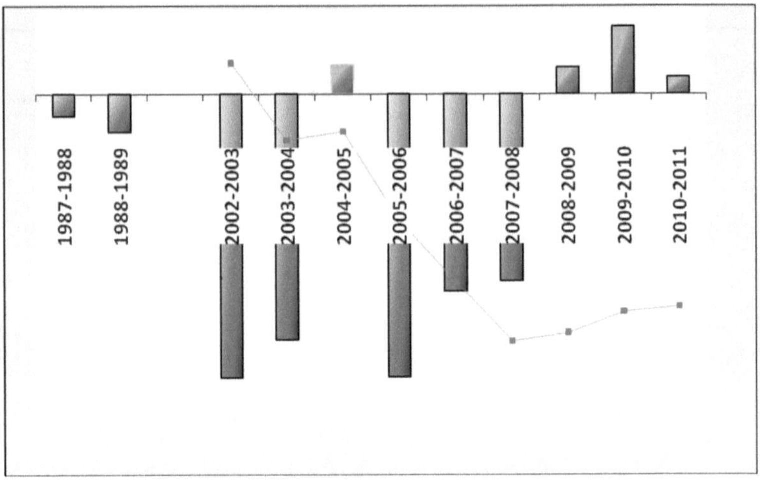

Fig. 3 Specific mass balance (m we) of Chhota Shigri glacier 2002-2011

Table 3 Mass Balance data of Chhota Shigri glacier

Year	*Sp Mass Balance (m we)*	*Cumulative mass balance (m we)*	*ELA (m amsl)*	*AAR*
1987-1988	–0.11	–0.11	4600	0.62
1988-1989	–0.19	–0.3	4840	
2002-2003	–1.42	–1.42	5170	0.31
2003-2004	–1.23	–2.64	5165	0.31
2004-2005	0.14	–2.50	4855	0.74
2005-2006	–1.41	–3.91	5185	0.29
2006-2007	–0.98	–4.89	5130	0.36
2007-2008	–0.93	–5.83	5120	0.38
2008-2009	0.13	–5.70	4950	0.63
2009-2010	0.33	–5.36	4890	0.70
2010-2011	0.08	–5.28	4973	0.59

Source: Dobhal et al., 1995, Wagnon et al., 2007, Vincent et al., 2013

References

Ageta, Y. and Pokhral, A.P. (1999). Characteristics of mass balance components of summer accumulation type glacier in the Nepal Himalaya. *Seppyo*, **45:** 81-105.

Berthier, E., Arnaud, Y., Kumar, R., Ahmad, S., Wagnon, P. and Chevallier, P. (2007). Remote sensing estimates of glacier mass balances in the Himachal Pradesh (Western Himalaya, India). *Remote Sens. Environ.*, **108:** 327-338.

Bhambri, R., Bolch, T., Chaujar, R.K. and Kulshreshtha, S.C. (2011). Glacier changes in the Garhwal Himalaya, India, from 1968 to 2006 based on remote sensing. *J. Glaciol.*, **57(203):** 543-556.

Bhutiyani, M.R., Kale, V.S. and Pawar, N.J. (2007). Long-term trends in maximum, minimum and mean annual air temperatures across the northwestern Himalaya during the twentieth century. *Climatic Change*, **85(1–2):** 159-177.

Bishop, M.P., Kargel, J.S., Kieffer, H.H., MacKinnon, D.J., Raupand, B.H. and Shroder, J.F. Jr. (2000). Remote Sensing science and technology for studying glacier processes in high Asia. *Ann.Glaciol.*, **31:** 164-170.

Bolch, T., Kulkarni, A.V., Kääb, A., Huggel, C., Paul, F., Cogley, J.G., Frey, H., Kargel, J.S., Fujita, K., Scheel, M., Bajracharya, S. and Stoffel, M. (2012). The State and Fate of Himalayan Glaciers. *Science*, **336:** 310-314.

Collins, D.N. and Hasnain, S.I. (1995). Runoff and sediment transport from glacierized basins at the Himalayan scale. International Association of Hydrological Sciences Publication 226 Symposium at Boulder 1995.

Dash, S.K., Jenamani, R.K., Kalsi, S.R. and Panda, S.K. (2007). Some evidence of climate change in twentieth-century India. *Climatic Change*, **85(3–4):** 299-321.

Dimri, A.P. and Ganju, A. (2007). Wintertime Seasonal Scale Simulation over Western Himalaya Using RegCM3. *Pure Appl. Geophys.*, **164:** 1733-1746. DOI 10.1007/s00024-007-0239-y

Dobhal, D.P., Kumar, S. and Mundepi, A.K. (1995). Morphology and glacier dynamics studies in monsoon–arid transition zone: An example from Chhota Shigri glacier, Himachal Himalaya, India. *Current Sci.*, **68(9):** 936-944.

Dutta, Shruti, Ramanathan, AL. and Linda Anurag (2012). Shrinking Glaciers in the Beas Basin observed through Remote Sensing techniques, 1972–2006, Himachal Pradesh, India. *J. Earth Syst. Sci.*, **121(5):** 1105-1112.

Giles, H. Brown (2002). Glacier meltwater hydrochemistry. *Applied Geochemistry*, **17:** 855-883.

Immerzeel, W.W, van Beeke, L.P.H. and Bierkens, M.F.P. (2010). Climate change will affect the Asian water towers. *Science*, **328:** 1382-1385.

Kaul, M.K. (1999). Inventory of Himalayan Glaciers. Geol. Surv. India, Spl. Pub. 34.

Kulkarni, A.V., Bahuguna, I.M., Rathore, B.P., Singh, S.K., Randhawa, S.S., Sood, R.K. and Dhar, S. (2007). Glacial retreat in Himalaya using Indian remote sensing satellite data. *Curr. Sci.*, **92(1)**: 69-74.

Paterson, W.S.B. (1994). The physics of glaciers. Third edition. Oxford, etc., Elsevier.

Shekhar, M.S., Chand, H., Kumar, S., Srinivasan, K. and Ganju, A. (2010). Climate-change studies in the western Himalaya. *Ann. Glaciol.*, **51(54):** 105-112.

Singh, Pratap and Jain, S.K. (2002). Snow and glacier melt in the Satluj River at Bhakra Dam in the western Himalayan region. *Hydrological Sciences*, **47(1):** February 93.

Singh, Pratap, Haritashya, Umesh K., Kumar, Naresh and Singh, Yatveer (2006). Hydrological characteristics of the Gangotri Glacier, central Himalayas, India. *Journal of Hydrology*, **327:** 55-67.

Singh, Virendra Bahadur, Ramanathan, AL., Jose, Pottakkal George, Linda, Anurag and Sharma, Parmanand (2013). Temporal Variation in the Major Ion Chemistry of Chhota Shigri Glacier Meltwater, Lahaul–Spiti Valley, Himachal Pradesh, India. *Natl. Acad. Sci. Lett.* ISSN 0250-541X. DOI 10.1007/s40009-013-0135-1.

Singh, Pratap, Ramasastri, K.S., Kumar, Naresh and Arora, Manohar (2000). Correlations between discharge and meteorological parameters and runoff forecasting from a highly glacierized Himalayan basin. *Hydrological Sciences*, **45(5):** 637.

Vincent, C., Ramanathan, AL., Wagnon, P., Dobhal, D.P., Linda, A., Berthier, E., Sharma, P., Arnaud, Y., Azam, M.F., Jose, P.G. and Gardelle, J. (2013). Balanced conditions or slight mass gain of glaciers in the Lahaul and Spiti region (northern India, Himalaya) during the nineties preceded recent mass loss. *The Cryosphere*, **7:** 569-582.

Wagnon, P., Linda, A., Arnaud, Y., Kumar, R., Sharma, P., Vincent, C., Pottakkal, J.G., Berthier, E., Ramanathan, AL., Hasnain, S.I. and Chevallier, P. (2007). Four years of mass balance on Chhota Shigri Glacier, Himachal Pradesh, India. A New Benchmark Glacier in the Western Himalaya. *J. Glaciol.*, **53(183):** 603-611.

Yadav, R.R., Park, W.K., Singh, J. and Dubey, B. (2004). Do the western Himalayas defy global warming? *Geophysical Research Letters*, **31(17)**. doi: 10.1029/2004GL020201.

Seasonal Variations and Flux of Arsenic in Gomati River, Ganga Alluvial Plain, Northern India

Dharmendra Kumar Jigyasu, Rohit Kuvar, Satyendra Singh, Sandeep Singh, Ashwini Kumar Chowdhary, and Munendra Singh

Introduction

Arsenic is a naturally occurring toxic metalloid and its contamination in the hydrological system has significantly received worldwide attention in the last three decades. It is widely distributed in all the components of the environment and varies by more than four orders of magnitude ranging from <0.5 μg/l to 5000 μg/l in the natural water systems. Natural systems, with climate as the controlling mechanism, play an important role in strong geochemical fractionation with quantitative elemental transportation (Raju, 2012a). Natural systems of tropical environment are, therefore, a major concern for environmental scientists (Dissanayake and Chandrajith, 1999; Smedley and Kinniburgh, 2002). The study of fluvial time series provides the key for the understanding of elemental mobilization that controls the dissolved elemental concentration in the various components of the hydrological system (Raju 2012a). Studies have shown that dissolved As concentration in river water varies at a great extent and is mainly dependent on geology, hydrology, climate as well as various anthropogenic activities (Raju 2012b; Masson et al., 2007; Elbaz-Poulichet et al., 2006; Pettine et al., 1997; McLaren and Kim, 1995). In northern India, the Ganga Alluvial Plain (GAP) is one of the most densely populated regions of the world. It is drained by several alluvial rivers and supports nearly 500 million people. The objective of this paper is to report

D.K. Jigyasu (✉) • R. Kuvar • M. Singh
Centre of Advanced Study in Geology, University of Lucknow, Lucknow 226007, India
e-mail: jigyasu2012@gmail.com

S. Singh
Government Girls P.G. College, Ghazipur 233001, India

S. Singh
Department of Earth Science, Indian Institute of Technology, Roorkee 247667, India

A.K. Chowdhary
Institute Instrumentation Centre, Indian Institute of Technology, Roorkee 247667, India

N.J. Raju et al. (eds.), *Management of Water, Energy and Bio-resources in the Era of Climate Change: Emerging Issues and Challenges*,
DOI 10.1007/978-3-319-05969-3_8

the seasonal variation and flux of dissolved As in the Gomati river and to understand the As mobilization for its eco-toxicological potentials in the GAP.

Study Area

The Gomati river (an alluvial river of the GAP) originates from the Gomath Tal and covers a distance of 900 km before its confluence with the Ganga river. It drains about 30,437 km^2 area, located between 25–29°N and 80–84°E in the interfluve region of the Ganga and Ghaghara rivers and encompasses an altitude range of 190 m to 60 m above mean sea level (Fig. 1). Geologically, the Gomati River Basin is made up of unconsolidated alluvial sediments derived from the Himalayan region. These alluvial sediments are made up of well sorted silt and silty fine sand, representing muddy and sandy interfluve deposits (Singh, 1996). Silt fraction of these alluvial sediments is predominantly composed of mica along with feldspar, quartz and mixed layer minerals in the GAP (Pal et al., 2011).

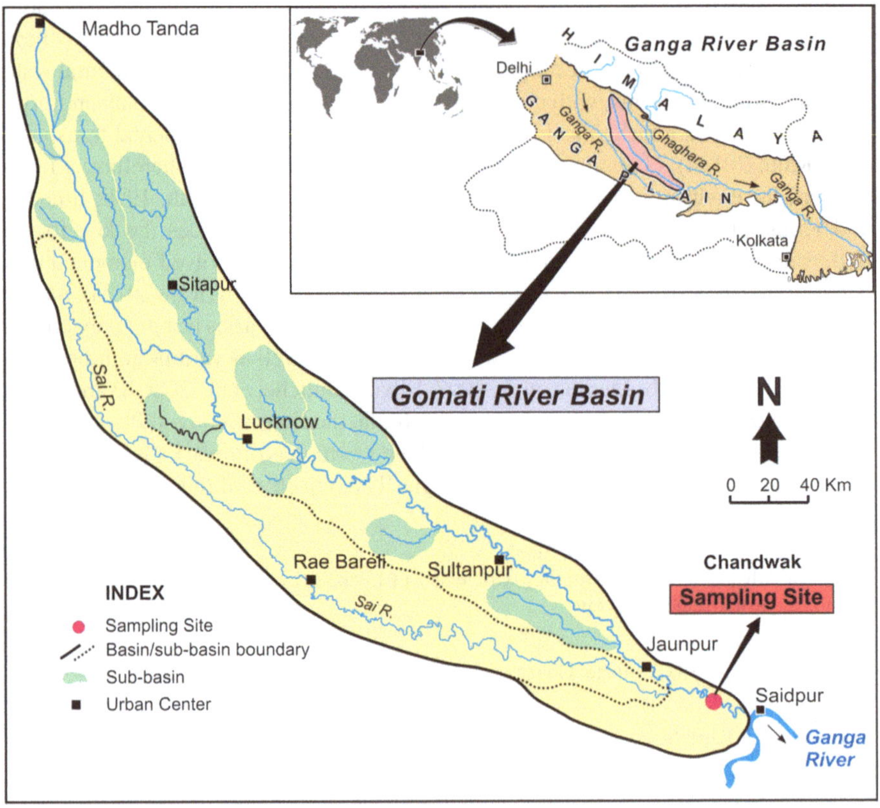

Fig. 1 Location map of the study area. The Gomati river drains the interfluve region of the Ganga and the Ghaghara rivers in the Ganga Alluvial Plain, northern India. Sampling site Chandwak was selected for the present study

The Gomati river basin experiences a warm and humid sub-tropical climate with four prominent seasons: the winter, the summer, the monsoon and the post-monsoon seasons (Dasgupta, 1984). The winter season is a cold dry season that extends from December to February, characterized by low temperature and low rainfall. The hot dry summer season extends from March to May and is marked by a continuous rise in temperature that reaches a maximum of 40 to 47 °C in May. The monsoon season is a warm wet season, extending from June to September. It is characterized by high rainfall due to the southwest monsoon system. Daily mean maximum temperature is about 30 °C. Nearly 70 – 80 % of the total annual rainfall occurs during this season. The Gomati river along with its tributaries and other water bodies including the low-lying areas in the regional surface of the GRB get flooded due to excessive monsoon rainfall. The post-monsoon season extends from October to November and is characterized by the decreasing trend of maximum (35 – 25 °C) and minimum (20 – 10 °C) temperatures along with moderate rainfall.

The Gomati river is a groundwater fed river. Its hydrology is strongly controlled by the seasons. The river discharge varies from 31.85 m^3/s in the summer season, to 721.32 m^3/s in the monsoon season. The first monsoon peak discharge (>600 m^3/s) occurs in the end of July and is ~20 times higher than the average discharge (30 m^3/s) during the summer season (Table 1). The $\delta^{18}O$ characterization of precipitation recorded at New Delhi shows the lower-most values in September (late monsoon) and November (post-monsoon) months (IAEA, 1981). Low isotopic characters ($\delta^{18}O = -10$ ‰) of the monsoon precipitation pass into the shallow groundwater and the Gomati river water as the monsoon precipitation is the dominant process of the recharge activity of the Gomati river basin in the GAP.

Methodology

Chandwak (25°35′ N, 83°00′ E) has been selected to represent the entire drainage basin and is located upstream of the confluence of the Gomati with the Ganga river at an elevation of 70 m above mean sea level (Fig. 1). Thirty-six river water samples were collected from the middle of the river channel during June 2009-May 2010 at an interval of ten days. The river water samples were collected in wide mouth 250 ml polypropylene bottles (© Tarsons) with airtight caps and were acidified in the field with HNO_3 (5 ml/l). Each sample bottle was tagged with appropriate label and details. All these water samples were carefully transported into the laboratory and stored at 4 °C in the laboratory till further chemical analysis. The water samples were filtered by using 0.45 μm cellulose filters. All the filtered samples were analysed by Induced Coupled Plasma-Mass Spectrophotometer at the Institute Instrumentation Centre, Indian Institute of Technology, Roorkee. Each sample was analysed in duplicate and mean values were taken as the result. All the samples were analysed in the laboratory following the standard protocols (APHA, 2002). For quality assurance, replicates and analytical blanks were also prepared and analysed to check the

Table 1 Sampling details and dissolved As concentration (μg/l) along with discharge and flux of the Gomati river at Chandwak

Sample code	Sampling date (dd.mm.yy)	Season	Dissolved As (μg/l)	Discharge (m^3/s)	Flux (kg/day)
GRW01	05.06.09	Monsoon	4.34	88.17	33.06
GRW02	14.06.09	Monsoon	2.41	95.86	19.96
GRW03	25.06.09	Monsoon	2.43	63.76	13.39
GRW04	05.07.09	Monsoon	3.80	66.03	21.68
GRW05	15.07.09	Monsoon	3.27	95.02	26.85
GRW06	25.07.09	Monsoon	1.306	11.73	68.71
GRW07	05.08.09	Monsoon	1.012	09.91	18.32
GRW08	16.08.09	Monsoon	2.112	27.13	41.41
GRW09	25.08.09	Monsoon	1.26	232.66	25.33
GRW10	05.09.09	Monsoon	2.40	409.92	85.00
GRW11	16.09.09	Monsoon	0.5371	1.83	32.60
GRW12	25.09.09	Monsoon	1.01	380.57	33.21
GRW13	05.10.09	Post-monsoon	1.28	721.32	79.77
GRW14	17.10.09	Post-monsoon	0.98	697.56	59.06
GRW15	25.10.09	Post-monsoon	0.83	280.80	20.14
GRW16	05.11.09	Post-monsoon	1.06	174.49	15.98
GRW17	15.11.09	Post-monsoon	1.61	91.73	12.76
GRW18	25.11.09	Post-monsoon	1.93	78.59	13.11
GRW19	05.12.09	Winter	1.61	110.45	15.36
GRW20	15.12.09	Winter	1.18	168.70	17.20
GRW21	25.12.09	Winter	1.37	102.34	12.11
GRW22	05.01.10	Winter	1.25	128.50	13.88
GRW23	15.01.10	Winter	1.28	164.37	18.18
GRW24	25.01.10	Winter	1.18	145.30	14.81
GRW25	05.02.10	Winter	1.27	61.42	6.74
GRW26	15.02.10	Winter	1.89	132.70	21.67
GRW27	25.02.10	Winter	1.57	98.49	13.36
GRW28	05.03.10	Summer	2.19	63.95	12.10
GRW29	15.03.10	Summer	2.59	65.11	14.57
GRW30	25.03.10	Summer	2.65	53.30	12.20
GRW31	05.04.10	Summer	3.31	60.17	17.21
GRW32	15.04.10	Summer	3.55	42.67	13.09
GRW33	25.04.10	Summer	4.71	36.59	14.89
GRW34	05.05.10	Summer	5.15	35.84	15.95
GRW35	15.05.10	Summer	5.18	31.85	14.25
GRW36	25.05.10	Summer	5.67	33.50	16.41

Note: The office of Middle Ganga Division, Central Water Commission provided the discharge data for the Gomati river at Maighat, Chandwak

reliability of the data. Analytical precision for all samples were within ±5 % in As determination. Dissolved As concentration and discharge data were used to estimate the dissolved flux by using commonly applied equations (Table 1).

Results

Dissolved As Concentration and Seasonal Variations

The dissolved As concentration in the river water varies from 0.53 to 5.67 μg/l, whereas baseline concentrations of As in the world's river waters were very low, about 0.1-0.8 μg/l (Smedley and Kinniburgh 2002). The dissolved As concentrations (<45 μm) varied by a factor ~6 in the Gomati river (Table 1). Figure 2 displays variations of dissolved As concentration in the Gomati river water during the monsoon, the post-monsoon, the winter and the summer seasons. The variations in dissolved As concentration have been already reported for the Garonne, the Dordogne and the Isle rivers in France and also for the Seine, Po and Waikato rivers (Masson et al., 2007; Elbaz-Poulichet et al., 2006; Pettine et al., 1997; McLaren and Kim, 1995). During the winter season, the dissolved As concentration in the Gomati river water was stable around 1.00 to 2.00 μg/l. In the summer season, As concentration continuously increases from 2.19 to 5.67 μg/l. This is the most significant characteristic of As variation in the Gomati river water which is due to increasing temperature in the summer season. High temperature leads to the enormous growth of microbial species within the river sediments which facilitates the mobilization of As from the river sediments to the Gomati river water (Oremland and Stolz, 2003; Islam et al., 2012).

During the monsoon and the post-monsoon seasons, dissolved As concentration showed cyclic decreasing trend from 4.34 to 0.53 μg/l. It is mainly dependent on the monsoon rainfall patterns and flood discharges that mainly control the hydrology of the river. Dissolved As concentrations, therefore, were affected by dilution in these seasons. These results suggest that the seasonal variations in dissolved As

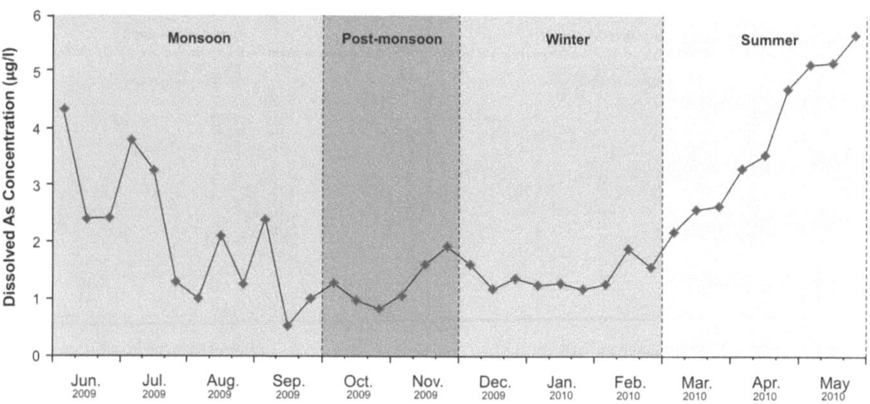

Fig. 2 Seasonal variation of dissolved As concentration (μg/l) in the Gomati river water at Chandwak during June 2009 to May 2010. Note the steady trend during the winter, continuous increasing trend during the summer and cyclic decreasing trend during the monsoon and the post-monsoon seasons

concentration show a single cyclic pattern controlled by the temperature during the winter and summer seasons, and the river's flood discharge during the monsoon and post-monsoon seasons.

Relationship between River Hydrology

The dissolved As concentrations in the Gomati river displayed the moderate negative correlation coefficient ($r^2 = 0.49$) with the river's flood discharges during the monsoon season (Fig. 3a). It can be summarized that dissolved As concentration in the river water truly depends on the river's flood discharge that is controlled by dilution factor through precipitation in the GAP during the monsoon season. During the post-monsoon and winter seasons, insignificant correlation coefficients were recorded (Fig. 3b, c). Significant correlation ($r^2 = 0.86$) has been observed between the river discharge and the dissolved As concentration during the summer season (Fig. 3d). This correlation can be linked with a decrease in the river's discharge along with rise in temperature that controls As mobilization in the river water.

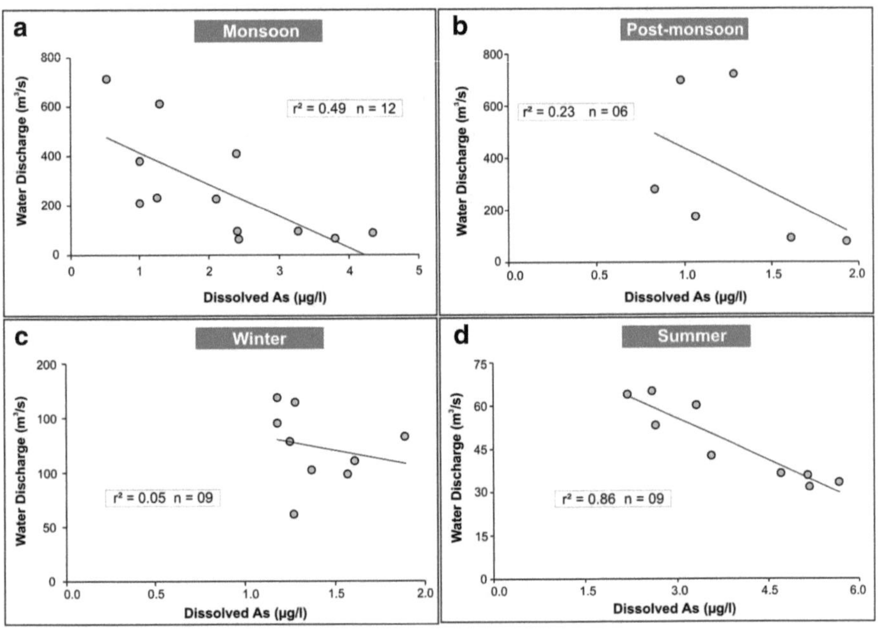

Fig. 3 Correlation plots of water discharge (m^3/s) and dissolved As concentration (µg/l) of the Gomati river in (**a**) the monsoon; (**b**) the post-monsoon; (**c**) the winter and (**d**) the summer seasons

Dissolved as Flux and Its Seasonal Variations

Figure 4a displays a bar diagram representing the monthly variations in the dissolved As flux of the Gomati river. The estimated total dissolved As flux of the Gomati river is 8.99×10^3 kg that is transported into the Ganga river during June, 2009-May, 2010. The fluxes of dissolved As of the Gomati river were recorded to be 4289 kg during the monsoon, 2028 kg during the post-monsoon, 1333 kg during the winter and 1335 kg during the summer seasons. The Gomati river contributes ~70 % of total dissolved As flux during the monsoon and post-monsoon seasons and ~30 % during the winter and summer seasons (Fig. 4b). This is due to the fact that Gomati river basin receives up to 80 % of the annual rainfall from June to September. The As mobilized from the top alluvial sediments of the GAP may be contributed to the dissolved As fluxes of the Gomati river during the monsoon and post-monsoon seasons, particularly after heavy precipitation events. This clearly indicates that the monsoonal rainfall and surface runoff contributes greater amount of As into the Gomati river.

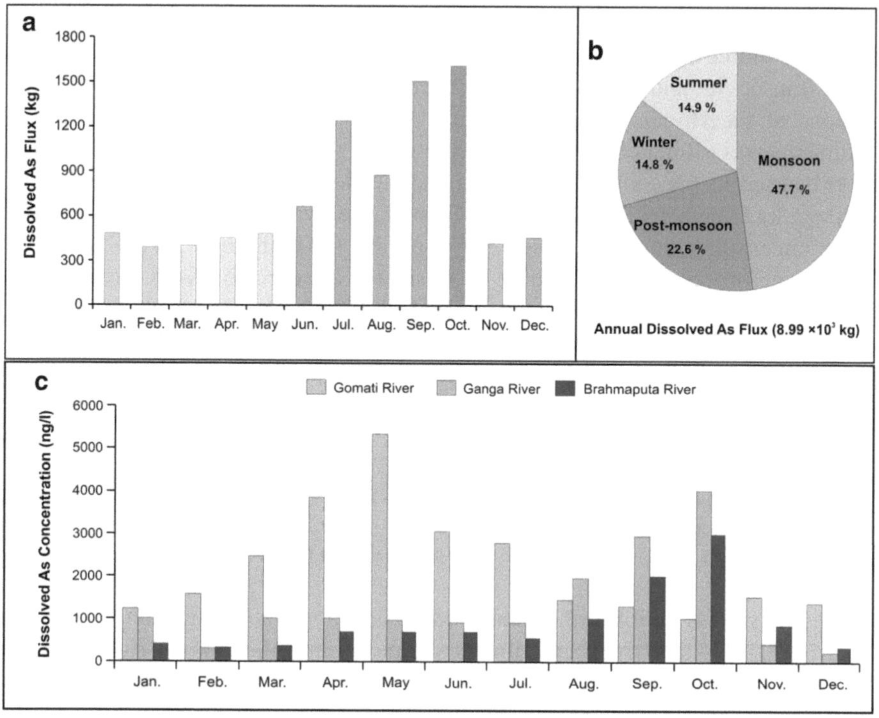

Fig. 4 (**a**) Bar diagram showing the monthly distribution of dissolved As flux of the Gomati river to the Ganga river; (**b**) Pie diagram showing the seasonal distribution of dissolved As flux of the Gomati river. Nearly 70 % of the total annual As flux (8.99×10^3 kg) is transported during the monsoon and post-monsoon seasons. (**c**) Bar diagram representing comparative variations in dissolved As concentration (ng/l) in the Gomati, the Ganga and the Brahmaputra rivers. [Data source: Islam et al., 2012]

Discussions

Comparative Study with the Ganga and Brahmaputra Rivers

The dissolved As concentrations showed seasonal variations ranging from 0.2 to 4.0 μg/l for the Ganga and 0.3 to 3.0 μg/l for the Brahmaputra rivers (Islam et al., 2012). Figure 4c displays the bar diagram showing monthly variations of dissolved As concentrations in the Gomati along with the Ganga and the Brahmaputra rivers. Except during the monsoon season, dissolved As concentrations in the Gomati river water is reported to be higher than the Ganga and the Brahmaputra rivers and are linked with the characteristics of the river hydrology and anthropogenic influences.

Empirical Modelling

Figure 5 displays an empirical model to explicate the seasonal variations of dissolved As concentration and flux of river in the GAP, northern India. The geogenic source of As in the GAP is most likely alluvial sediments forming mineral mica (muscovite and biotite), derived from the granitic and metamorphic source regions of the Himalaya (Breit, 2000). Average As concentrations in the GAP sediments was estimated to be about 10.44 mg/kg (Singh et al., 2012). The Gomati river bed sediments contain 1.36 mg/kg and the Gomati river suspended sediments 5.30 mg/kg (Singh et al., 2012). In the Gomati river water, dissolved As is released from unconsolidated alluvial sediments of the GAP through natural chemical weathering process. Mineral biotite is considered as source of As in the unconsolidated sediments of the GAP (Seddique et al., 2008; Chakraborty et al., 2011; Singh et al., 2013). Biotite weathers very easily and may be converted to kaolinite. In the process of kaolinization of biotite, the weathering reaction requires chemical changes by addition of H^+ and release of K^+, Al^{3+}, Mg^{2+}, Fe^{2+}(As), Ti^{4+}, Si^{4+} and H_2O (Dong et al., 1998).

$$
\begin{aligned}
&K_{0.65}(Al_{1.10}Ti_{0.15}Fe_{0.50}\,Mg_{0.55})(Si_{3.20}Al_{0.80})O_{10}(OH)_2 + 8H^+ \\
&\quad = (Al_{1.71}Fe_{0.31}Ti_{0.03})Si_2O_5(OH)_4 + 0.65\,K^+ + 0.19Al^{3+} + 0.55\,Mg^{2+} \\
&\qquad + 0.19Fe^{2+} + 0.12Ti^{4+} + 1.2Si^{4+} + 3H_2O
\end{aligned}
$$

Reductive dissolution is the established geochemical mechanism that mobilizes As from the Gomati river and alluvial sediments into the Gomati river water. It occurs in the presence of Fe containing silicate minerals, which break down under the influence of decaying organic matter. This chemical process is driven by the decay of organic matter (Ravenscroft et al., 2009). Microbes (such as cyanobacteria, algae and lichens etc.) are essentially ubiquitous in sediments and inhabit on the surface of minerals. These microbes dissolve silicate minerals through the corrosive action of metabolic products, such as NH_3, HNO_3 and CO_2

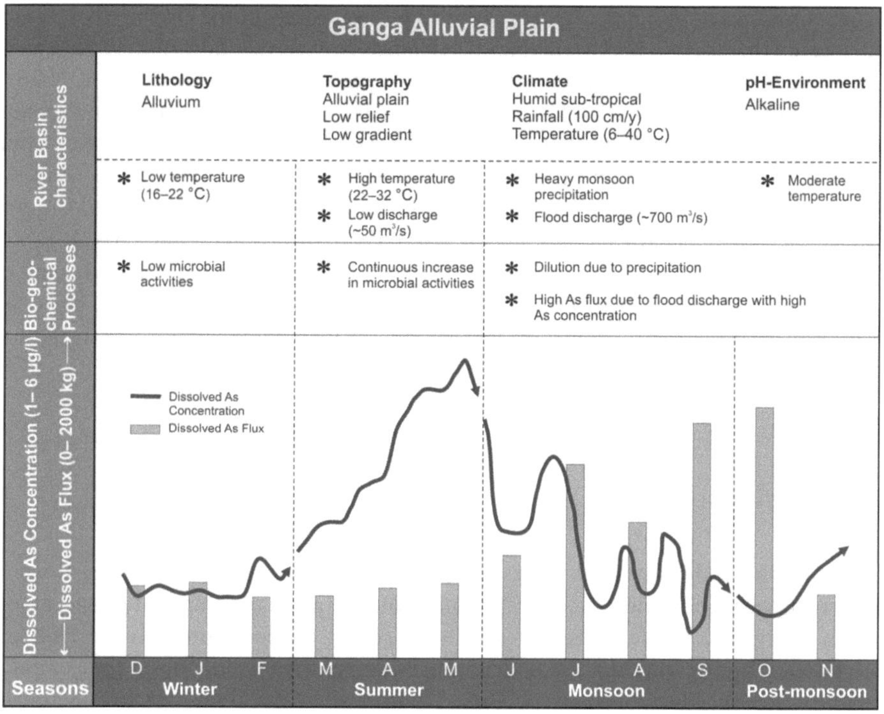

Fig. 5 Empirical model showing the characteristics and biogeochemical processes involved in the seasonal variation of dissolved As concentration and flux of river in the Ganga Alluvial Plain, northern India

forming H_2CO_3 in water etc. These metabolic products, execrated by micro-organisms that promote the chemical weathering of primary minerals, result in the displacement of cationic components from crystal lattice due to breaking of Si-O and Al-O bonds. Microorganisms, therefore, directly or indirectly, induce mineral disaggregation, hydration, dissolution and secondary mineral formation (Banfield et al., 1999; Ehrlich and Newman, 2009). The Gomati River receives organic matter rich untreated urban effluents from major urban centres located on its banks such as Lucknow, Sultanpur, Jaunpur etc. Singh et al. (2010) recently reported the downstream increase in the dissolved As concentration from 1.3 to 9.6 µg/l in the Gomati river during the winter season. The microbiological activity, therefore, plays a significant role in the mobilization of As from the sediments and dissolved As concentrations in the river water during the winter and summer seasons.

During the monsoon season, rainwater becomes either groundwater or floodwater after interacting with mica dominated silty alluvial sediments of the GAP after the 3-month long summer season. Robert et al. (2011) recently reported that monsoon derived floodwaters in the Ganga Delta region recorded high As concentrations (~500 μg/l). The recent $\delta^{18}O$ isotopic study of the Gomati river basin displayed that there is an instantaneous exchange between shallow groundwater and river water through precipitation during the monsoon season (Singh et al., 2013). The duration and intensity of precipitation is accountable for the river hydrology and also governs the dilution of dissolved As concentration due to the rapid exchange of the monsoon precipitation into the river water through the shallow groundwater. It can also be justified through similar oxygen isotopic compositions of the shallow groundwater and the Gomati river water (Singh et al., 2013). From the above studies, it can be concluded that heavy monsoon precipitation, instantaneous interactions of shallow groundwater and the Gomati river water, high As concentrations in flood waters are responsible for the significant dissolved As flux of the Gomati river during the monsoon and post-monsoon seasons.

Conclusions

The present study reveals that dissolved As in rivers of the GAP show significant seasonal variations controlled by the range and duration of temperature and monsoon precipitation. Nevertheless, all dissolved As concentrations in the Gomati river were higher than the Ganga and the Brahmaputra Rivers during the summer and winter seasons that can be linked with hydrological and biogeochemical factors. The Gomati River substantially contributes the dissolved As flux to the Ganga river. Anthropogenically induced organic matter rich untreated effluents draining into rivers of the GAP govern the fate and mobility of As in the fluvial environment of the GAP during the summer season. Our findings may have important implications for the designing of research and monitoring programmes in As-affected areas of the GAP as well as other sub-tropical regions of the world.

Acknowledgements Authors would like to thank Prof. Indra Bir Singh, University of Lucknow for constant encouragement and valuable discussions. This study was financially supported by the University Grants Commission (UGC), New Delhi under Rajiv Gandhi National Fellowship scheme to DKJ (No. 16-1731(SC)/2010 (SA-III), under Junior Research Fellowship to RK (No. 23154 Dt. 11/08/2010) and minor research project to SS (No. 8-29153)2011(MRP/NRCB). The Central Water Commission is gratefully acknowledged for providing discharge data of the Gomati river. Generous assistance by Vinay Singh during the field sampling work and by Nupur Srivastava in ICP-MS analysis is greatly appreciated. We thank Dr. Ratan Kar for comments and improvements in the early draft. We thank the anonymous reviewer for comments that substantially improved the paper.

References

APHA (2002). Standard methods for the examination of water and wastewater. 20th ed. American Public Health Association, Washington, DC.

Banfield, J.F., Barker, W.W., Welch, S.A. and Taunton, A. (1999). Biological impact on mineral dissolution: Application of the lichen model to understanding mineral weathering in the rhizosphere. *Proceedings of National Academy of Science* USA, **96:** 3404-3411.

Breit, G.N. (2000). Arsenic cycling in eastern Bangladesh: The role of phyllosilicates. Proceedings of the Annual Meeting of the Geological Society of America, A-192.

Chakraborty, S., Bardelli, F., Mullet, M., Greneche, J.M., Varma, S., Ehrhardt, J.J., Banerjee, D. and Charlet, L. (2011). Spectroscopic studies of arsenic retention onto biotite. *Chemical Geology*, **281:** 83-92. Doi: 10.1016/j.chemgeo.2010.11.030.

Dasgupta, S.P. (1984). The Ganga Basin, Part II. Central Board for Prevention and Control of Water Pollution, New Delhi.

Dissanayake, C.B. and Chandrajith, R. (1999). Medical geochemistry of tropical environments. *Earth Science Review*, **47:** 219-258. Doi:10.1016/S0012-8252(99)00033-1.

Dong, H., Peacor, D.R. and Murphy, S.F. (1998). TEM study of progressive alteration of igneous biotite to kaolinite throughout a weathered soil. *Geochimica et Cosmochimica Acta*, **62:** 1881-1887. doi:10.1016/S0016-7037(98)00096-9.

Ehrlich, H.L. and Newman, D.K. (2009). Geomicrobiology. CRC Press, Boca Raton.

Elbaz-Poulichet, F., Seidel, J-L., Casiot, C. and Tusseau-Vuillemin, M-H. (2006). Short-term variability of dissolved trace element concentrations in the Marne and Seine Rivers near Paris. *Science of Total Environment*, **367:** 278-287. doi.org/10.1016/j.scitotenv.2005.11.009.

IAEA (1981). Statistical Treatment of Environmental Isotope Data in Precipitation. International Atomic Energy Agency, Vienna, Austria. *Technical Report Series* **206**. Accessible at http://naweb.iaea.org. October 2012.

Islam, S.M.N., Rahman, S.H., Chowdhury, D.A., Rahman, M.M. and Tareq, S.M. (2012). Seasonal variations of arsenic in the Ganges and Brahmaputra River, Bangladesh. *Journal of Scientific Research*, **4(1):** 65-75. doi:10.3329/jsr.v4i1.7820.

Masson, M., Schäfer, J., Blanc, G. and Pierre, A. (2007). Seasonal variations and annual fluxes of arsenic in the Garonne, Dordogne and Isle Rivers, France. *Science of Total Environment*, **373:** 196-207. doi:10.1016/j.scitotenv.2006.10.039.

McLaren, S.J. and Kim, N.D. (1995). Evidence for a seasonal fluctuation of arsenic in New Zealand's longest river and the effect of treatment on concentrations in drinking water. *Environmental Pollution*, **90:** 67-73. doi.org/10.1016/0269-7491(94)00092-R.

Oremland, R.S. and Stolz, J.F. (2003). The Ecology of Arsenic. *Science*, **300:** 939-944. doi:10.1126/science.1081903.

Pal, D.K., Bhattacharyya, T., Sinha, R., Srivastava, P., Dasgupta, A.S., Chandran, P., Ray, S.K. and Nimje, A. (2011). Clay minerals record from late quaternary drill cores of the Ganga Plains and their implications for provenance and climate change in the Himalayan foreland. *Palaeogeography Palaeoclimatology Palaeoecology,* **356–357:** 27-37. doi:10.1016/j.palaeo.2011.05.009.

Pettine, M., Mastroianni, D., Camusso, M., Guzzi, L. and Martinotti, W. (1997). Distribution of As, Cr and V species in the Po–Adriatic mixing area (Italy). *Marine Chemistry*, **58:** 335-349. doi.org/10.1016/S0304-4203(97)00060-1.

Raju, N.J. (2012a). Arsenic exposure through groundwater in the middle Ganga plain in the Varanasi environs, India: A future threat. *Journal of Geological Society of India*, **79:** 302-314.

Raju, N.J. (2012b). Evaluation of hydrogeochemical processes in the Pleistocene aquifers of Middle Ganga plain, Uttar Pradesh, India. *Environmental Earth Sciences*, **65(4):** 1291-1308.

Ravenscroft, P., Brammer, H. and Richards, K. (2009). Arsenic pollution: A global synthesis. Wiley-Blackwell, United Kingdom.

Robert, L.C., Hug, S.J., Voegelin, A., Dittmar, J., Kertzschmar, R., Wehrli, B., Saha, G.C., Badruzzaman, A.B.M. and Ali, M.A. (2011). Arsenic Dynamics in Porewater of an Intermittently Irrigated Paddy Field in Bangladesh. *Environmental Science and Technology*, **45:** 971-976. doi.org/10.1021/es102882q.

Seddique, A.A., Masuda, H., Mitamura, M., Shinoda, K., Yamanaka, T., Itai, T., Maruoka, T., Uesugi, K., Ahmed, K.M. and Biswas, D.K. (2008). Arsenic release from biotite into a

Holocene groundwater aquifer in Bangladesh. *Applied Geochemistry*, **23:** 2236-2248. doi:10.1016/j.apgeochem.2008.03.007.

Singh, I.B. (1996). Geological Evolution of Ganga Plain – An overview. *Journal of the Palaeontological Society of India*, **41:** 99-137.

Singh, M., Kumar, S., Kumar, B., Singh, S. and Singh, I.B. (2013). Investigation on the hydrodynamics of Ganga Alluvial Plain using environmental isotopes: A case study of Gomati River Basin, northern India. *Hydrogeology Journal*, **21:** 687-700. doi: 10.1007/s10040-013-0958-3.

Singh, M., Singh, A.K., Swati, Srivastava, N., Singh, S. and Chowdhary, A.K. (2010). Arsenic mobility in fluvial environment of the Ganga Plain, northern India. *Environmental Earth Science*, **59:** 1703-1715. doi: 10.1007/s12665-009-0152-z.

Singh, M., Srivastava, A., Shinde, A.D., Acharya, R., Reddy, A.V.R. and Singh, I.B. (2012). Study of arsenic (As) mobilization in the Ganga Alluvial Plain using neutron activation analysis. *Journal of Radioanalytical and Nuclear Chemistry*, **294:** 241-246. doi:10.1007/s10967-1592-y.

Smedley, P.L. and Kinniburgh, D.G. (2002). A review of the source, behavior and distribution of arsenic in natural waters. *Applied Geochemistry*, **17:** 517-568. doi:10.1016/S0883-2927(02)00018-5.

Stable Isotopic Signatures for Hydrogeochemical Characterisation of Ground Water from Pondicherry to Nagapattinam, Tamil Nadu

S. Chidambaram, K. Tirumelesh, M.V. Prasanna, R. Thilagavathi, S. Pethaperumal, G. Johnson Babu, and P. Paramaguru

Introduction

Groundwater recharge is a critical hydrological parameter. Identification of recharge zones and recharge estimates are essential to water resources management (Scanlon and Cook, 2002; Raju et al., 2006). In recent years, the overgrowing population and climate change are putting water resources under pressure all over the world. Sustainable management of aquifers to meet human and ecosystem needs will require protection of recharge areas and their augmentation (Reddy et al., 2000).

A precise knowledge in distribution of isotopic composition about the local precipitation and its relationship to the local environmental condition is essential for hydrological studies on a regional scale yielding vital information on the mean elevation of recharge area for aquifers, hydrological flow, etc. (Longinelli and Selmo, 2003). Precipitation and stable isotope values are primarily a function of prevailing atmospheric conditions such as relative humidity, temperature etc. during rainfall. Environmental isotopes have been used to understand the moisture

S. Chidambaram (✉) • R. Thilagavathi • G.J. Babu
Department of Earth Sciences, Annamalai University, Annamalai Nagar, 608002 India
e-mail: chidambaram_s@rediffmail.com

K. Tirumelesh
SOE, IAD, HIS, BARC, Mumbai, India

M.V. Prasanna
Department of Applied Geology, School of Engineering & Science, Curtin University, Sarawak Malaysia, CDT 250, 98009 Miri, Malaysia

S. Pethaperumal
Hydrogeologist, State Groundwater Department, Pondicherry, India

P. Paramaguru
Mohamed Sathak Engineering College, Kelakarai, Tamil Nadu, India

N.J. Raju et al. (eds.), *Management of Water, Energy and Bio-resources in the Era of Climate Change: Emerging Issues and Challenges*,
DOI 10.1007/978-3-319-05969-3_9

source of precipitation, precipitation recharge to ground water and also to understand the relative contribution of precipitation in different seasons (Chidambaram et al., 2009). Many investigators studied the ^{2}H and ^{18}O variations in precipitation to characterize the precipitation isotopically (Dansgaard, 1964; Yurtsever and Gat, 1981; Rozanski, 1982; Nativ and Riggio, 1990; Gedzelman and Lawrence, 1990; Krishnamurthy and Bhattacharya, 1991; Gat, 1996; Araguas-Araguas et al., 1998; Despande et al., 2003; Saravana Kumar et al., 2009).

Groundwater tracers have been widely used to identify the areas contributing to groundwater recharge. Specifically, stable isotopes of hydrogen (^{2}H) and oxygen (^{18}O) have been used as conservative groundwater tracers because values remain constant as long as there are no phase changes or fractionation along the flow path (Clark and Fritz, 1997). The interest for the hydrological framework of the coastal aquifers of Nagapattinam region can be explained by the extensive exploitation of these aquifers for irrigation and growing population water needs. Both shallow and deep groundwaters are being harnessed in this region leading to undesirable effects such as seawater intrusion and up coning of deep seated saline waters. There are cases of sea water intrusion in parts of this region (Chidambaram et al., 2010). The present study involves hydrogeochemical and stable isotopic characters of deep and shallow aquifers during premonsoon and postmonsoon. It also aims to investigate the δD-$\delta^{18}O$ relationships and to understand the hydrogeochemical variation of ground water with respect to depth.

Study Area

The area chosen for study falls along the coastal stretch of Pondicherry to Nagapattinam region, which is situated on the eastern margin of Tamil Nadu. It includes the taluks of Cuddalore, Chidambaram, Sirkazhi, Mayiladuthurai, Tharangambadi, Nagapattinam, Kilvelur, Thirukkuvalai and Vedaranyam. The area has also been extended into the Karaikkal Union Territory so as to get a complete overview along the coast. The subsurface water in the study area are being tapped by the industries along the coastal regions. Apart from the industrial threat, paddy cultivation is also practiced in many parts of this region extracting sufficient amount of water from the aquifers. Moreover this region has salt pans and several distributary channels, which serve as back-water during non-monsoonal periods. The normal annual rainfall of the study area is 1230 mm. The major part of the study area is covered by black clay soils (matured) as per the classification of soil survey and land use shows an isolated patch of brown clay loam soil in the area bordering the north-western boundary of the Karaikal Region. The study area forms part of Cauvery river basin and is drained by a network of rivers like the Cauvery, the Kollidam, the Arasalar, the Vettar and the Ponnaiyar. All the rivers are almost ephemeral in nature. At the estuaries, all the rivers are filled with the back-waters. The general descriptions of the various geological formations occurring in the study area are briefed in Table 1. Location map of the study area is shown in Fig. 1.

Table 1 Stratigraphic succession of geological formations in study area

ERA	Period	Formation	Lithology
Quaternary	Recent to pliestocene	Alluvium	Soils, coastal and river sand
Tertiary	Pliocene	Podakkal formation	Sand and clays
	Mio-Pliocene	Cuddalore formation	Sand, Lignite, clay

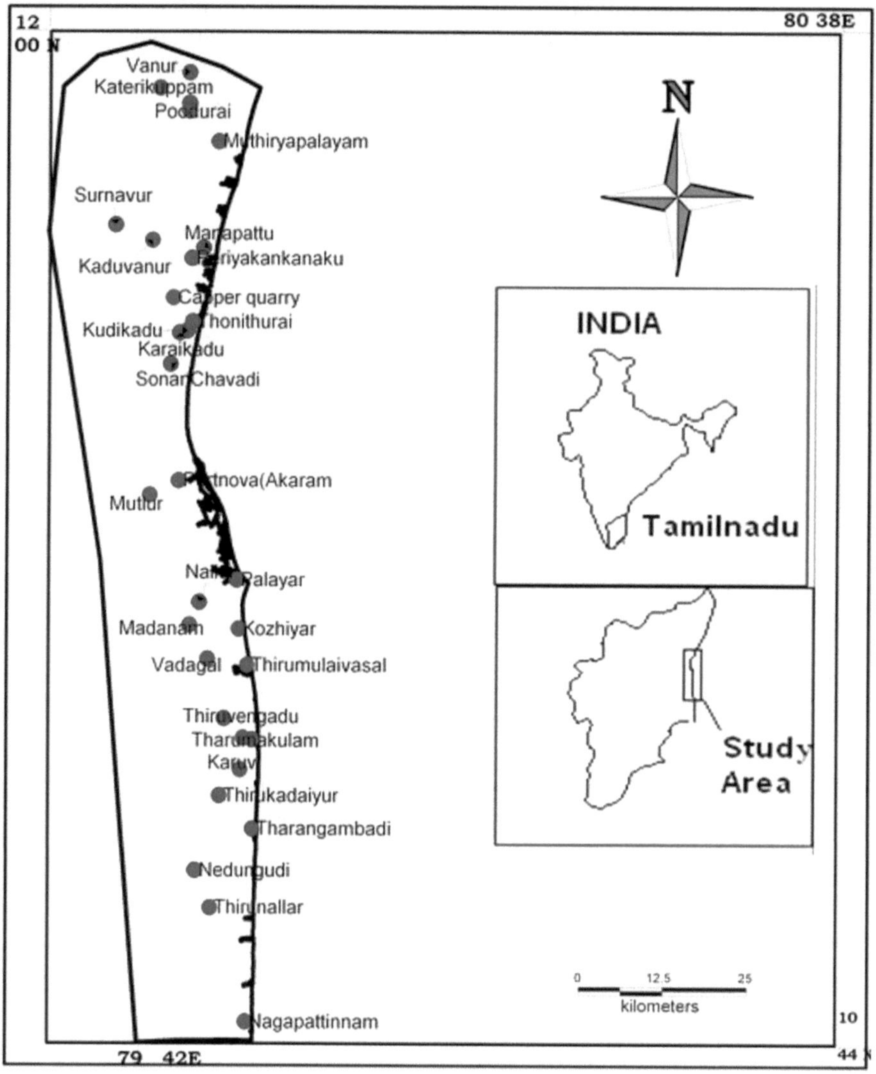

Fig. 1 Location map of the study area

Sampling and Methodology

Stable isotopic studies ($\delta^{18}O$ and δD) were carried out for 62 samples from different locations of the study area during PRM in June month and POM in January month by adapting standard procedures (IAEA) for isotopes and the samples were analysed for major cation and anions by using standard procedures (APHA, 1998). Sixteen samples from shallow groundwater representing pre-monsoon (SPRM) and 16 representing post-monsoon seasons (SPOM) were collected and analysed. Similarly, 15 samples from deep groundwater during pre-monsoon (DPRM) and 15 from post-monsoon (DPOM) seasons were collected. Samples of shallow and deep groundwater locations are shown in Table 2. The shallow and deep aquifers were considered based on the intermittent clay layers approximately at an average depth of 15-20 m. Environmental stable isotopes and radioisotopes (^2H and ^{18}O) have also been measured for hydrogen and oxygen-18 isotope analyses. The results are expressed in δ values (IAEA), which are in per mil (‰) and are computed as:

$$\delta^2H\left(\text{or}\delta^{18}O\right) = \left(\frac{R_{\text{sample}} - R_{\text{standard}}}{R_{\text{standard}}}\right)10^3$$

where $R = \frac{^2H}{^1H} \text{or} \frac{^{18}O}{^{16}O}$

The precision of measurement of δ^2H was ± 0.5‰ and that of $\delta^{18}O$ was ± 0.1‰.

Table 2 The classification of groundwater samples with respect to depth

S.No	Location name	Deep (metre)	Location name	Shallow (metre)
1	Periyagangkuppam	140	Karaikadu	4.55
2	Capper Quarry	215	Sonanchavadi	3.64
3	Vanur	180	Portonovo	3.03
4	Katerikuppam	180	B. Mutlur	3.94
5	Karasur	185	Thirumullaivasal	4.55
6	Poodurai	30	Vadagal	8.18
7	Muthiryapalayam	85	Thiruvenkadu	5.45
8	Manapattu	215	Tharmakulam	7.58
9	Kaduvanur	46	Karuvi	8.79
10	Surnavur	70	Thirukadaiyur	6.67
11	Thaikkal Thurarai	54.55	Tharankampadi	5.45
12	Kudikadu	39.39	Kozhaiyar	3.03
13	Tharmakulam (Bore)	37.88	Palayar	2.73
14	ThiruNallar	21.21	Madhanam	7.58
15	Nedunkadu	24.24	Nagapattinam	12.12
			Nallur	6.67

Results and Discussion

The δ^2H-$\delta^{18}O$ Relationship in Ground Water

The stable δD and $\delta^{18}O$ isotopes are ideal tracers for determining the recharge areas and flow path of ground water, because they are from the water molecules and susceptible to physical processes such as mixing and evaporation (Coplen, 1993). Groundwater samples of shallow zone isotopes enrichment of $\delta^{18}O$ and δ^2H values ranges between –6.4 and –1.4‰ and –43.1 to –10.3‰ during PRM respectively and in POM $\delta^{18}O$ ranges from –5.9 to –0.3‰ and δ^2H from –41.1 to –11.3‰ (Table 3). In the case of deep samples the $\delta^{18}O$ and δ^2H during PRM ranges between –6.4 to –2.3‰ and –41.5 to –16.2‰, and during POM the range is from –7.6 to +0.5‰ and –55.6 to –1.8‰ respectively. The slopes and intercepts of the various regression lines are lower than the GMWL and the LMWLs for which the precipitation data are available. The $\delta^{18}O$ and δD of the groundwater samples of the region (Fig. 2) was plotted to correlate with the SM (Summer Monsoon), NEM (North East Monsoon) and SWM (South West Monsoon) local meteoric water lines (LMWL).

A wide variation is noted in shallow groundwater samples of both POM and PRM. Few of the samples are found to be contributed from the GMWL and SM (LMWL); further few samples were observed to fall in line with SM. The stable isotope signatures of $\delta^{18}O$ and δD values obtained here were also correlated with derived local meteoric water line of SWM precipitation (Chidambaram et al., 2009).

$$\delta D = 7.89\,\delta^{18}O + 10.38\ (\mathrm{SWM})$$

Various studies have also developed LMWLs for several Indian regions (Deshpande et al., 2003; Unnikrishnan Warrier et al., 2010). The LMWL provides a baseline for groundwater investigations of a region, which differs from the global line owing to the variations in climatic and geographic parameters. The isotopic composition of ground water has resulted due to the recharge by local precipitation from both SWM and SM.

Some deep groundwater samples in POM clearly reflect a recharge due to the NEM monsoon without any significant influence of the SM and SWM. Few samples reflect the influence of a saturated soil that limited rainwater percolation during the second monsoon stage (Negrel et al., 2011). On the contrary, the few shallow PRM sample points also show more depleted values when compared to SM and SWM, suggesting that they are more influenced by the NEM monsoon. Groundwater samples were divided into three groups as: (group A) the samples are depleted in heavy isotopes, and that these waters are isotopically not altered during PRM and POM, these sample plots near to the Global meteoric water line and LMWL; (group B) the samples are lighter during PRM and become heavier during the POM; and (group C) show samples with lighter isotopes during POM and heavier during PRM.

Most of the shallow groundwater samples are observed in group A and followed by group B. The samples of group B are represented at Karaikadu, Nagapattinam, Madhanam, Thamarakulam and Thiruvenkadu due to the fact that ground water

Table 3 Maximum, minimum and average of geochemical parameters and isotopes

		pH	EC (µS/cm)	Ca (mg/l)	Mg (mg/l)	Na (mg/l)	K (mg/l)	Cl (mg/l)	HCO_3 (mg/l)	NO_3 (mg/l)	SO_4 (mg/l)	$\delta^{18}O$ (‰)	δ D (‰)
DPRM	Max	7.50	2308.00	264.00	70.20	493.00	16.30	397.70	540.00	60.36	570.00	−2.35	−16.22
(n = 15)	Min	6.00	248.00	11.30	7.95	12.70	0.60	17.00	87.50	0.01	8.20	−6.44	−41.52
	Avg	6.92	1135.87	56.53	23.07	145.07	4.98	151.39	299.08	9.92	77.14	−5.06	−31.21
SPRM	Max	7.80	6910.00	209.00	111.00	1330.00	154.50	1823.00	567.50	147.00	604.48	−1.43	−10.31
(n = 16)	Min	6.50	585.00	19.80	9.68	54.40	2.81	61.72	160.00	0.70	8.00	−6.45	−43.16
	Avg	7.11	1988.63	64.45	39.03	271.67	37.92	397.01	314.88	32.72	139.74	−4.21	-28.56
DPOM	Max	7.54	2444.00	155.00	53.40	410.00	21.60	441.00	732.00	24.60	186.50	0.54	−1.88
(n = 15)	Min	5.50	262.00	11.50	8.70	7.90	1.50	23.10	97.60	0.14	8.60	−7.70	−55.63
	Avg	6.81	998.13	55.67	25.41	127.53	6.62	132.73	322.89	5.54	46.57	−4.19	−31.36
SPOM	Max	7.60	6522.00	516.00	164.00	672.00	250.00	1567.00	683.20	198.00	954.00	−0.34	−11.39
(n = 16)	Min	6.55	230.00	23.80	12.50	19.20	1.50	12.90	87.50	0.05	12.50	−5.95	−41.18
	Avg	7.03	2305.13	119.22	48.44	278.97	52.60	480.81	344.81	45.22	166.54	−4.04	−30.94

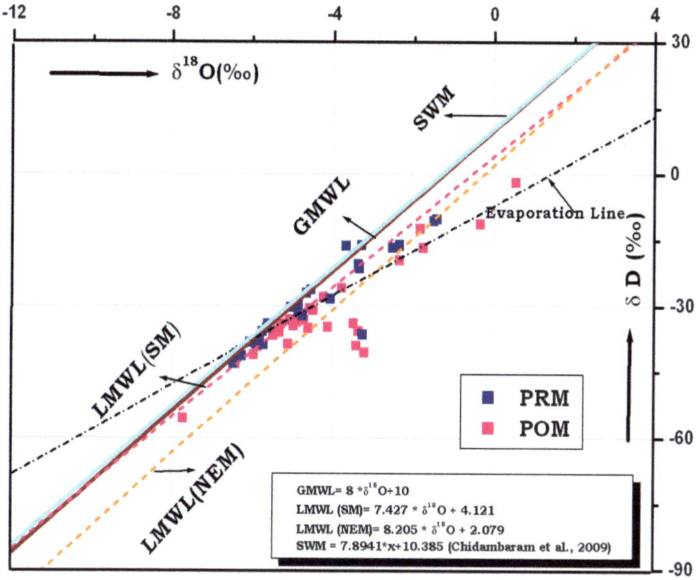

Fig. 2 δD-δ^{18}O relationship of ground water at different seasons

may be recharged by seepage from surface water, such as rivers. Most of the recharge is from seepage, the ground water should reflect the mean isotopic composition of the river or the lake instead of local precipitation (Aggarwal et al., 2000). In this case samples with heavier isotopes during POM is nearby the river and water may be considerably enriched in heavy isotopes due to the impact of evaporation (Fig. 3) in the surface water.

The sample of group C reflects the fact that enrichment is limited by direct isotopic exchange with atmospheric moisture. Thus this enrichment is higher here and evaporation is more intense with respect to the total volume of water, that is enclosed by rivers in this area (Aggarwal et al., 2000) and the samples like Tharangapadi, Palayar, Karuvi, Kozhaiyar, Vadagal, B. Mutlur belong to the group C which have the lighter isotopes during POM that became heavier during PRM. Due to evaporation the concentration of dissolved salts increases in water as a function of its initial composition, the intensity of evaporation and atmospheric humidity. The direct meteoritic water recharge dilutes total ionic concentration and EC of the samples which are reduced after monsoon reflecting the same. This phenomenon occurs at the surface and during water infiltration to deep saturated zones (Clark and Fritz, 1997) and also may be due to seawater intrusion. Shallow groundwater samples during POM at B. Mutlur, Thirukadaiyur, Palayar and Nallur during PRM fall right below the LMWL line indicate the lighter isotopes may also have contributions from anthropogenic activities like aquaculture and salt pan activities in the study area (Aggarwal et al., 2000).

In deep waters most of the PRM and POM season samples show a clear contribution from SM source indicated by their alignment along the LMWL of

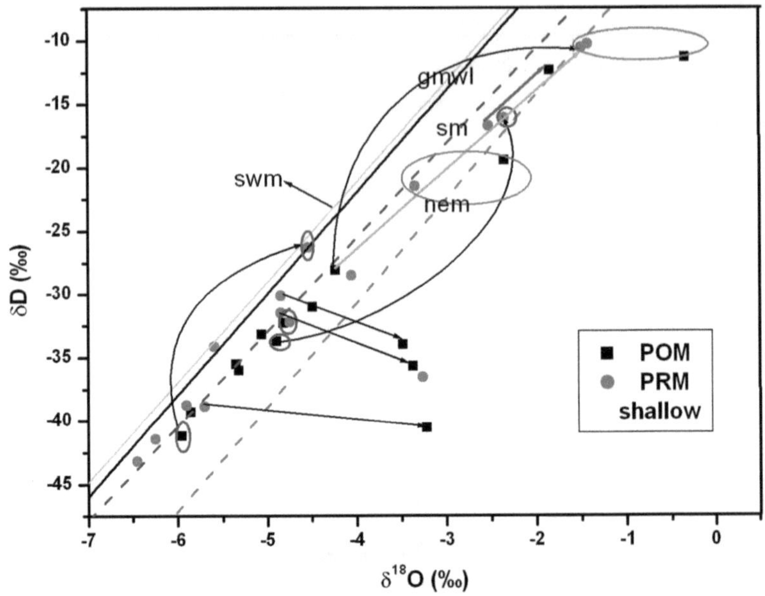

Fig. 3 δD-δ^{18}O relationship of shallow groundwater at different seasons

SM and few of the PRM samples fall along the SWM (LMWL) which is recharged from the local regional precipitation of SWM (Fig. 4). Few of the POM samples correlate with NEM (LMWL) which indicates that isotopes are derived from the local recharge by NEM monsoon. Most of the samples in deep groundwater in group A followed by Kudikadu, Capper quary, Vanur and Taikalthonithurai belong to the group B and Surnavur, Kaduvanur and Thamaraikulam samples belong to the group C. Groundwater in Kanduvanur and Capper quary during POM observed below the NEM may be due to the anthropogenic factor; heavier isotopes are noted in Thirunallar, Nendukadu, Kudikadu and Nagapattinam in POM, which along the coastal region may be due to saltwater intrusion or dissolution of salts precipitation along the pore spaces.

The δ^{18O}-Deutrium – Excess Relationship

The d-excess value may also be impacted by evaporation of the precipitation, either as it falls through the air (Gat, 1996) or as it sits in the rain collector. D-excess is a useful parameter to distinguish the vapour sources. It is calculated by the following equation:

$$\text{d-excess} = \delta^2\text{H} - 8 \times \delta^{18}\text{O}$$

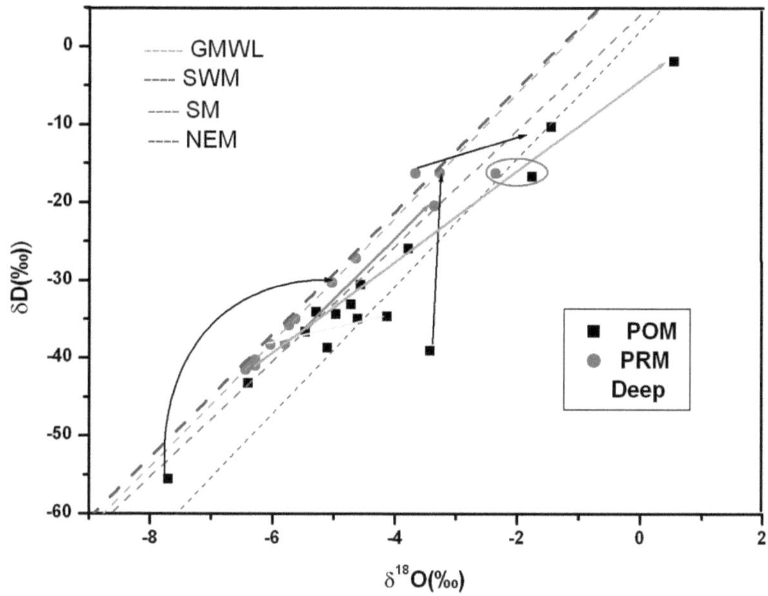

Fig. 4 δD-δ^{18}O relationship of deep groundwater at different seasons

The d-excess values of the shallow groundwater during PRM and POM are varying from −10.50‰ to +10.594‰ and −14.85‰ to 7.4‰ respectively. In deep groundwater majority of the samples during PRM and POM show d-excess variation between 2.5 and 13.0‰ and from −11.70 to 8.1‰ respectively. Values of the d-excess factor (d − excess = δ^2H − 8 × δ^{18}O) around +5‰ indicates that ground water has undergone evaporation prior to infiltration (Caro et al., 2009). In these regions, in PRM deep and shallow groundwater samples are enriched with respect to the precipitation in δ^{18}O and show (Dalai, 2002; Gupta, 2003) lower values of d-excess. This indicates that there is little significance to kinetic evaporation of the precipitated water before groundwater recharge.

Increase of d-excess for ground water in PRM and POM season suggested a new source whose d-excess (>10.5%) was characterized by precipitation or storm runoff (Liu et al., 2010) and the decline in d-excess in POM indicated the recharge of severe-evaporated sources either by lateral recharge or by surface runoff retained by silt arresters. High d-excess values in PRM shallow and deep samples indicate that more evaporated moisture has been added to the atmosphere (Gat and Matsui, 1991), and low values are associated with samples fractionated by evaporation. In Nallur, samples show the negative d-excess values during POM; most of the samples are showing same trend as PRM (Fig. 5). Shallow groundwater from Mutlur, Thirukadaiyur, Palayar, Madhanam and Thamarakulam, and deep groundwater samples from Kaduvanur and Capperquary show the negative d-excess values with depleted δ^{18}O, where 'd-excess' values of less than 5, however, suggest

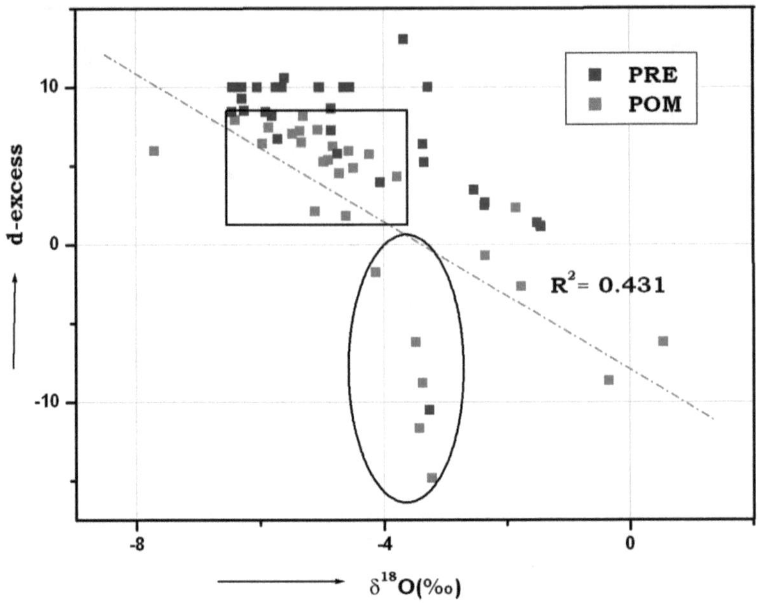

Fig. 5 $\delta^{18}O$ and d-excess relationship of ground water at different seasons

significant eva-poration of rainwater, leaving the residual groundwater (Negrel et al., 2011). The deep groundwater samples of Kudikadu show the positive $\delta^{18}O$ with negative d-excess value.

Stable Isotopes ($\delta^{18}O$, δD) vs. Chloride

A comparison of the deuterium and chloride data provides greater understanding of ground water-surface water interaction processes in the study area. There are few significant variation of isotopic characters with respect to Cl. The chloride-$\delta^{18}O$ plot (Fig. 6) which indicates that major mechanism functioning in the study area are (i) recharge from precipitation, (ii) recharge may be from tank/river (evaporated water) and (iii) recharge of the mixed sea water and the saline back waters. The chloride-deuterium plot suggests that some of the groundwater samples from SPRM and SPOM along the coast have high chloride and enriched in δD (Fig. 7) indicating that these areas rarely receive recharge from meteoric water and recharge of the evaporated waters from a different source nearby or due to the evaporation taking place along the coast. Higher concentration of chloride was attributed to the leaching of secondary salts from the formation (Chidambaram et al., 2007) or due to sea water intrusion. It has been reported that few samples from DPOM are characterized by low chloride and relatively enriched δD signatures, indicating frequent recharge by river waters (Prasanna et al., 2008). Certain groundwater

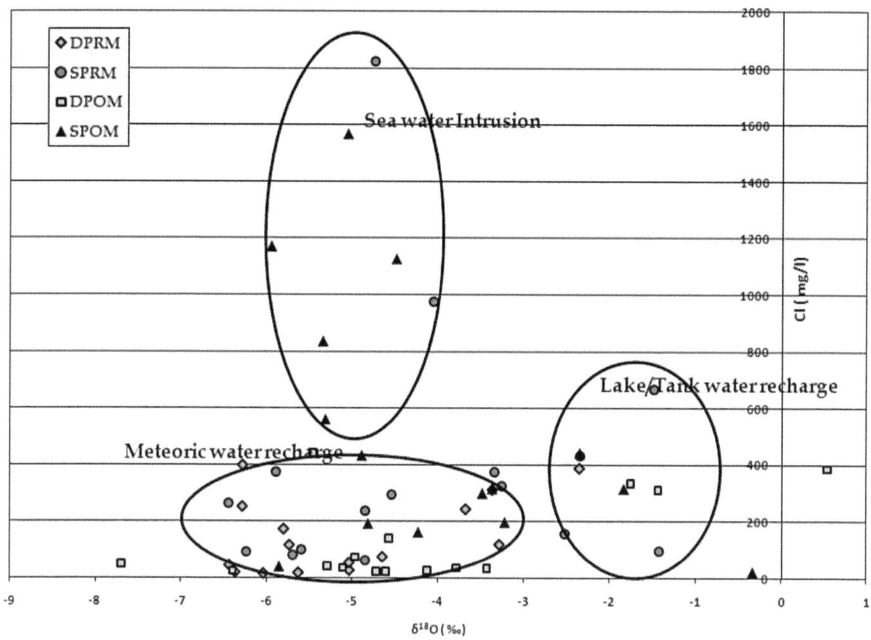

Fig. 6 Plot for chloride versus $\delta^{18}O$ data of groundwater samples

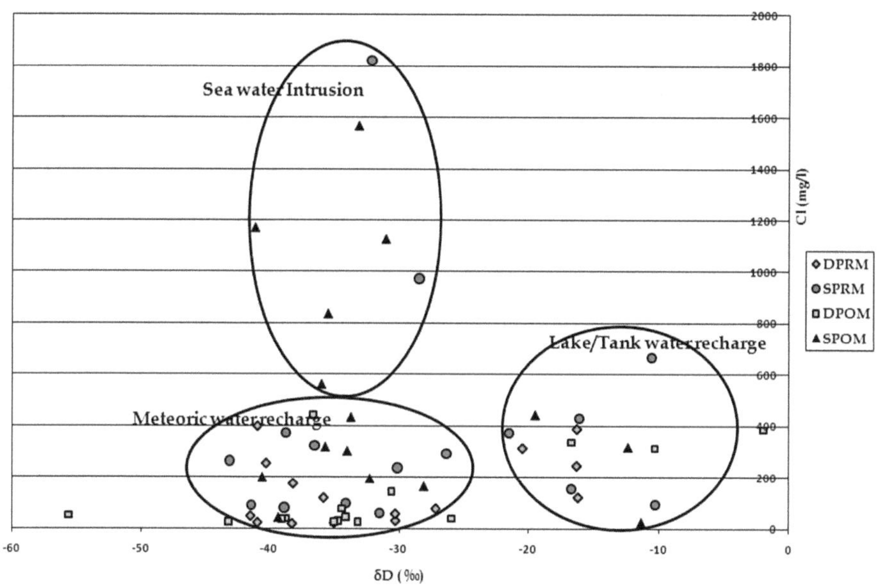

Fig. 7 Plot for chloride versus δD data of groundwater samples

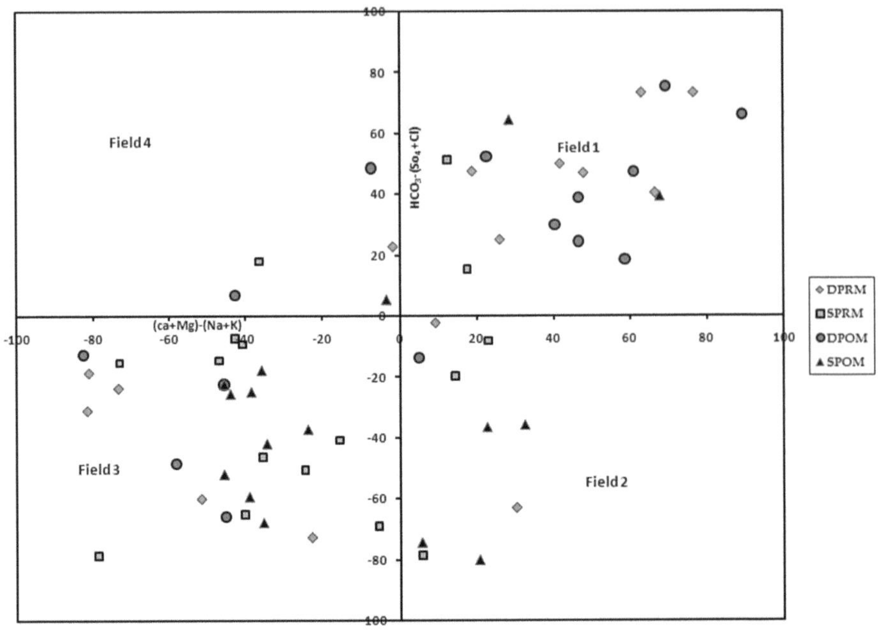

Fig. 8 Chadda's geochemical process evolution plot

samples of both the depths during PRM and POM show very high Cl^- and high deuterium is due to the dominance of seawater/brackish water and some samples from shallow PRM are represented with low Cl^- with low δD signatures shows the recharge of fresh water from less evaporated source.

These processes can also be easily discerned by Cl^- versus $\delta^{18}O$. The distribution of Cl^- variation with respect to $\delta^{18}O$ in samples (Fig. 6) shows the dominance in some of the SPRM and SPOM of seawater/brackish water. Few of the samples from DPOM, SPRM and SPOM show the dominance of river/tank water recharge. Some samples show high Cl^- and moderate enrichment in $\delta^{18}O$ indicating contribution of saline and evaporated surface water bodies such as backwater. Most of the samples are dominated by the meteoric water in both the seasons of the deep aquifers.

Compositional Variations of Major Ions

The maximum/minimum and average is shown in Table 3. The resultant diagram is exhibited in (Fig. 8): Field 1 (Recharging water); Field 2 (Reverse ion-exchange); Field 3 (Na^+-Cl^-) waters are typical sea water mixing and are mostly constrained to the coastal areas or water with higher residence time; and Field 4 (Na^+-HCO_3^-) waters possibly represent Base Exchange reactions (Chadha, 1999). The plot shows that majority of samples from deep groundwater during PRM and POM fall in

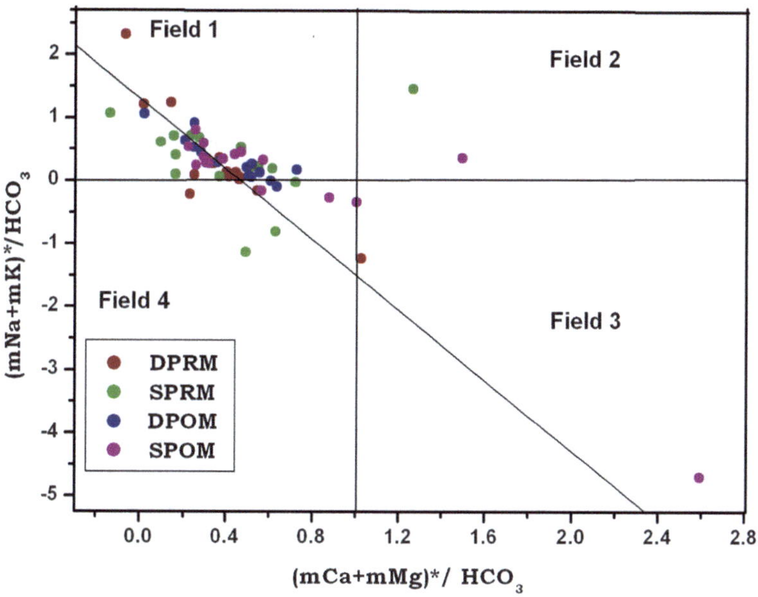

Fig. 9 The ratio between the (Ca + Mg)*/HCO_3 and (Na + K)*/HCO_3

Field 1 (Fig. 8) indicating recharging waters which is evidenced from the $\delta^{18}O$ and δD shows that the POM samples are influenced by the LMWL and the PRM samples follows the SM water line SPRM and SPOM samples are represented in Field 3 indicating the dominance of Na^+ and Cl^- for sea water intrusion. This is also clearly evidenced by stable isotopes followed by these very few representations which are also noted in Field 2, indicating the characteristics of reverse ion exchange.

Stoichiometry of the Waters

Stoichiometry of these waters shows that a considerable amount of $[Ca^{2+} + Mg^{2+}]$ is balanced by SO_4^{2-}. Assuming that SO_4^{2-} is balanced only by Ca^{2+} and Mg^{2+}, the balance $[Ca^{2+} + Mg^{2+}]^*$, which is given by subtracting $SO_4{}^{2-}$ valences from the total $[Ca^{2+} + Mg^{2+}]$ equivalences, is accounted for by weathering of carbonate and/or silicate rocks by carbonic acid. Covariation of $[Na^+ + K^+]^*/[HCO_3^-]$ versus $[Ca^{2+} + Mg^{2+}]^*/[HCO_3^-]$ in these waters indicate (Fig. 9) the relative contributions of carbonate and silicate weathering process in waters. In Fig. 9, most of the waters plot near the intersection of these two lines. The ground water from the study area of SPOM, SPRM, DPRM and DPOM seasons plots in the (Field 1) first quadrant. This suggests that weathering of silicate minerals or saline intrusions are the significant contributors for the water chemistry of this region. On the other hand, in Karuvi during SPRM and Nallur during SPOM (Field 2) second quadrant, showing excess of both [Ca^{2+} + Mg^{2+}] and excess [Na^{+} + K^{+}] over [HCO_3^{-}], which may not be ascribed to weathering either of silicate, or of carbonate and of evaporate minerals, but probably to anthropogenic inputs, are responsible for the excess of ions in these waters. Few shallow samples and few deep samples during both seasons are in Field 4 representing Ca–Mg–HCO_3 water type, with low Mg + -Ca with the increasing Na + K.

Conclusions

The study on the stable isotopes of oxygen and hydrogen of groundwater samples collected during PRM and POM shows that POM samples are generally more depleted than PRM samples. In deep aquifers of both PRM and POM, clear contribution from SM and SWM local meteoric source represents the recharging sources. Shallow groundwater samples during PRM and POM results in lighter isotopes along with the contributions from anthropogenic activities and from salt water intrusion in the study area. High d-excess values in PRM shallow and deep samples indicate that more evaporated recharge and low d-excess values in POM represent the fractionation. Shallow samples in both the seasons show the saline water contribution and deep samples show the recharge by local precipitation and by seepage from surface water. The depth-wise and seasonal variations of the hydrogeochemical process are reflected in the isotopes signature.

Acknowledgement The authors wish to express thanks to Board of Research in Nuclear Sciences (BRNS) for providing necessary financial support to carry out this study and author R.T. wishes to express thanks to DST for providing the Inspire fellowship (DST/Inspire Fellowship/2010/[220] 18 March, 2011).

References

Aggarwal, P.K. et al. (2000). A report on isotope hydrology of groundwater in Bangladesh: Implications for characterization and mitigation of arsenic in ground water. IAEA–TC Project (BGD/8/016), IAEA, Vienna.

APHA (1998). Standard methods for the examination of water and wastewater. 19th edn. APHA, Washington, DC.

Araguas-Araguas, L., Froehlich, K. and Rozanski, K. (1998). Stable isotope composition of precipitation over Southeast Asia. *Journal of Geophysical Research*, **103, D22:** 28,721-28,742.

Chadha, D.K. (1999). A proposed new diagram for geochemical classification of natural waters and interpretation of chemical data. *Journal of Hydrogeology*, **7(5):** 431-439.

Chidambaram, S., Prasanna, M.V., Ramanathan, AL., Vasu, K., Hameed, S., Warrier, U.K., Srinivasamoorthy, K., Manivannan, R., Tirumalesh, K., Anandhan., P. and Johnsonbabu, G. (2009). A study on the factors affecting the stable isotopic composition in precipitation of Tamil Nadu, India. *Journal of Hydrological Processes*, **23(12):** 1792-1800.

Chidambaram, S., Vijayakumar, V., Srinivasamoorthy, K., Anandhan, P., Prasanna, M.V. and Vasudeven, S. (2007). A Study on variation in Ionic Composition of Aqueous System in Different Lithounits Around Perambalur Region, Tamil Nadu. *Journal Geological Society of India*, **70:** 1061-1069.

Chidambaram, S., Ramanathan, AL., Prasanna, M.V., Karmegam, U., Dheivanayagi, V., Ramesh, R., Johnsonbabu, G., Premchander, B. and Manikandan, S. (2010). Study on the hydrogeochemical characteristics in groundwater, post- and pre-tsunami scenario, from Portnova to Pumpuhar, southeast coast of India. *Environmental Monitoring Assessment*, **169:** 553-568.

Clark, I.D. and Fritz, P. (1997). Environmental isotopes in hydrogeology. Lewis Publishers, New York.

Coplen, T.B. (1993). Uses of environmental isotopes. *In:* Alley, W.M. (ed.), Regional groundwater quality. Van Nostrand Reinhold, New York.

Dalai, T.K., Bhattacharya, S.K. and Krishnaswami, S. (2002). Stable isotopes in the source water of the Yamuna and its tributaries: Seasonal and altitudinal variations and relation to major cations. *Hydrological Process*, **271:** 3345-3364.

Dansgaard, W. (1964). Stable isotopes in precipitation. *Tellus*, **16(4):** 436-468.

Deshpande, R.D., Bhattacharya, S.K., Jani, R.A. and Gupta, S.K. (2003). Distribution of oxygen and hydrogen isotopes in shallow groundwaters from Southern India: Influence of a dual monsoon system. *Journal of Hydrology*, **271:** 226-239.

Gat, J.R. (1996). Oxygen and hydrogen isotopes in the hydrological cycle. *Annual Review of Earth and Planetary Sciences*. **24:** 225-262.

Gat, J.R. (1980). The isotopes of hydrogen and oxygen in precipitation. *In:* P. Fritz and J. Ch. Fontes (eds), Handbook of Environmental Isotope Geochemistry. Elsevier Scientific Pub. Co., Amsterdam.

Gat, J.R. and Matsui, E. (1991). Atmospheric water balance in the Amazon Basin: An isotopic evapotranspiration model. *Journal of Geophysical Research*, **96:** 13179-13188.

Gedzelman, S.D. and Lawrence, J.R. (1989). The isotopic composition of precipitation from two extra tropical cyclones. American Meteorological Society.

Gupta, S.K., Deshpande, R.D., Bhattacharya, S.K. and Jani, R.A. (2003). Groundwater $\delta^{18}O$ and δD from Central Indian Peninsula: Influence of Arabian Sea and Bay of Bengal branches of summer monsoon. **303(1-4):** 38-55.

IAEA (International Atomic Energy Agency) (1992). Statistical Treatment of data on environmental isotopes in precipitation. Technical reports series 331.

Krishnamurthy, R.V. and Bhattacharya, S.K. (1991). Stable oxygen and hydrogen isotope ratio in shallow ground waters from Northern India and a study of the role of evapotranspiration in the Indian monsoon. *In:* Taylor, H.P., O'Neil, J.R. and Kaplan, I.R. (eds), Stable Isotope Geochemistry: A Tribute to Samuel Epstein. Spec. Publ. No. 3, The Geochem. Society, San Antonio, Texas.

Longinelli, A. and Selmo, E. (2003). Isotopic composition of precipitation in Italy: A first overall map. *Journal of Hydrology*, **270:** 75-88.

Nativ, R. and Riggio, R. (1990). Precipitation in the southern high plains: Meteorological and isotopic features. *Journal of Geophyisical Research*, **95:** 22,559-22,564.

Negrel, Ph., Pauwels, H., Dewandel, B.J., Gandolfi, M., Mascré, C. and Ahmed, S. (2011). Understanding Groundwater Systems and Their Functioning through the Study of Stable Water Isotopes in a Hard-Rock Aquifer (Maheshwaram Watershed, India). *Journal of Hydrology*, **397(1–2):** 55-70.

Prasanna, M.V., Chidambaram, S., Pethaperumal, S., Srinivasamoorthy, K., John Peter, A., Anandhan, P. and Vasanthavigar, M. (2008). Integrated geophysical and chemical study in the lower sub-basin of Gadilam River, Tamilnadu, India. *Environmental Geosciences*, **15(4):** 145-152.

Raju, N.J., Reddy, T.V.K. and Muniratnam, P. (2006). Subsurface dams to harvest rainwater – A case of Swarnamukhi River basin, Southern India. *Hydrogeology Journal*, **14:** 526-531.

Reddy, M.R., Raju, N.J., Reddy, Y.V. and Reddy, T.V.K. (2000). Water resources development and management in the Cuddapah district, Andhra Pradesh, India. *Environmental Geology*, **39:** 342-352.

Rozanski, K., Araguas, A.L. and Gonfiantini, R. (1993). Isotopic patterns in modern global precipitation. *Geophysical Monograph.*, **78:** 1-37.

Rozanski, K., Sonntag, G. and Munnich, K.O. (1982). Factors controlling the stable isotopic composition of European precipitation. *Tellus*, **34:** 142-150.

Saravana Kumar, U., Suman, S., Navada, S.V. and Deodhar, A.S. (2009). Environmental isotopes investigation on recharge processes and hydrodynamics of the coastal sedimentary aquifers of Tiruvadanai, Tamil Nadu State, India. *Journal Hydrology*, **364:** 23-39.

Scanlon, B.R. and Cook, P.G. (2002). Theme issue on groundwater recharge. *Hydrogeology Journal*, **10:** 3-4.

Unnikrishnan Warrier, C., Praveen Babu, M., Manjula, P., Velayudhan, K.T., Shahul Hameed, A. and Vasu, K. (2010). Isotopic characterisation of dual monsoon precipitation evidence from Kerala, India. *Current. Science*, **98:** 1487-1495.

Xin Liu, Xianfang Song, Yinghua Zhang, Jun Xia, Xuecheng Zhang, Jingjie Yu, Di Long, Fadong Li and Bing Zhang (2010). Spatio-temporal variations of δ^2H and δ^{18}O in precipitation and shallow groundwater in the Hilly Loess Region of the Loess Plateau, China. *Environmental Earth Sciences*. DOI 10.1007/s12665-010-0785-y.

Yurtsever, Y. and Gat, J.R. (1981). Atmospheric waters. *In:* Stable isotope hydrology: Deuterium and oxygen-18 in the water cycle. *Technical Report Series*, **210:** 103-142.

Assessment of Hydrochemical Evolution of Ground Water through Silica Geothermometry in a Part of Ganga Basin

Rashid Umar and Zameer Ahmad Shah

Introduction

The hydrogeochemical processes and hydrogeochemistry of the ground water vary spatially and temporally, depending on the geology and chemical characteristics of the aquifer. Fresh groundwaters flowing through different aquifers may be identified and differentiated by their characteristic salinity levels and ionic ratios (Rosenthal, 1987). Changes in chemical characteristics of ground water in different aquifers over space and time often serve as an important technique in deciphering a geochemical model of the hydrological system (Cheboterev, 1955; Hem, 1959; Back and Hanshaw, 1965; Gibbs, 1970; Srinivasamoorthy, 2005; Srinivasamoorthy et al., 2008; Dehnavi et al., 2011). An understanding of geochemical evolution of ground water is important for a sustainable development of water resources for any region; in this connection, an attempt was made to assess the hydrochemical characteristics and chemical alteration of ground water in the study area.

The chemistry of ground water is mainly controlled by the geochemical reactions that take place along the direction of flow. It is necessary to identify these geochemical reactions in the aquifer in order to assess the distribution of major ion chemistry of the region (Fisher and Mullican, 1997; Rosen and Jones 1998; Mohan et al., 2000; Satyanarayanan et al., 2007; Reddy et al., 2010; Reddy and Kumar, 2010; Tirumalesh et al., 2010; Raju et al., 2011). Studying the seasonal and temporal variation of various ions in ground water helps to identify the major processes controlling the composition of ground water in a particular area (Laluraj and Gopinath, 2006; Raju, 2006; Rao, 2006; Andrade and Stigter, 2011; Rouxel et al., 2011; Senthilkumar and Elango, 2013).

R. Umar (✉) • Z.A. Shah
Department of Geology, Aligarh Muslim University, Aligarh 202002, India
e-mail: rashidumar@rediffmail.com

N.J. Raju et al. (eds.), *Management of Water, Energy and Bio-resources in the Era of Climate Change: Emerging Issues and Challenges*,
DOI 10.1007/978-3-319-05969-3_10

Agricultural land-use practices, population growth, rapid industrialization and unplanned urbanization in the Ganga Plain exert a major influence on groundwater quality in many ways (Ansari et al., 2000; Umar, 2006). A number of studies have already been carried out on the quality and chemical characteristics of ground water of Ganga basin (Umar and Ahmad, 2000; Ravenscraft et al., 2005; Umar et al., 2006, 2007; Saha, 2009; Saha et al., 2009; Joshi et al., 2009; Ghose et al., 2009; Dhakyanaika and Kumara, 2010; Raju et al., 2009; Raju, 2012). The present study was performed to assess the chemistry of the groundwater system and to identify the relative roles of water–sediment interaction and anthropogenic impacts.

Each groundwater system in the area has a unique chemistry, acquired as a result of chemical alteration of meteoric water recharging the system (Back, 1966; Drever, 1982). The chemical alteration of the rain water depends on several factors such as soil-water interaction, dissolution of mineral species and anthropogenic activities (Faure, 1998; Subba Rao, 2001; Umar and Ahmed, 2007). The study of a relatively large number of groundwater samples from a given area offers clues to various chemical alterations which the meteoric groundwater undergoes, before acquiring distinct chemical characteristics.

The aim of this work is to provide an unequivocal evidence for involvement of anthropogenic processes in ion acquisition through silica geothermometry.

Study Area

The study area is north-western part of Unnao district, sandwiched between south-easterly flowing rivers Sai and Ganga (Fig. 1). The climate of the study area is semi-arid type. The area lies in the Central Ganga Plain, consisting mostly of alluvial sediments of Ganga basin, which have been classified into Older Alluvium and Newer Alluvium. The major part of the study area is occupied by the Older Alluvial litho-unit which is mainly constituted of the cyclic sedimentary formations of oxidized clay, silt and grey to brown sand occasionally mixed with kankar and ferro-magnesium nodules (Kumar, 2005).

The top clay layer is persistent throughout the area varying in thickness from 5 to 35 m bgl. Underlying this is a more porous granular zone and this granular zone is intervened more frequently by clay layers in north than in the south of the study area. The ground water occurs in the pore spaces of unconsolidated alluvial material in the zone of saturation. The near-surface clay and beds of $CaCO_3$ concretions (locally called *kankar*) supports many open wells where ground water occurs under shallow water table condition (Jha et al., 2009). The Sarda canal passing through this area helps re-charging shallow aquifers in some parts besides rainfall during monsoon seasons.

The depth to water table in the study area ranges between 2.92 to 15.25 m bgl and 3.07 to 14.23 m bgl during pre-monsoon and post-monsoon period of 2010, respectively. Overall, the general groundwater flow direction is from northeast to southwest with some local diversions. The hydraulic gradient varies from 0.68 to 1.95 m/km, the average gradient is 1.12 m/km.

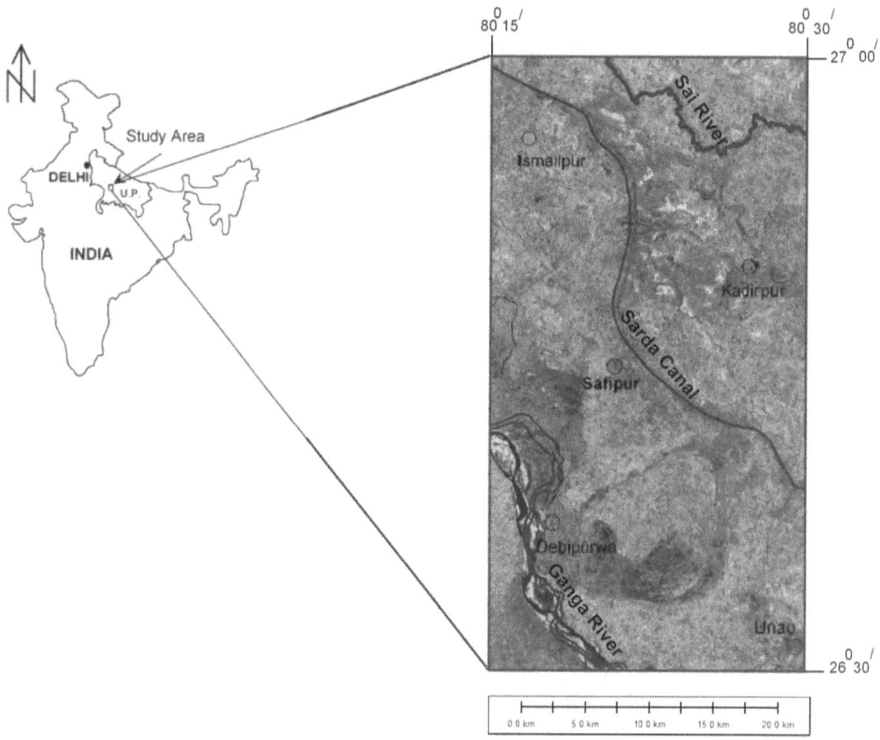

Fig. 1 Location map of the study area

Hydrochemical Aspects

In the present study, 37 groundwater samples were collected for physico-chemical and 20 among them were also selected for silica analysis in two successive pre- and post-monsoon seasons corresponding to June 2010 and November 2010. Hereafter, these sampling periods will be referred as pre- and post-monsoon samples. Hydrochemical data reveals that ground water has pH 6 to 7.6 during pre-monsoon period and 6.8 to 8.5 in post-monsoon period. The ground water in the study area is thus mildly acidic to slightly alkaline in nature. The area is characterized by higher TDS values, showing average values of 982 and 1548 mg/l during pre-monsoon and post-monsoon period, respectively. EC values ranges between 600-1600 μS/cm during pre-monsoon and 400-1400 μS/cm in post-monsoon period. Average hardness values of 202 and 280 mg/l were recorded in pre-monsoon and post-monsoon, respectively.

So far as major ion chemistry is concerned, the area under investigation exhibits an increasing trend in most of the ions towards post-monsoon period. Average Na values of 158 and 238 mg/l were recorded during pre-monsoon and post-monsoon periods, respectively. Likewise K also show an increasing average values, from

27 mg/l in pre-monsoon to 61 mg/l in post-monsoon period, respectively. The area is characterized by a prominent Ca deficiency, showing average values of 59 and 72 mg/l in pre-monsoon and post-monsoon periods, respectively. Mg concentration has remained more or less consistent ranging between average values of 41 and 43 mg/l in pre-monsoon to post-monsoon, respectively.

Discussion

General Alteration Trends

Applying simple logic, it can be inferred that water-rock interaction alone cannot account for the acquisition of solutes and observed chemical changes during the period of study. The questions that need to be answered are that what are those different processes in addition to water-rock interaction which have resulted in the observed chemistry and what is the relative significance of these processes.

The natural processes which may cause chemical changes in ground water are various modes of water-rock interaction involving mainly precipitation and dissolution of chemical species. All of the processes which may account for any kind of change in groundwater system may be various forms of anthropogenic activities.

To assess alteration trends in groundwater chemistry, help has primarily been taken from L-L diagrams. Had there been a clear representative of local meteoric water (LMW), its chemical alteration trends could have been easily deciphered. However, as mentioned above meteoric signatures of all the samples are highly obliterated, the approach that has been adopted is to take some representative samples and assess chemical changes in these during the period of study. These samples (samples 11 and 19, Table 1) shift their position on the L-L plot from pre-monsoon to post-monsoon, thus providing some qualitative estimates of the chemical changes that have occurred during the period of the study (Fig. 2).

Samples 11 and 19 depict reverse trends on the L-L plot (Fig. 2) for pre- and post-monsoon sampling seasons. While the former which plots on the periphery of the meteoric water field in pre-monsoon, qualifying to be the nearest representative of LMW, shifts diagonally when sampled in post-monsoon. This diagonal shift depicts relative enrichment of alkalis and $Cl + SO_4$.

On the other hand, sample 19 shows a more pronounced diagonal shift, but it is reverse in comparison to that of sample 11. While sample tends to get enriched in alkalis and $Cl + SO_4$ in post-monsoon, sample 19 shows relative enrichment of Ca + Mg and HCO_3 during the same period to an extent that it is the only sample in the entire pre-monsoon and post-monsoon population which, though fortuitously, plots within the field of LMW.

The observations on samples 11 and 19 give very significant clues to the processes that might have been operating in the area to account for solute acquisition mechanisms and observed chemical changes.

Table 1 Silica values and the temperatures derived for groundwater samples

		(Pre-monsoon)				*(Post-monsoon)*			
S/No.	*Location*	SiO_2	*Cl*	*TDS*	*T°C*	SiO_2	*Cl*	*TDS*	*T°C*
1	Jagdeshpur	24.5	120.0	986.0	39.5	28.1	128.0	761.0	45.2
2	Argurpur	48.0	159.0	875.0	69.8	22.0	369.0	2384.0	35.1
3	Shadipur	31.5	204.0	774.0	50.2	41.0	355.0	1798.0	62.2
4	Dolawa	48.0	240.0	1107.0	69.8	20.0	256.0	1211.0	31.4
5	Utmanpur	26.0	230.0	1240.0	42.0	22.0	142.0	851.0	35.1
6	Safipur	27.0	119.0	959.0	43.5	30.0	199.0	1204.0	48.1
7	M. Brahmnan	37.0	383.0	1399.0	57.4	34.0	227.0	2141.0	53.6
8	Barithana	47.3	196.0	838.0	69.1	63.5	128.0	1452.0	84.3
9	Munda	19.3	85.0	652.0	29.9	39.5	182.0	1570.0	60.5
10	Jamalpur	34.5	94.0	801.0	54.2	39.3	176.0	1762.0	60.2
11	Sarmba	26.0	127.0	1224.0	42.0	21.0	71.0	1163.0	33.3
12	Hansakhera	21.0	116.0	924.0	33.3	18.0	151.0	1446.0	27.3
13	Barikhera	18.0	437.0	1170.0	27.3	30.0	200.0	1456.0	48.1
14	Mahipatkhera	49.6	128.0	824.0	71.4	59.0	256.0	1599.0	80.4
15	Muminpur	29.0	259.0	1186.0	46.6	23.0	199.0	1759.0	36.9
16	Dostinagar	31.0	227.0	874.0	49.5	37.0	142.0	1538.0	57.4
17	K. Tadwa	31.0	207.0	1253.0	49.5	28.1	440.0	2481.0	45.2
18	S. Sarrai	23.5	173.0	649.0	37.8	29.5	133.0	1588.0	47.3
19	Kanikamau	39.0	295.0	1265.0	59.9	34.0	113.0	1579.0	53.6
20	Dewarakalla	21.5	139.0	634.0	34.2	21.5	335.0	1222.0	54.9

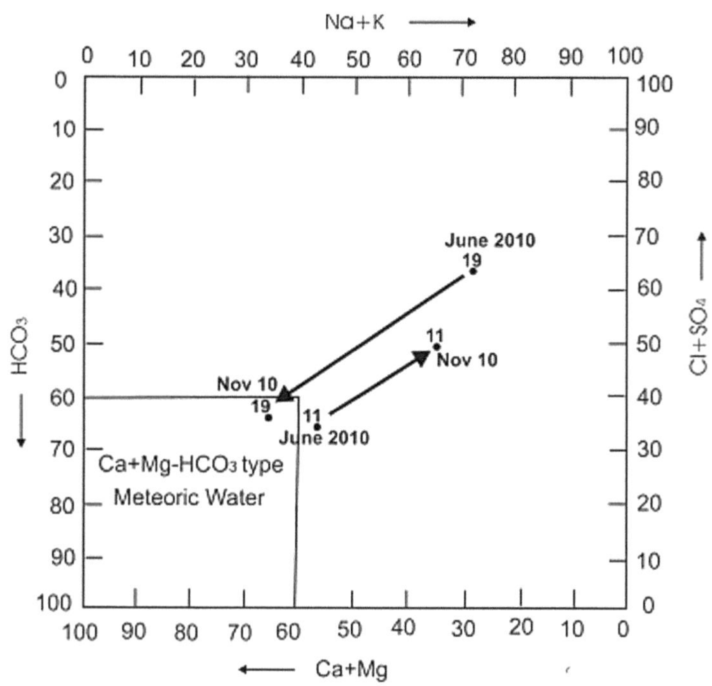

Fig. 2 Groundwater chemistry alteration

Table 2 Aqueous ionic species (mg/l) of sample 11

	Sarmba	
Ionic species in mg/l	*Pre-monsoon*	*Post-monsoon*
K-Cl	72	22
Na-Cl	154	101
Na-HCO_3	245	475
Mg-$(HCO_3)_2$	359	169
Ca-$(HCO_3)_2$	211	
Ca-SO_4	134	201
K-NO_3		78
Surplus SO_4		121
Relative abundance	Mg-$(HCO_3)_2$ > Na-HCO_3 >	Na-HCO_3 > Ca-SO_4
	Ca-$(HCO_3)_2$ > Na-Cl >	> Mg-$(HCO_3)_2$ > Na-Cl
	Ca-SO_4 > K-Cl	> K-NO_3 > K-Cl

- The whole of the natural processes such as water-rock interaction, dissolution and precipitation of minerals does not seem to be highly significant. In case the governing mechanism was geogenic processes, chemical alteration would have been more consistent and regulated and not like that observed in samples 11 and 19, showing different trends for different periods of sampling.
- If that is the case, one has to look for various anthropogenic influences, such as agricultural activities and tanneries.
- Trends contrary to one another shown by samples 11 and 19 depict that seasonal variations and other not so well defined activities at this stage tend to control the chemical signatures of ground water in the area.

Ionic Complexes and Possible Solute Acquisition Processes

On the basis of law of chemical combination, the main ionic complexes that are likely to occur in the two representative samples (samples 11 and 19) have been determined. The differences in the acquisition of ionic species in pre- and post-monsoon provide a clue that ground water in the study area has undergone a series of reactions and processes in the intervening period of two sampling seasons.

As depicted by Fig. 2 and indicated in Table 2, following observations are made in the behaviour of sample 11 during the two sampling periods:

1. The bulk of the chemistry in pre-monsoon is constituted of bicarbonate complexes of Mg, Na and Ca.
2. Bicarbonates of Ca and Mg (possible meteoric signature) put together overwhelmingly dominate over Na-HCO_3 in pre-monsoon.
3. Other complexes which may be present in pre-monsoon samples are Na-Cl, Ca-SO_4 and K-Cl.

Table 3 Aqueous ionic species (mg/l) of sample 19

	Kanikamau	
Ionic species in mg/l	*Pre-monsoon*	*Post-monsoon*
K-NO_3	120	13
K-Cl	193	87
Na-Cl	338	119
Na-HCO_3	252	420
Mg-$(HCO_3)_2$	170	320
Mg-SO_4	45	292
Ca-SO_4	170	
Ca-$(HCO_3)_2$		320
Relative abundance	Na-Cl > Na-HCO_3 > K-Cl >	Na-HCO_3 > Ca-$(HCO_3)_2$
	Mg-$(HCO_3)_2$ = Ca-SO_4 >	= Mg-$(HCO_3)_2$ > Mg-SO_4 >
	K-NO_3 > Mg-SO_4	Na-Cl > K-Cl > K-NO_3

4. In the post-monsoon period, clear indication of relative abundance of alkalis and Cl + SO_4 compared to pre-monsoon samples as also indicated by Fig. 2.
5. The most dominant ionic complex in post-monsoon period is Na-HCO_3 which is overwhelmingly abundant compared to other quantitatively significant complexes of sulphate, bicarbonate and chloride.
6. The most significant change that has occurred in post-monsoon period is the addition of sulphate.

Observations with respect to the behaviour of sample 19 (Fig. 2 and Table 3) are given below:

1. In pre-monsoon sample, Na-Cl is the most dominant complex followed by Na-HCO_3.
2. Due to the anomalously high K content, K-Cl figures out as a dominant ionic species.
3. Bicarbonate complexes of Na and Cl are significant and Ca and Mg tend to form sulphate complexes.
4. In the post-monsoon period, the scenario has totally changed and bicarbonate species of Na, Ca and Mg constitute more than 70 % of the bulk chemistry.
5. Abundance of bicarbonate and tendency of Ca and Mg to form complexes with it, which together are more abundant than Na-HCO_3 results in getting the sample plotted in LMW field giving a fortuitous impression of the sample being representative of local meteoric recharge.

Silica Geothermometry

The physico-chemical analytical results of temperature, TDS, Cl and silica in groundwater samples are presented in Table 1. The main purpose of this approach is to make a distinction between water of deeper origin and shallower origin, which can provide a significant clue either to geogenic or anthropogenic involvement in evolving groundwater quality. Relatively rich literature exists on relationship of silica with parameters, such as temperature, regional heat flow values and depth (Khan and Umar, 2010; Swansberg et al., 1978; Stewart et al., 2007).

In pre-monsoon, when the ambient temperature was in excess of 30 °C, a silica temperature of around 70 °C would mean a temperature of about 40 °C over and above the average air temperature which, in turn, would correspond to a depth of about 1300 m taking into consideration an average thermal gradient value of 30 °C/km (GSI, 1991). Other samples with average silica values of 43 °C, as mentioned earlier, would in the same manner correspond to a depth of about 600 m. An important noteworthy point in both the time periods is that a prominent high temperature occurs in the vicinity of Barithana and Mahipatkhera area, which provide sturdy evidence that ground water in these two areas is emanating from a depth of >1300 m.

Relationship of SiO_2 with Other Chemical Species

Cl acts as an inert element in practically all natural water-rock environments, i.e. once in solution, it stays there both at ambient and elevated temperatures (Ellis, 1970). It is not adsorbed to any marked degree on mineral surfaces (Hem, 1970) and usually does not enter common rock-forming minerals due to the large size of its ion. Keeping this in consideration, Cl values have been related to silica concentrations for both the time periods (Figs. 3a and 3b) to evaluate the role of

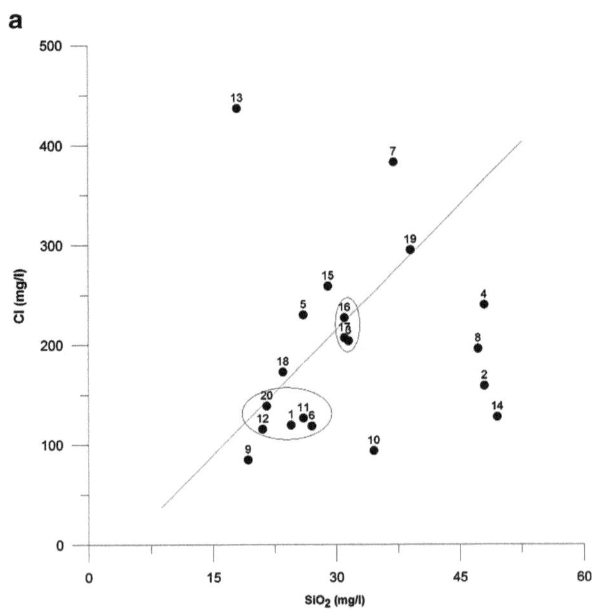

Fig. 3a SiO_2 versus Cl plot (pre-monsoon)

Fig. 3b SiO_2 versus Cl plot (post-monsoon)

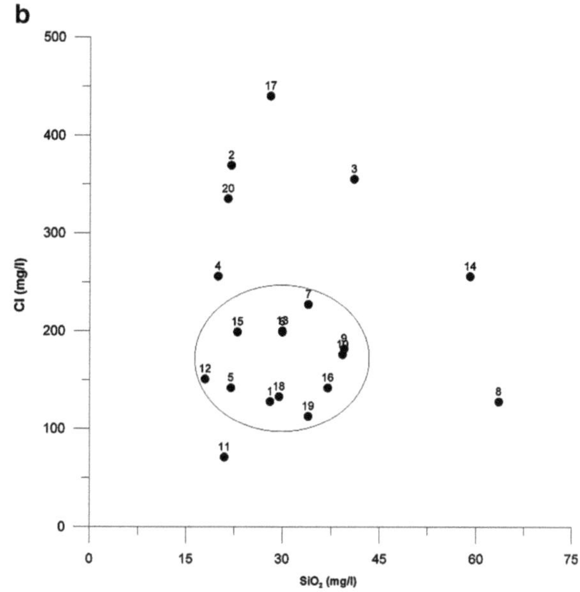

anthropogenic and natural activities in ion acquisition. In SiO_2-Cl plot of pre-monsoon period (Fig. 3a), eight samples tend to cluster in two distinct groups. One group is constituted by samples 3, 16 and 17 characterized by silica content around 31 mg/l and high Cl content ranging between 204 and 227 mg/l. Another group involving samples 1, 6, 11, 12 and 20 exhibits comparatively lower silica content of 21 to 27 mg/l and Cl varying between 116 to 129 mg/l, respectively. These two groups show an average chalcedony temperature of about 50 and 42 °C. Both the groups show relatively high Cl concentrations. The only possible explanation for the high Cl concentrations takes into account sources other than geogenic. The high values of Cl in both the time periods support this assertion that Cl is contributed to the system from more than one source. These two distinct groups seem to have attained their silica level at temperatures around 50 and 40 °C corresponding to a depth of 350 to 700 m at which some reactions or processes involving solid phase cannot be ruled out.

The SiO_2-Cl plot of post-monsoon period (Fig. 3b) does not provide any significant clue as all samples plot without any preferred trend. This may be due to the availability of the new recharge water which might have caused dilution to the extent that any specific signatures might have been obliterated. Twelve out of 20 samples plot in a cluster in which there are clear indications that for nearly identical Cl content of about 150 mg/l, there is large variation in SiO_2 values from 18 to 37 mg/l. If SiO_2 values are taken to have derived through solid-liquid interaction, Cl values seem to be insensitive to such processes and may have been derived through non-geogenic processes.

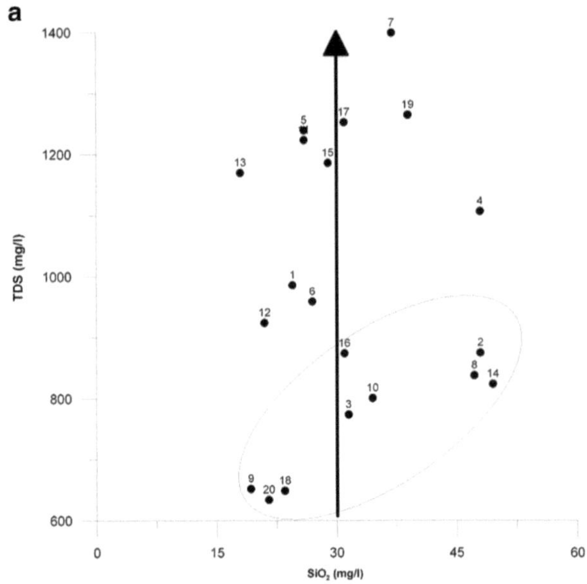

Fig. 4a SiO_2 versus TDS plot (pre-monsoon)

Some trends that may be discovered from the SiO_2-TDS plot of pre-monsoon period (Fig. 4a) are:

1. A group of nine samples show a positive correlation between SiO_2 and TDS implying some contribution of water-rock interaction in acquisition of species.
2. For nearly identical SiO_2 values of about 30 mg/l, TDS varies from <800 to >1200 mg/l. This suggests that bulk of TDS has been acquired by processes other than those responsible for acquisition of SiO_2.

SiO_2-TDS plot for the post-monsoon period (Fig. 4b) also depicts the trend of highly variable TDS values for similar SiO_2 concentration, as observed in pre-monsoon samples. This tends to suggest that relative increase in TDS as observed in post-monsoon period has been brought by processes in which water-rock interaction has played relatively insignificant role.

Conclusion

To provide a concise summary, it may be mentioned that the study area is characterized by a groundwater system in which primordial meteoric signatures have been obliterated. Groundwater chemistry seems to have a very limited role of water-rock interaction, and if there is any, it is masked by the influences of anthropogenic activities. TDS tends to increase in the post-monsoon period due to monsoonal cultivation activities, influences of industrial pollution and dissolution of surface salts deposited in the pre-monsoon period by recharging meteoric water. Silica analysis have revealed that ground water

Fig. 4b SiO_2 versus TDS plot (post-monsoon)

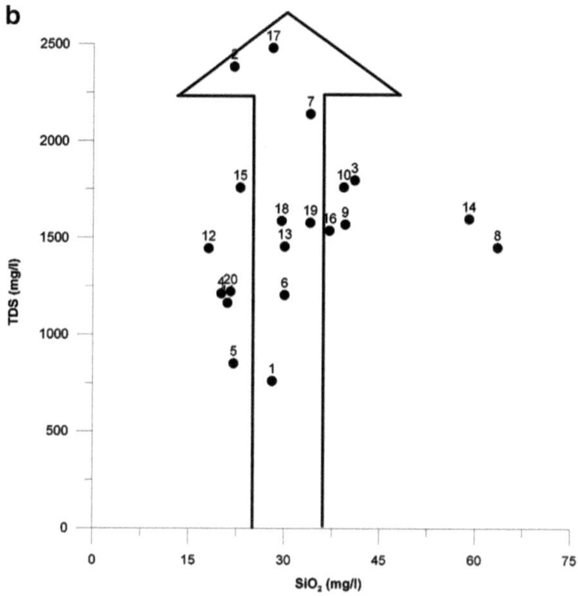

could have emanated from a depth range of 600 to 1300 m bgl. Plots of silica with TDS and Cl suggest that ground water has acquired various chemical species not predominantly through water-rock interaction.

Acknowledgements The financial assistance received by second author in the form of Project Assistant Fellow from DST under project number NoSR/S4/ES-544/2010(G) is acknowledged. The authors express their gratitude to the reviewers of the manuscript; their suggestions have improved the manuscript substantially. The authors are also thankful to the Chairman, Department of Geology, Aligarh Muslim University, Aligarh for providing the basic facilities.

References

Andrade, A.I.A.S.S. and Stigter, T.Y. (2011). Hydrochemical controls on shallow alluvial groundwater under agricultural land: Case study in central Portugal. *Environ. Earth Sci.*, **63(4):** 809-825.

Ansari, A.A., Singh, I.B. and Tobschal (2000). Role of monsoon rain on concentrations and dispersion pattern of metal pollution in sediments and soils of the Ganga Plain, India. *Environmental Geology*, **39(3-4):** 221-237.

Back, W. (1966). Hydrochemical facies and groundwater flow patterns in the northern part of the Atlantic Coastal Plain. USGS Prof. Paper 498-A.

Back, W. and Hanshaw, B. (eds) (1965). Chemical geohydrology advances in hydroscience. Academic Press.

Cheboterev, II. (1955). Metamorphism of natural waters in the crust of weathering-I. *Geochim. Cosmochim. Acta.*, **8(1–2):** 22-48.

Dehnavi, A.G., Sarikhani, R. and Nagaraju, D. (2011). Hydrochemical and rock-water interaction studies in east of Kurdistan, N-W of Iran. *Inter. Jour. of Envir. Sciences and Research*, **1:** 16-22.

Dhakyanaika, K. and Kumara, P. (2010). Effects of pollution in River Krishni on hand pump water quality. *Journal of Engineering and Science Technology Revised*, **3(1):** 14-22.

Drever, J.I. (1982). The geochemistry of natural waters. Prentice-Hall, Englewood Cliffs, NJ.

Ellis, A.J. (1970). Quantitative interpretation of chemical characteristics of hydrothermal systems. *Geothermics*, **2:** 516-527.

Faure, G. (1998). Principles and applications of geochemistry. 2nd edn. Prentice Hall, Englewood Cliffs, New Jersey.

Fisher, R.S. and Mullican, III. W.F. (1997). Hydrochemical evolution of Na-SO_4 and Na-Cl groundwater beneath the northern Chihuahuan desert, Trans-Pecos, Texas, USA. *Hydrogeol. J.*, **5(2):** 4-16.

Geological Survey of India (1991). Geothermal Atlas of India. Geological Survey of India special publication.

Ghose, N.C., Saha, D. and Gupta, A. (2009). Synthetic detergents (Surfactants) and organochlorine pesticide signatures in surface water and groundwater of greater Kolkata, India. *Journal Water Resources Protection*, **4:** 290-298.

Gibbs, R.J. (1970). Mechanism controlling world water chemistry. *Science*, **17:** 1088-1090.

Hem, J.D. (1959). Study and interpretation of the chemical characteristics of natural water. *USGS Water Supply Paper*, **2254:** 117-120.

Hem, J.D. (1970). Study and interpretation of the chemical characteristics of natural water. *USGS Water Supply Paper*, **1473**.

Jha, S.K., Nayak, A.K. and Sharma, Y.K. (2009). Fluoride occurrence and assessment of exposure dose of fluoride in shallow aquifers of Makur, Unnao district, Uttar Pradesh, India. *Environ Monit Assess*, **156:** 561-566, DOI 10.1007/s10661-008-0505-1.

Joshi, D.M., Kumar, A. and Agrawal, N. (2009). Assessment of the irrigation water quality of river Ganga in Haridwar District. *Rasayan Journal of Chemical*, **2(2):** 285-292.

Khan, M.M.A. and Umar, R. (2010). Significance of silica analysis in groundwater in parts of Central Ganga Plain, Uttar Pradesh, India. *Cur. Sci.*, **98(9):** 1237-1240.

Kumar, G. (2005). Geology of Uttar Pradesh and Uttranchal. *Geol. Soc. India*, Bangalore, pp. 267-291.

Laluraj, C.M. and Gopinath, G. (2006). Assessment on seasonal variation of groundwater quality of Phreatic aquifers – A river basin system. *Environ. Monit. Assess.*, **117:** 45-57.

Mohan, R., Singh, A.K., Tripathi, J.K. and Chowdhary, G.C. (2000). Hydrochemistry and quality assessment of groundwater in Naini industrial area, Allahabad district, Uttar Pradesh. *J. Geol. Soc. India*, **55:** 77-89.

Raju, N.J. (2006). Seasonal evolution of hydro-geochemical parameters using correlation and regression analysis. *Curr. Sci.*, **91(6):** 820-826.

Raju, N.J. (2012). Arsenic Exposure through Groundwater in the Middle Ganga Plain in the Varanasi Environs, India: A Future Threat. *Jour. Geol. Soc. India*, **79:** 302-314.

Raju, N.J., Ram, P. and Dey, S. (2009). Groundwater Quality in the Lower Varuna River Basin, Varanasi District, Uttar Pradesh. *Jour. Geol. Soc. India*, **73:** 178-192.

Raju, N.J., Shukla, U.K. and Ram, P. (2011). Hydrochemistry for the assessment of groundwater quality in Varanasi: A fast urbanizing center in Uttar Pradesh, India. *Environ. Mont. Assess.*, **173:** 279-300.

Rao, N.S. (2006). Seasonal variation of groundwater quality in a part of Guntur District, Andhra Pradesh, India. *Environ. Geol.*, **49:** 413-429.

Ravenscraft, P., Burgess, W.G., Ahmed, K.M., Burren, M. and Perrin, J. (2005). Arsenic in groundwater of the Bengal Basin, Bangladesh: Distribution, field relations, and hydrogeological setting. *Hydrogeology Journal*, **13:** 727-751.

Reddy, A.G.S. and Kumar, K.N. (2010). Identification of hydrogeochemical processes in groundwater using major ion chemistry: A case study of Penna-Chitravathi river basins in southern India. *Environ. Monit. Assess.*, **170:** 365-382.

Reddy, A.G.S., Reddy, D.V., Rao, P.N. and Prasad, K.M. (2010). Hydrogeochemical characterization of fluoride rich groundwater of Wailapalli watershed, Nalgonda district, Andhra Pradesh, India. *Environ. Monit. Assess.*, **171:** 561-577.

Rosen, M. and Jones, S. (1998). Controls on chemical composition of groundwater from alluvial aquifers in the Wanaka and Wakatipu basins, central Otago, New Zealand. *Hydrogeol. J.*, **6:** 264-281.

Rosenthal, E. (1987). Chemical composition of rainfall and groundwater in recharge areas of the Bet shean-harod multiple aquifer system, Israel. *Journal of Hydrology*, **89:** 329-352.

Rouxel, M., Molenat, J., Ruiz, L., Legout, C., Faucheux, M. and Gascuel-Odoux, C. (2011). Seasonal and spatial variation in groundwater quality along the hill slope of an agricultural research catchment (Western France). *Hydrol. Process.*, **25(6):** 831-841.

Saha, D. (2009). Arsenic groundwater contamination in parts of middle Ganga plain, Bihar. *Current Science*, **97(6):** 753-755.

Saha, D., Sarangam, S.S., Dwivedi, S.N. and Bhartariya, K.G. (2009). Evaluation of hydrogeochemical processes in arsenic-contaminated alluvial aquifers in parts of Mid Ganga Basin, Bihar, Eastern India. *Environmental Earth Science*. doi:10.1007/s12665-009-0392-y.

Satyanarayanan, M., Balaram, V., Hussin, M.S.A., Jemaili, M.A.R.A., Rao, T.G., Mathur, R., Dasaram, B. and Ramesh, S.L. (2007). Assessment of groundwater quality in a structurally deformed granitic terrain in Hyderabad, India. *Environ. Monit. Assess.*, **131:** 117-121.

Senthilkumar, M. and Elango, L. (2013). Geochemical processes controlling the groundwater quality in lower Palar river basin, southern India. *J. Earth syst. Sci.*, **122:** 419-432.

Srinivasamoorthy, K. (2005). Hydrogeochemistry of groundwater in Salem district, Tamil Nadu, India. Unpublished Ph.D. Thesis, Annamalai University.

Srinivasamoorthy, K., Chidambaram, S., Prasanna, M.V., Vasanthavihar, M., John, P. and Anandhan, P. (2008). Identification of major sources controlling groundwater chemistry from a hard rock terrain – A case study from Mettur taluk, Salem district, Tamil Nadu, India. *Jour. Earth. Syst. Sci.*, **117(1):** 49-58.

Stewart, M.K., Mehlhorn, J. and Elliott, S. (2007). Hydrometric and natural tracer (oxygen-18, silica, tritium and sulphur hexafluoride) evidence for a dominant groundwater contribution to pukemanga stream, New Zealand. *Hyrol. Process.*, **21:** 3340-3356.

Subba Rao, N. (2001). Geochemistry of groundwaters in parts of Guntur district, Andhra Pradesh, India. *Environ. Geol.*, **41:** 552-562.

Swansberg, Chandler, A. and Morgan, P. (1978). The linear relation between temperatures based on the silica content of groundwater and regional heat flow: A new heat flow map of the United States. *Jour. of Pure and Applied Geophysics*, **117:** 1-12.

Tirumalesh, K., Shivanna, K., Sriraman, A.K. and Tyagi, A.K. (2010). Assessment of quality and geochemical processes occurring in groundwaters near central air conditioning plant site in Trombay, Maharashtra, India. *Environ. Monit. Assess.*, **163(1–4):** 171-184.

Umar, R. (2006). Hydrogeological environment and groundwater occurrences of the alluvial aquifers in parts of the Central Ganga Plain, Uttar Pradesh, India. *Hydrogeology Journal*, **14:** 969-978.

Umar, R. and Ahmed, I. (2007). Hydrochemical characteristics of groundwater in parts of Krishni-Yamuna Basin, Muzaffarnagar district, UP. *Jour. Geol. Soc. India*, **69:** 989-995.

Umar, R., Alam, F. and Ahmed, I. (2007). Groundwater quality characteristic and its suitability for drinking and agricultural uses of Hindon-Yamuna sub-basin in parts of western Uttar Pradesh, India. *Indian Journal of Geochemistry*, **22(2):** 223-236.

Umar, R. and Ahmad, M.S. (2000). Groundwater quality in parts of Central Ganga Basin, India. *Environmental Geology*, **39(6):** 673-678.

Umar, R., Khan, M.M.A. and Absar, A. (2006). Groundwater hydrochemistry of a sugarcane cultivation belt in parts of Muzaffarnagar district, Uttar Pradesh, India. *Environmental Geology*, **49:** 999-1008.

Electrical Resistivity Survey for Groundwater Investigation at Sumbli of Jammu District (J&K)

Birendra Pratap and Hari Dev

Introduction

The ground water is the major resource for meeting the water demand of the population for drinking, agriculture and industries throughout the year. Management of the groundwater resource is most essential for achieving goal of good health and prosperity (Yadav et al., 2003; Birendra Pratap, 2006). It is essential for increasing agricultural income, improving urban governance and strengthening the social fabric. The resources management is a multi-disciplinary task. Geophysical electrical resistivity survey is one of the tools for assessment of groundwater resources and management. The occurrence, movement and control of ground water, particularly in hard rock areas, are governed by topography, lithology, structures like fractures, faults and nature of weathering (Raju et al., 1996; Raju and Reddy, 1998).

The concerted efforts were to evolve strategies for maximizing the availability of hazard-free water resources. Therefore, there is a need to focus on a more holistic groundwater management for resources evolution and sustainable use. It may appear strongly that water is going to be scarcer and scarcer day by day, if this resource is not properly conserved and managed effectively. A world-wide water crisis is not unlikely in the near-future. The signals are already being received in many countries including India.

The Kandiand Sirowal area of Jammu and Kashmir State is characterized by shallow soils full of boulders with almost negligible natural water resources (Birendra Pratap et al., 2009). This area is facing acute shortage of water which in turn affects the productivity of the land also. The average annual rainfall of the area is 1100 mm. About 85 % of the total rainfall is received during monsoon

B. Pratap (✉) • H. Dev
Central Soil and Materials Research Station, New Delhi, India
e-mail: bpbhu2002@yahoo.co.in

N.J. Raju et al. (eds.), *Management of Water, Energy and Bio-resources in the Era of Climate Change: Emerging Issues and Challenges*,
DOI 10.1007/978-3-319-05969-3_11

period. The study area is located at 25 km south-east from Jammu, near Khara Madhana, on Jammu-Purmandal road via Ratnu Chak. The study area falls in the Survey of India Toposheet no. 43 L/14 and is located at the North latitudes 32°41'38" and East longitudes 74°58'56". The survey area is represented by hilly rugged topography and lies in outer plain physiographic unit, in the foothills of Siwalik ranges. Altitude of the area varies from 470 to 520 m amsl. The Sumbli Khad, a northern tributary of the Balawal Nalla, drains the area. The general slope of the area is in north–south direction.

Electrical resistivity survey was carried out in and around Sumbli village, Block Purmandal, Tehsil Samba, District Jammu (J&K), which included the hydrogeological traversing of the area. The hydrogeological survey was carried out to identify the suitable site/s having the potential groundwater zone/s to augment the water supply to the village Sumbli and nearby villages by construction of tube well or percolation well. Resistivity survey was also conducted to delineate subsurface geological formation, their hydrological characteristics, thickness of productive groundwater potential zones and to estimate the thickness of alluvium formations overlying the Siwalik formation.

The study indicated that the water shortage is affected by various geological characteristics and structures, porous and permeable rock of Siwalik formations, hilly rugged topography and deep infiltration due to presence of big boulders with sand up to greater depth. As a result, shallow aquifers are not able to support productivity/crop for even one season. Due to unhindered infiltration, water percolates to a depth of more than 100 m. The aquifers in this area cannot be easily identified and utilized.

Geology and Hydrogeology

Geologically, the surveyed area is underlain by Siwalik formation, comprising formations from recent and sub-recent to Miocene, in age. These deposits comprise a horizontally layered thin sequence of pebbles, cobbles, boulders, sandstones and coarse sands intercalated with clays. The geological succession of the area is given in Table 1.

Ground water in the area occurs within the pore spaces of the terrace sediments under water table conditions. The formation is predominantly underlain by the clay due to which the groundwater potential of the area is very poor. One tube-well was

Table 1 The geological succession of the area

Sub group	*Stage*	*Lithology*	*Age*
Upper Siwalik	Dhok Pathan stage	Sandstone, shale with isolated nodules	Lower Pliocene
Middle Siwalik	Nagri stage	Sandstone and shale	Upper Miocene

Table 2 The details of dug wells inventoried in the area

S. No.	*Location of Year of Remarks*			*Depth (m bgl)*		*Dia (m)*	
Lined	*dug well*	*SWL (m bgl)*		*construction*			
1.	Diwan Lalu	15.50	15.70	4.03	Yes	1812 (1866 Vikrami Filled with Shah well garbage Samvat)	
2.	Old well among	6.81	7.50	2.15	Yes	March, 1989	Clay exposed at two wells 6.81 m
3.	New well among two wells	5.45	7.55	1.95	Yes	May, 2006	–

drilled by CGWB under Exploratory Drilling Programme in village Khada Maidana in the year 1991 with the drilled depth of 305.37 m bgl and constructed upto the depth of 293 m bgl. The static water level was 46.17 m bgl. The main lithology encountered in the exploratory well was found to be sandstone and shale. Dug wells in the area mainly tap the ground water at the contact of gravel and clay zones at shallow depths. The details of dug wells inventoried in the area are given in Table 2.

Electrical Resisitivity Survey

Electrical resistivity surveys are usually designed to measure the electrical resistivity of sub-surface materials by making measurements at the earth surface. An electrical field is imposed on the study area by a pair of electrodes at varying spacing expanding symmetrically from a central point, while measuring the surface expression of the resulting potential field with additional pair of electrodes at the appropriate spacing. For an array of current electrodes A and B, and potential electrodes M and N, the apparent resistivity, ρ_a, is expressed by the equation:

$$\rho_a = 2\pi R \times \frac{[\{(AB2) + (MN2)\} \times \{(AB2) - (MN2)\}]}{MN}$$

where R = resistance $\{R = \Delta V/I\}$, I is the current introduced in the earth and ΔV is the potential difference between the potential electrodes.

A total of seven wenner resistivity profiling and thirteen vertical electrical sounding (VES) with schlumberger electrode configuration were carried out. To achieve the above objectives, wenner resistivity profiling and vertical electrical soundings were conducted by using the DDR–4MP resistivity meter, manufactured by IGIS, Hyderabad. The soundings were conducted with the maximum current electrode separation (AB) ranging from 100 m to 220 m. Current electrode separation were expanded in steps of 2 m up to $AB/2 = 20$ m and then in steps of 5 m up to $AB/2 = 50$ m and then in steps of 10 m from $AB/2 = 50$ m onwards, with appropriate *MN* separation. The electrode spread azimuth was kept approximately north–south for all the soundings, in order to avoid variations in measured resistivity due to formation anisotropy.

The values of apparent resistivity (ρ_a: product of resistance and geometric factor) in ohm-m were plotted against the related half-current electrode separation on double logarithmic scale paper of moduli 62.5 mm. The curves were carefully smoothened for the interpretation. Preliminary quantitative interpretations of VES curves were attempted by semi-empirical 'Auxiliary Point' method with the help of two-layer master curves and auxiliary point charts of Orellana-Mooney (1966). The interpreted results give the resistivity of different layers and the depth of various interfaces underneath. The data was also processed and interpreted using software IPI2Win to verify the manually interpreted results. Any deviation of the computed curves from the related field curves was modified keeping in view of the local geology to arrive at a realistic model. The location of wenner profile and VES points are given in Fig. 1.

Interpretation and Results

The interpretation of results include seven wenner resistivity profiles (WRP) for deciding to carry out VES at lowest apparent resistivity; assessment of layer parameters using VES curves; preparing pseudo geo-electrical cross section using the apparent resistivity values (directly received from the resistivity equipment) and respective current electrode separations; and preparing of geo-electrical cross-section using the layer parameters (resistivity and thickness) obtained from interpretation of VES curves.

Wenner Resistivity Profile (WRP)

Seven wenner resistivity profiles (WRP) were carried out in the study area. At Sheesham trees field, three WRPs were conducted on the same profile line, with current electrode separation of 10 m, 20 m and 5 m for WRP1, WRP2 and WRP3, respectively, keeping the station interval of 5 m in all WRPs. The apparent resistivity values with respect to the profile line distances were plotted on graph paper for qualitative analysis (Fig. 2) for deciding to carry out VES at lowest apparent resistivity. Rise in apparent resistivity value, from the distance of 70 m, in all WRPs was observed. From the WRP3 curve, it was inferred that there is high resistivity zone (boulder formation) in the distance range of 90 m to 120 m, in profile line. VES1 to VES3 and VES5 to VES7 were conducted at the locations of low resistivity apparent values.

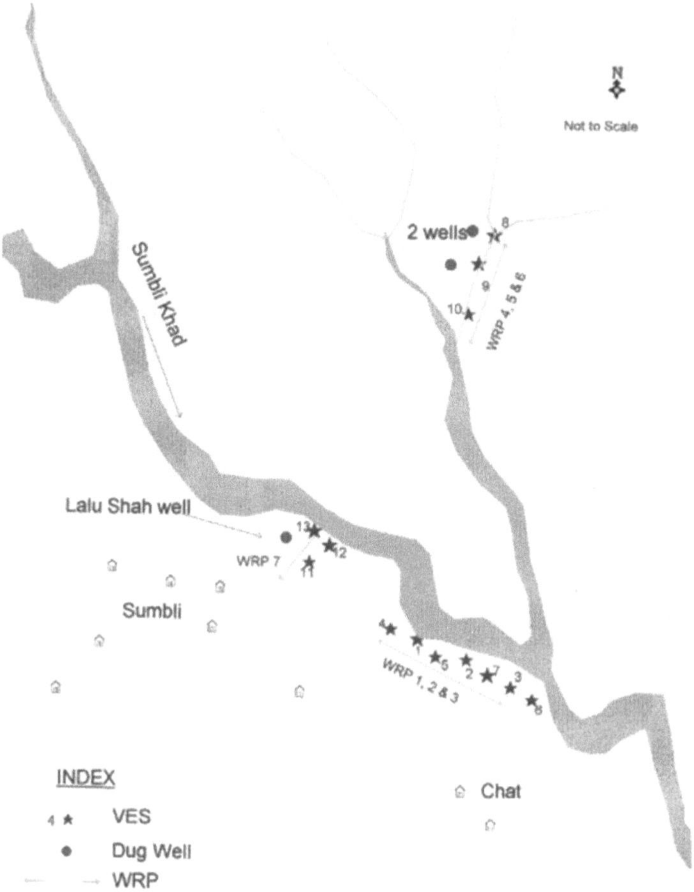

Fig. 1 Location map of WRP and VES at Sumbli

At two dug wells location, three WRPs were conducted. WRP4 and WRP5 are on the same line while WRP6 was conducted across the WRP4 and WRP5's profile line. All the WRPs were conducted with electrode separation of 10 m and station interval of 5 m. These WRPs show that there is a low resistivity zone in the distance range of 70 m to 105 m on profile line (Fig. 3) where VES 8, 9 and 10 were conducted.

Only one WRP was conducted at Lalu Shah well (Fig. 4), with electrode separation of 10 m and station interval of 5 m and the lowest apparent resistivity value (17.81 Ohm-m) was found at the distance of 65 m on profile line. At this low resistivity point, VES13 was conducted with *AB*/2 of 60 m. It is evident from the figure that there is a low resistivity area from the distance of 50 m onwards in profile line.

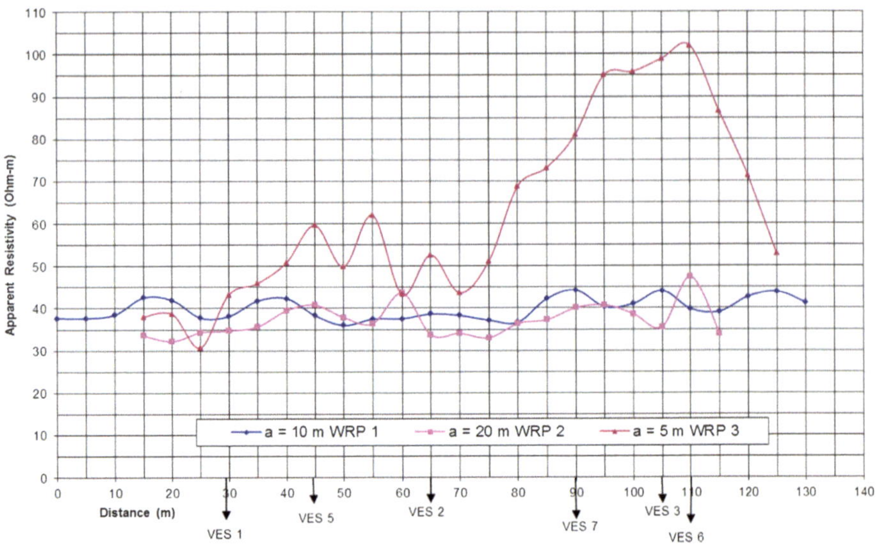

Fig. 2 Wenner Resistivity Profiles (WRP1 to WRP3) at Sumbli, Tehsil Samba, District Jammu (at Sheesham Trees Field)

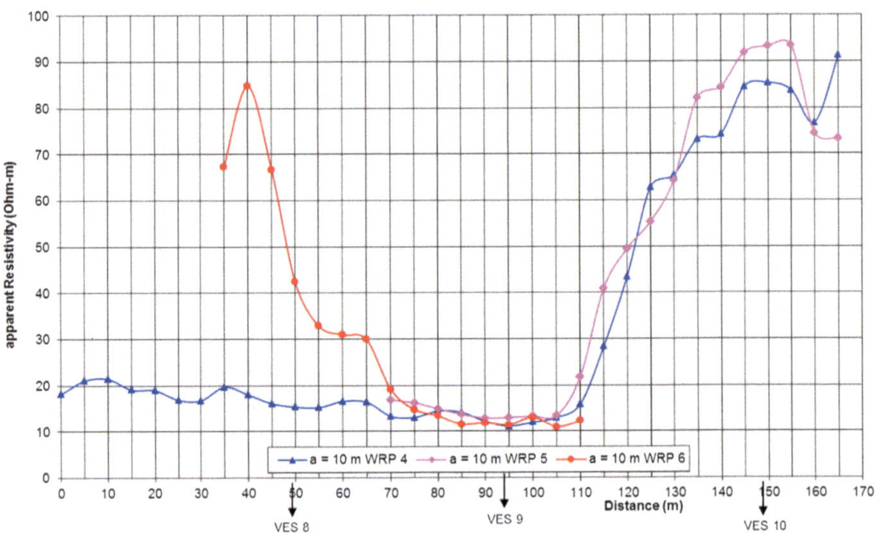

Fig. 3 Wenner Resistivity Profiles (WRP1 to WRP3) at Sumbli, Tehsil Samba, District Jammu (near Two Dug Wells)

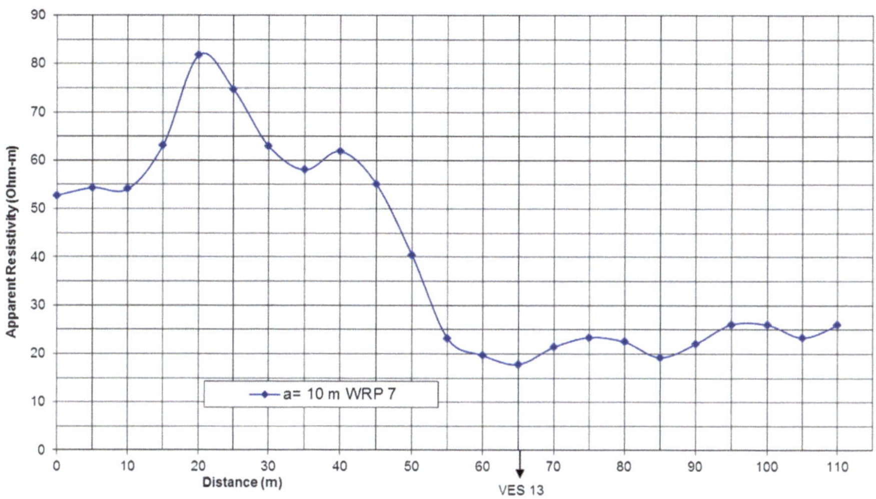

Fig. 4 Wenner Resistivity Profiles (WRP1 to WRP3) at Sumbli, Tehsil Samba, District Jammu (at Lalu Shah Well)

VES Curves

The interpreted true resistivity of the field VES curves indicates 8 to 13 sub-stratum geo-electrical layers. The eight geo-electrical layers have been identified in VES7 conducted at Sheesham trees field whereas thirteen geo-electrical layers have been found in VES9 and VES11, conducted near two dug wells and Lalu Shah well respectively. The resistivity sounding curves obtained in the area are of Q or KH or KQ type. The resistivity of the first layer ranges from 42 to 720 Ohm-m indicating moist to dry topsoil found at VES1 and VES7. The thickness of first layer ranges from 0.45 to 1.2 m. The second layer resistivity ranges from 5 to 1680 Ohm-m, with thickness ranging from 0.08 to 13.23 m. The maximum resistivity value of 1680 Ohm-m is interpreted at VES1, indicating boulders and the minimum resistivity value of 1 Ohm-m is interpreted at VES10, which can be inferred as clay. The true resistivity values, thickness and depth range are given in Table 3. The VES curves are given in Figs. 5(a) to 5(d).

Pseudo Geo-electrical Cross-section

The pseudo geo-electrical cross-section was drawn, using the apparent resistivity values (directly received from the resistivity equipment) and respective current electrode separations. The pseudo geo-electrical cross-section shows the lateral distribution of apparent resistivity along a particular survey line. The first pseudo geo-electrical cross-section was prepared on the basis of apparent resistivity data

Table 3 The interpreted results of VES data

					True resistivity $\rho_1, \rho_2, \rho_3, \ldots$ (Ohm-m)								
VES No.					Depth $h_1, h_2, h_3, \ldots$ (m bgl)								
1	42	1680	600	285	17.5	46	25	70	22				
	0.45	0.53	1.02	3.66	4.17	6.77	26.51	33.41	–				
2	500	75	16.4	55.5	73.5	4.2	58.5	2.35	140				
	1.05	3.89	5.21	11.81	21.01	23.21	39.31	43.41	–				
3	350	1225	186	23	57	10.4	266	2.95	360	3.01	150		
	0.64	1.12	2.8	4.96	17.26	18.96	22.76	25.76	31.2	37.3	–		
4	100	500	320	125	17	14.1	22.5	17.6	28.5				
	0.57	0.75	2.44	4.66	5.07	11.6	27.58	43.18	–				
5	220	550	400	27	35	57	33.6	68	20.5	150			
	0.51	0.86	1.61	2.58	8.83	11.12	27.62	36.37	52.57	–			
6	550	825	17	35	61.5	4.06	145	22	77.5	7.6			
	0.9	1.98	2.18	6.14	22.34	24.54	31.74	47.34	60.34	–			
7	720	288	23.1	73.5	3.3	56.25	8.55	35.2					
	0.9	2.34	3.66	10.33	11.33	37.58	41.28	–					
8	92	184	40.8	2.25	10.8	28	1.0	300	16	85	1.2		
	1.0	1.48	4.93	5.37	8.82	14.14	15.54	17.19	31.49	41.24	–		
9	300	90	5.5	92	1	138	1	140	1.2	120	1	320	1
	0.88	1.58	11.32	13.52	14.47	16.71	19.61	21.11	24.71	32.51	35.71	41.65	
10	200	5	62.4	124	72	3200	46	2.65	13.2	840	1.95		
	0.9	1.03	4.17	8.57	13.07	13.2	29.9	31.76	54.16	57.84	–		
11	560	840	544	15	54	23	77	1.45	820	1	900	1.35	880
	1.0	1.6	2.4	2.63	4.32	27.72	32.72	34.42	35.47	37.87	38.85	41.3	–
12	61	79	4.1	64	1.4	1480	2.6	700	2.4	700	2		
	1.2	14.43	15.63	23.38	24.58	25.58	28.3	30.05	32.3	34.14	–		
13	640	128	3.5	360	1.15	400	1.35	61	18.5				
	1.05	2.0	10.36	10.91	11.61	12.36	13.5	41.5	–				

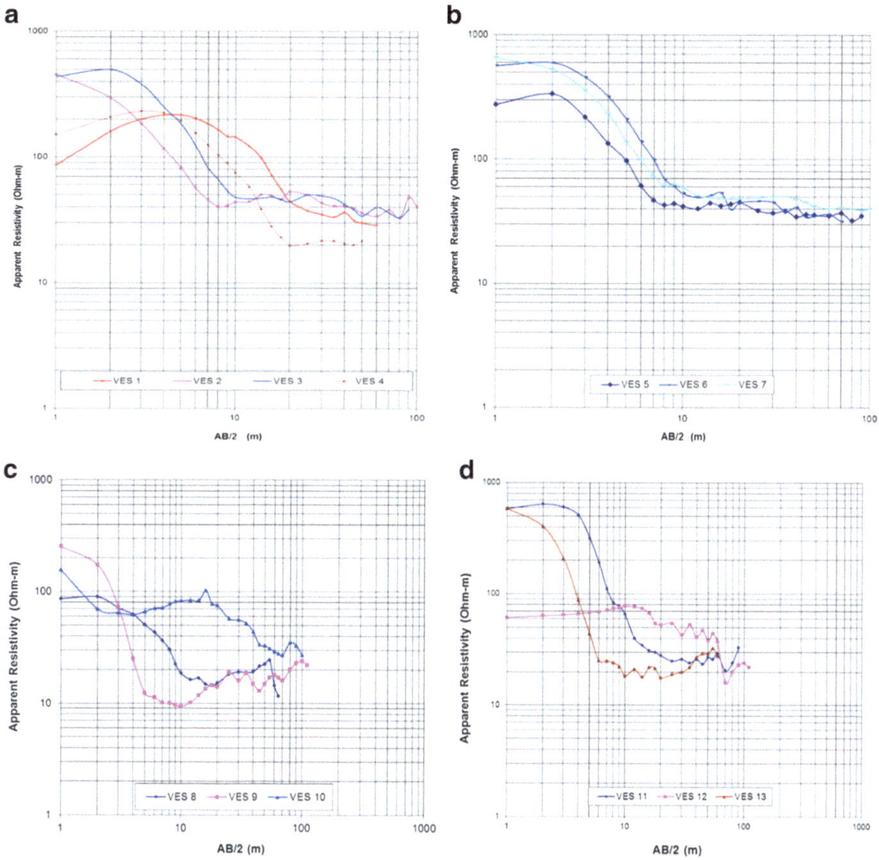

Fig. 5 Field VES curves at Sumbli. (**a**) VES Curves 1–4 (**b**) VES Curves 5–7 (**c**) VES Curves 8–10 (**d**) VES Curves11-13

obtained from VES1 to VES7 near the Sheesham trees field (Fig. 6). The cross-section shows the higher apparent resistivity values (100–300 ohm-m) at shallow depths up to 10 m, along the cross-section line. An isolated patch of low apparent resistivity value (20 Ohm-m) was found below the depth of 38 m onwards, at VES4 and VES1. All VES shows low apparent resistivity values (less than 40 Ohm-m) below the depth range of 15 m to 50 m.

The second pseudo geo-electrical section is prepared on the basis of apparent resistivity of VES8, VES9 and VES10, conducted near two dug wells as shown in Fig. 7. In this pseudo section, the higher apparent resistivity values more than 90 Ohm-m was found up to 10 m depth, representing surface layer. The 20 Ohm-m

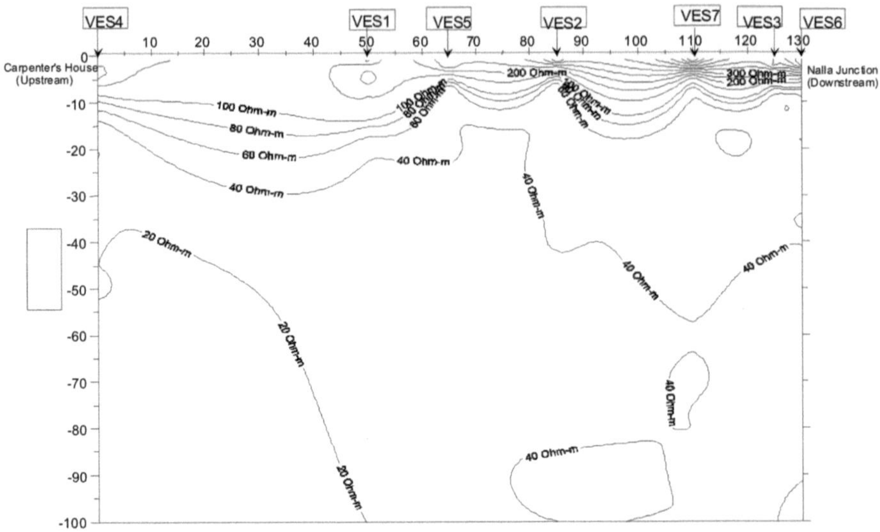

Fig. 6 Pseudo geo-electrical cross-section at Sumbli Rakh, Tehsil Samba, District Jammu (at Sheesham trees field along the Nalla)

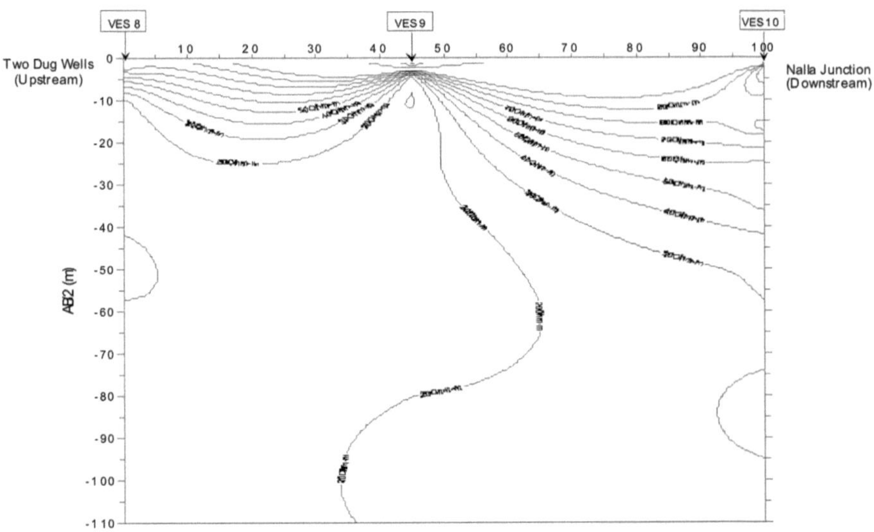

Fig. 7 Pseudo geo-electrical cross-section at Sumbli, Tehsil Samba, District Jammu (near Two Dug Wells location)

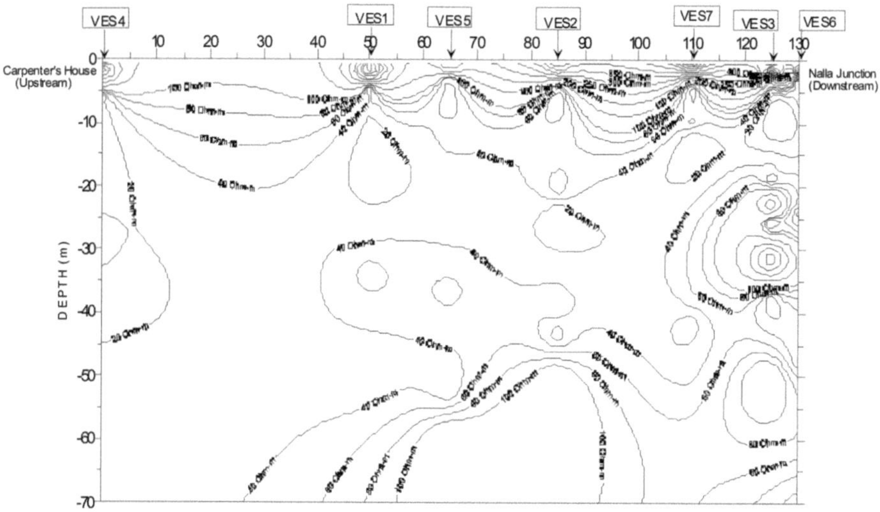

Fig. 8 Geo-electrical cross-section at Sumbli, Tehsil Samba, District Jammu (at Sheesham trees field along the Nalla)

apparent resistivity layer starts at depth of 10 m bgl in VES8 while it appears at 5 m bgl in VES9 and 55 m bgl in VES10.

Geo-electrical Cross-sections

The geo-electrical cross-sections were prepared using layer parameters (resistivity and thickness) obtained from the interpretation of VES curves. The chief objective of the preparation of geo-electrical sections was to show the distribution of true resistivity values i.e. different formations in the survey area along the survey line. The first geoelectrical cross-section was prepared with the help of seven VES points as shown in Fig. 8 which represents the surface layer containing the resistivity ranging from 80 to 400 Ohm-m upto a depth of 10 m. This variation in resistivity values is due to variation in moisture content present in the top surface layer. The second layer shows the five small isolated patches of low resistivity i.e. 20 Ohm-m. The subsequent layers contain the resistivity values from 40 to 100 Ohm-m from 5 m depth onwards.

The second geo-electrical cross-section was prepared on the basis of VES8, VES9 and VES10 (Fig. 9). The topmost layer contains resistivity value in the range of 60–550 Ohm-m. The variation of resistivity value is due to the variation in moisture content present in the surface layer. The resistivity value of less than

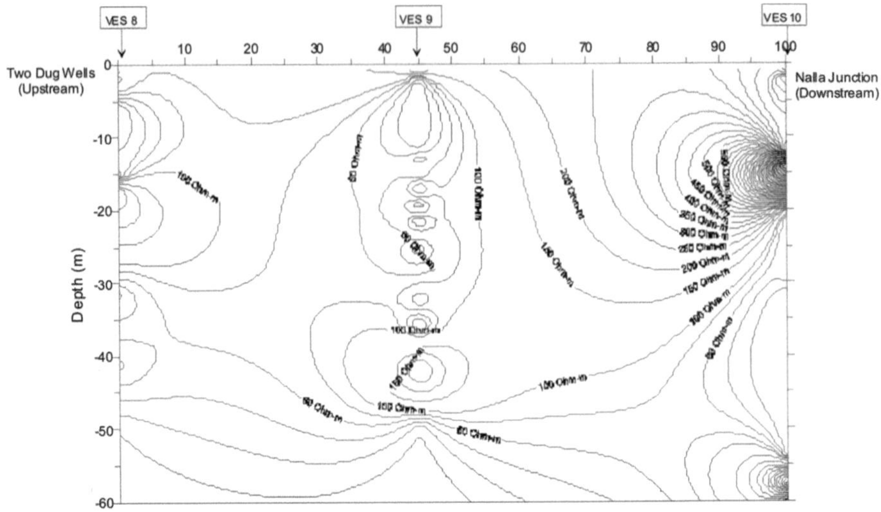

Fig. 9 Geo-electrical cross-section at Sumbli Rakh, Tehsil Samba, District Jammu (near Two Dug Wells location)

60 Ohm-m was found below the depth of 30 m bgl, 50 m bgl and 20 m bgl for VES8, VES9 and VES10, respectively.

Conclusions

From this study the following conclusions can be inferred:

- Based on the analysis of the field VES curves, interpreted layer parameters and study of geo-electrical cross-sections, it is deciphered that there is a possibility of encountering a groundwater potential zone at VES9 point.
- The identified prominent expected groundwater feasible or major lithology change zones are in the depth ranges of 14.4 to 16.7 m bgl, 19.6 to 21.1 m bgl and 24.7 to 32.5 m bgl.
- Study indicated presence of hard sticky clay layers in between groundwater potential layers.
- At VES9 point, potential groundwater zone may be encountered in dug well down to the depth of 20 m bgl or a bore hole down to the depth of 35 m bgl.
- The resistivity values in the range of 120 ± 30 Ohm-m indicates the gravel or sand mixed formation, saturated/partially saturated with water.

References

Birendra Pratap (2006). Some problems of ground water and its management using Geophysical techniques around Jodhpur city of Rajasthan, India. Ph.D. Thesis, Banaras Hindu University, Varanasi.

Birendra Pratap and Singh, K.P. (2009). Estimation of ground water quality from electrical logs in Kashmir Valley, Jammu and Kashmir. Proc. Workshop on "Ground Water Scenario and Quality Issues in J&K State", 2-3 March, 2009.

Raju, N.J., Reddy, T.V.K. and Nayudu, P.T. (1996). Electrical resistivity surveys for groundwater in the upper Gunjanaeru catchment, Cuddapah district, Andhra Pradesh. *Journal of Geological Society of India*, **47(6):** 705-716.

Raju, N.J. and Reddy, T.V.K. (1998). Fracture pattern and electrical resistivity studies for groundwater exploration. *Environmental Geology*, **34(2/3):** 175-182.

Orellana, E. and Mooney, H.M. (1966). Master tables and curves for vertical electrical sounding over layer media. Madrid Interciencia. p. 150 and 66 tables.

Yadav, G.S., Birendra Pratap and Singh, Shashikant (2003). Ground water characteristics in parts of Jodhpur city and adjoining area, Rajasthan. Proc. National Seminar on Status of Environmental Management in Mining Industry, Department of Civil Engineering, IT, BHU.

Isotopic-Chemical Framework of Groundwater Aquifer to Study the Pollution Dynamics at Delhi, India

Shilpi Saxena, J.P. Shrivastava, M.S. Rao, and Bhishm Kumar

Introduction

Indiscriminate disposal of anthropogenic wastes and leaching of pollutants from these resulted in an ever-increasing threat to the quality of ground water resource base. Large-scale groundwater withdrawal for domestic, commercial, and industrial purposes leads to widespread decline of ground water table. For protection of groundwater from pollution (Singhal et al., 2003 and Singh et al., 1997), it is a matter of concern for the planners and decision makers to clearly characterize the groundwater renewal, quality of water and causes of its deterioration, sources of pollution, trace the movement of pollutants and containment of spreading from known sources. The characteristics of pollutants level and transport in groundwater are associated with variations in one or two parameters at one scale and several parameters at another scale. In order to understand the groundwater system (Glynn et al., 2005) and its hydro chemical and pollutant transport study for the Najafgarh drain basin area, a detailed study has been undertaken about the major ions and isotopic signatures of groundwater of the Najafgarh drain basin area of Delhi (Shivanna et al., 1998).

S. Saxena (✉) • J.P. Shrivastava
Geochemical Laboratory, Department of Geology, University of Delhi, Delhi 110007, India
e-mail: docs.shilpi@gmail.com

M.S. Rao
Nuclear Hydrology Laboratory, National Institute of Hydrology,
Roorkee, Uttarakhand 247 667, India

B. Kumar
Isotope Hydrology Section, International Atomic Energy Agency, Vienna, Austria

N.J. Raju et al. (eds.), *Management of Water, Energy and Bio-resources in the Era of Climate Change: Emerging Issues and Challenges*,
DOI 10.1007/978-3-319-05969-3_12

Description of the Study Area

The present study area is the Najafgarh Drain Basin, where it starts from South West of Delhi near Delhi-Gurgoan border at Chhawla village and extends through Qutub Vihar, Goyala Khurd, Najafgarh, Nangli Dairy, Kakrola, Vikaspuri, Raghuveer Nagar, Meera Bagh, Moti Nagar, Shakti Nagar, Inderlok, Roop Nagar, Kinsgway Camp, Hakikat Nagar, Timarpur, Nehru Vihar and Mukherjee Nagar. The Najafgarh drain basin area in Delhi region is about 832 sq. km and most of the industrial units of Delhi including few major medium scales are situated in this drain basin area. The other important drain that passes through Delhi is the River Sohibi. It enters in Delhi near Dhansa where Dhansa Bund (regulator) is constructed. The floodwater of river Sohibi is drained through Najafgarh drain into the river Yamuna (CPCB, 1995). On its way to the river Yamuna, the Najafgarh drain receives discharge from other drains like Palam Drain (Ganda Nala), Mungeshpur Drain, Mundka Drain, Nangloi Drain, Western Yamuna Canal (Delhi Tail Distributary) and finally from Supplementary drain just before its culmination into the river Yamuna near Wazirabad Barrage.

There are 21 drains that conveys sewage and rainwater from different parts of the city to the river Yamuna. Out of these drains, Najafgarh Drain, Okhla Drain and Shahdara drain are the major drains. Its sewered catchment area is around 374 sq.km. Because of severe constraints in the sewerage system, drainage basin carries a very large quantity of raw sewage. In addition, it receives treated effluent from various STPs and fresh water from tail-end of the Western Yamuna Canal.

Najafgarh drain basin in East Delhi has been ranked eleventh in the country's highly polluted 88 industrial clusters. The Najafgarh drain basin, which is also the biggest polluter to Yamuna, tops the list of thirty-three "critically" polluted industrial hubs, scoring 70-80% in the Comprehensive Environmental Pollution Index (CEPI). And the drain has scored 79.54 points, which is attributed to huge effluents and sewage in its drainage basin. The location of sampling points in Najafgarh drain basin area is given in Fig. 1.

Physiography of the Study Area

The Delhi region is a part of the Indo-Gangetic Alluvial Plains, situated at an elevation ranging between 198 and 220 m above mean sea level (msl), transacted by a quartzite rocky ridge (maximum elevation of 318.52 m above msl) extending roughly from north-east to south-east. The ridge forms the principal watershed of the area and acts as a groundwater divide between the western and eastern parts of NCT, Delhi. Physiographically, the NCT, Delhi can be divided into four parts: (1) the Delhi Ridge, (2) the alluvial plain to the east of the Delhi Ridge; (3) the alluvial plain to the west of the Delhi Ridge; and (4) the closed basin of Chatttarpur (CGWB, 2006). The territory is made up of quartzite ridge tending from north-northeast (NNE) to south-southwest (SSW). The alluvial plain on the east

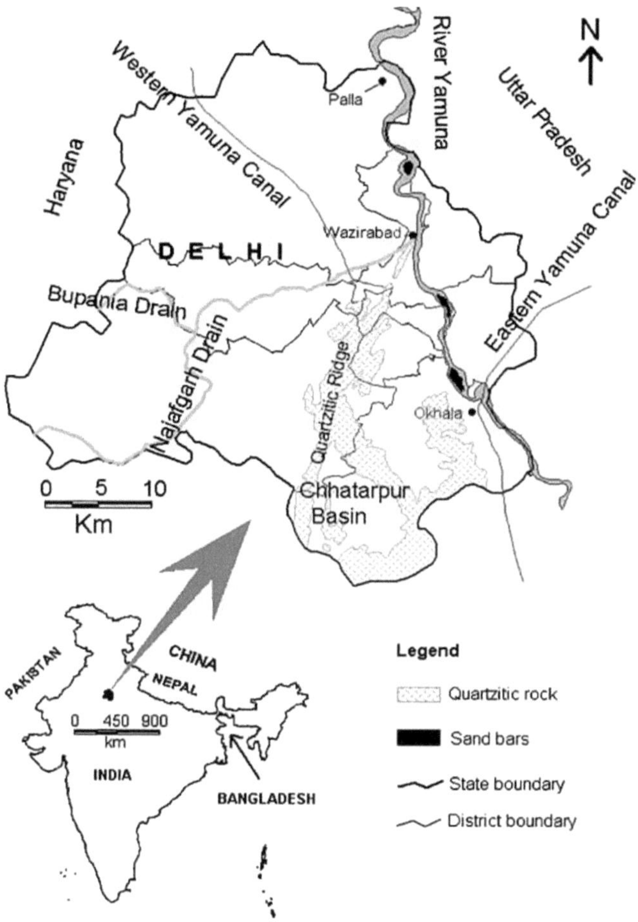

Fig. 1 Map of Delhi. (*Source:* CGWB, 2006)

and sand dunes at the west of the ridge are observed. The territory is drained by the river Yamuna which enters from the north eastern direction and flows through the south eastern area. The quartzite ridge is the prolongation of the Aravalli group of the rock and enters from the southern borders and ends in the north of Delhi on the west bank of Yamuna. The ridge forms the principal watershed in the area, the eastern drainage passing into the Yamuna and the western into the Najafgarh basin.

Hydrogeology of the Study Area

Geological formation of the north eastern area of Najafgarh drain basin (Tyagi et al., 1997) are generally composed of unconsolidated silts, sand and nodular calcareous material in the vicinity of the river Yamuna. These have more permeable

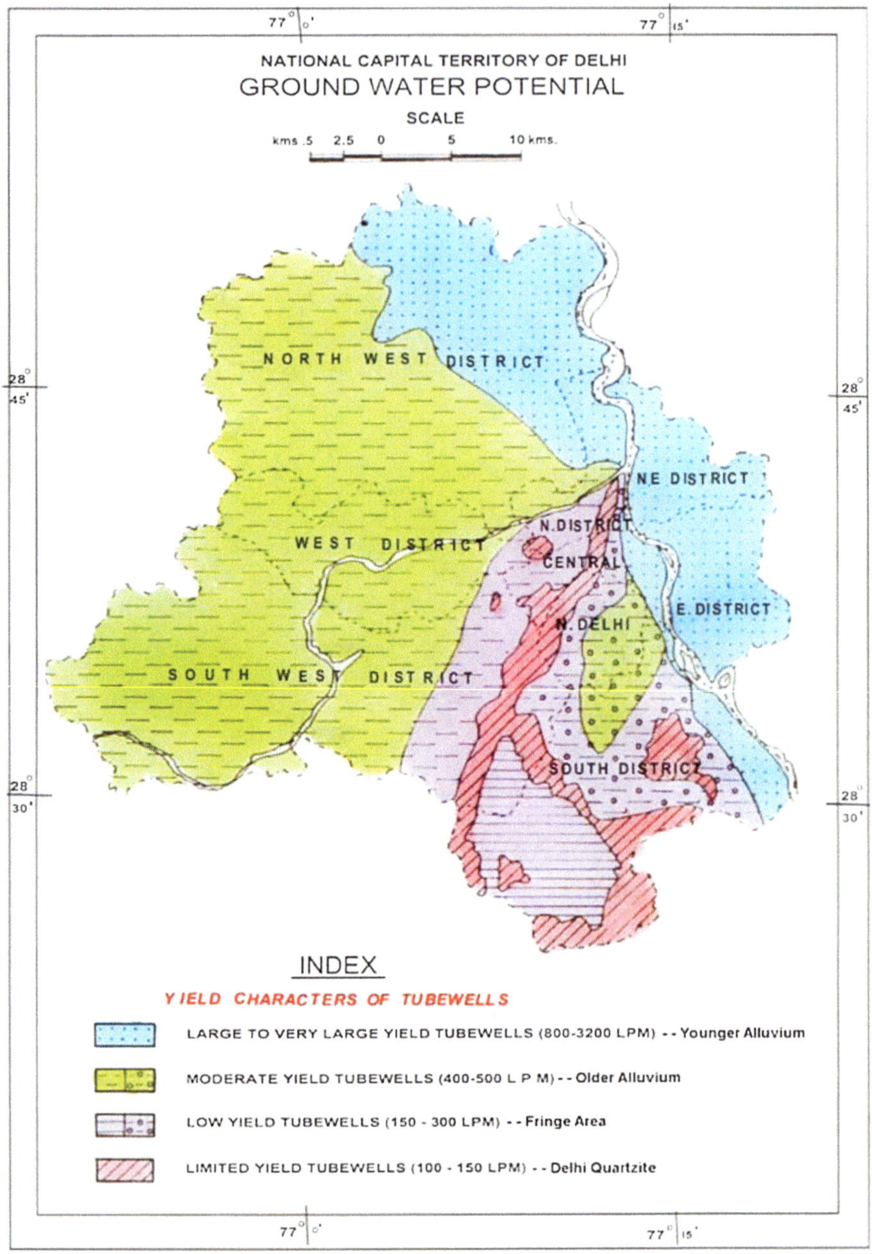

Fig. 2 The groundwater potential of the different formations in NCT Delhi. (*Source:* CGWB, 2006)

horizons than the older alluviums. The younger alluviums of this zone Alipur block have moderate to fairly good potential of ground water. The ground water availability in the territory is controlled by the hydrogeological situation characterized by occurrence of alluvial formation and quartzite hard rocks.

The alluvial deposits are of Quaternary age. The newer alluvium belongs to recent age and is referred to the sediments deposited in the flood plains of Yamuna River. These sediments range in texture from clay/silt mixed with tiny mica flakes to medium/coarse sand and gravel. Newer alluvium, in general, is characterised by absence of kankar. The older alluvium consists of sediments deposited as a result of past cycles of sedimentation of Pleistocene age and occurs extensively in the alluvial plains of the territory. This is comprised of inter bedded, lenticular and inter-fingering deposits of clay, silt and sand ranging in size from very fine to very coarse with occasional gravels. The kankar or secondary carbonates of lime occur with clay/silt deposits and sometimes as hard/compact pans. Older alluvium is predominantly clayey in nature in major parts of the territory except nearly closed alluvial basin of Chattarpur where the alluvial formation is derived from the weathered quartzite rocks. The hard rock formations mainly the Alwar quartzite of Delhi System, exposed in the area belong to Pre-Cambrian age. The quartzites are pinkish to grey in colour, hard, compact, highly jointed/fractured and weathered. These occur with interbeds of mica-schist and are intruded locally by pegmatites and quartz veins. The strike of these rocks varies northeast southwest to north-northeast south-southwest with steep dips towards southeast and east except for some local variations due to folding. Quartzites are ferruginous and gritty types on weathering and subsequent disintegration give rise to coarse sand (Badarpur sands).

Material and Methods

Field Investigations

The open land dumping of solid waste, discharge of liquid waste of both industrial and domestic origin and improperly lined drains lead to high groundwater contamination through leaching and seepage and this phenomenon is further accelerated due to gradient slope of this basin. Keeping these points in view, we have selected 17 different sampling stations throughout the stretch of the basin. The sampling coordinates were noted on a global positioning system (GPS III, Garmin). This was further supported by topographic sheets made available from Survey of India, Dehradun, Uttrakhand.

For the analysis of hydro chemical and isotopic (oxygen-18 and deuterium) analysis we have collected 102 groundwater samples using hand-pumps at a shallow depth in range of 10.66 to 36.57 m and using jet pump at a deeper depth ranging from 60 to 80 m in close proximity to the drain reach, during the year 2010-2012, from 17 different sampling stations for October-November 2010-12 (post-monsoon) and May-June 2011-12 (pre-monsoon) and analyzed for 17 different parameters covering major ions and stable isotopes by following the "Standard Methods of Examination of Water and Wastewater" (APHA-AWWA-WEF, 2005) and for isotopes analysis "Standard methods of (IAEA) International Atomic Energy Agency, Vienna", Austria.

Samples were taken only from those pumps which are actively in use. Sufficient water is flushed out before sampling to ensure that aquifer water and not the stagnant water is sampled. All the samples so collected were packed in a well sealed HDPE bottle and stored at low temperature to avoid any evaporation to occur before they get analyzed for isotopic ratios. Electrical conductivity, pH and temperature of samples were measured at sampling locations at the time of sampling. Four litre sample for chemical analysis and 125 ml for isotopes analysis from each sampling station were collected and preserved with the sulphuric acid for the analysis of major ions. Figure 3 shows X-Y coordinate or the map of plots used to simply display the sampling station locations throughout the Najafgarh drain basin area, Delhi.

Chemical Techniques

All groundwater samples were collected and filtered using 0.45 μm Millipore membranes. The samples were brought to the laboratory and stored at temperatures below 4 °C. Major ions were determined by using a Spectronic20,

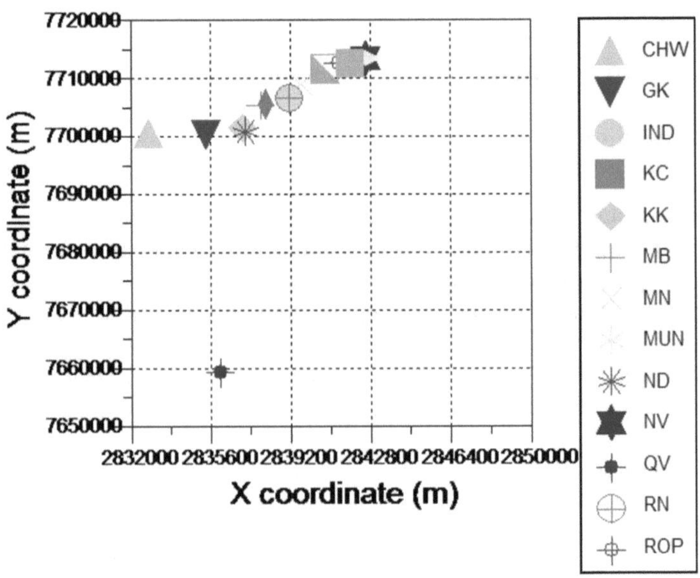

Fig. 3 X-Y plot for sample location based on GPS reading

Flame photometer (CL-378) and titration methods. Analysis of the chemical parameters were done by following the "Standard Methods of Examination of Water and Wastewater" (APHA, 2005). Apart from the collection and analysis of water samples, additional information related to geology, climate, soil type, and land use of the area were collected from the Central Ground Water Board (CGWB), Indian Meteorology Department (IMD) and Central Pollution Control Board (CPCB), New Delhi.

Nuclear Techniques

Measurement of $\delta^{18}O$

Oxygen analysis was carried out using CO_2 equilibration method (Epstain and Mayeda, 1953). 400 μl of sample water and secondary standard sample water were taken in 3.5 ml vials in a batch of 60 vials. The secondary standards used in the batch were pre-calibrated using the primary standards SLAP, VSMOW and GISP and also using the pre-analyzed samples procured from BARC, Mumbai and IAEA, Vienna. The batch samples were evacuated and equilibrated with Tank CO_2 gas (which is taken as calibrated reference gas) for seven hours at 40 °C (±0.1 °C. Isotope exchange between sample water oxygen and oxygen of gas CO_2 occurs during the equilibrium process. The gas after the equilibrium attains an isotope ratio of oxygen depending upon the sample water oxygen isotopic ratio and the temperature of equilibrium.

The isotopic exchange can be described as:

$$C^{16}O_{2(g)} + H_2{}^{18}O_{(l)}k = C^{18}O^{16}O_{(g)} + H_2{}^{16}O_{(l)}$$

where k is the isotopic exchange rate constant. The equilibrium isotopic fractionation factor $\alpha = (^{18}O/^{16}O)\ g/(^{18}O/^{16}O)\ g = 1.0412$ at 25 °C (Hut, 1987).

The temperature dependency is 0.2‰/°C. The equilibrated CO_2 gas is dehydrated using a water trap for 3 min at –72 °C. The equilibrated gas is then passed into the mass spectrometer for isotope ratio analysis. In the mass spectrometer, the gas is ionized, and its isotopes are dispersed according to their mass to charge ratio using combination of electric and magnetic field. The ionic currents of the separated isotopic beams (44, 45 and 46 in the case of CO_2) are measured using Faraday cups. After each measurement of isotopic ratio in a sample (46/44) S and in Tank-CO_2 Gas (46/44) R, δ in the sample is estimated as:

$$\begin{aligned}\delta &= \left[(44/46)_S - (44/46)_R/(44/46)_R\right] \times 10^3‰ \\ &= \left[(44/46)_S/(44/46)_R - 1\right] \times 10^3‰\end{aligned}$$

where $44 = {}^{12}C^{16}O^{16}O$ and $46 = {}^{12}C^{16}O^{18}O$. Abbreviations: R stands for reference (Tank Gas) and S for sample (including that for secondary standard).

In the measurement, Tank Gas (Reference) and also the sample gas is broken into 10 equal parts and the ratios (44/46) are measured on each of these sub-portion. As the instrument alternately switches between sample and Tank Gas, in the ten switch-overs ten δs get estimated. These ten δs are averaged after removing outliers using $\pm 1\sigma$ criterion and the mean δ is quoted as the δ for the sample. The estimated δ is then converted into $\delta^{18}O$ by introducing isotopic correction factor for the small presence of ^{17}O (Gat and Gonifiantini, 1981). The precision of the estimated $\delta^{18}O$ is within ±0.1‰. For statistical consistency two aliquots are taken for each sample. The $\delta^{18}O$ of sample is taken from the average $\delta^{18}O$ of these two aliquots. The secondary standards provides an overall check on the stability of the system during the batch analysis, long-term stability of the system and in calibrating the Tank Gas at the time of its replacement.

Measurement of δD

Measurement of D/H ratio in sample is done similar to oxygen isotope analysis (Fritz et al., 1980) with a difference that the equilibration is carried out with hydrogen gas in the presence of platinum (Pt catalyst, marketed as Hokko Beads) in place of CO_2 gas. The catalyst Hokko beads are made of 1 wt % Pt coated on 1 mm diameter styrene divinyl benzene porous resin that provides hydrophobic support to the platinum. These are re-usable by simply activating them by washing in distilled water and drying at 80-90 °C.

The exchange reaction takes place as:

$$HD_{(g)} + H_2O_{(l)}k \qquad H_{2(g)} + HDO_{(l)}$$

where k is the isotopic exchange rate constant. The equilibration is carried out for 180 minutes at 40 °C.

The fractionation factor $\alpha = (D/H)_{water}/(D/H)_{gas} = 3.81$ at 25 °C.

The reaction is very sensitive to temperature (6.3‰ per °C and precise temperature control is required (±0.05 °C corresponds to 0.3‰ error).

Result and Discussion

The chemical and isotopic study of ground water of Najafgarh drain basin area was carried out for 17 parameters covering major ions and stable isotopes study. The results of all chemical and isotopic study for 102 groundwater samples of 17 sampling stations were modelled with a Hydro Chemical Model AQUACHEM v. 2012.1 given below.

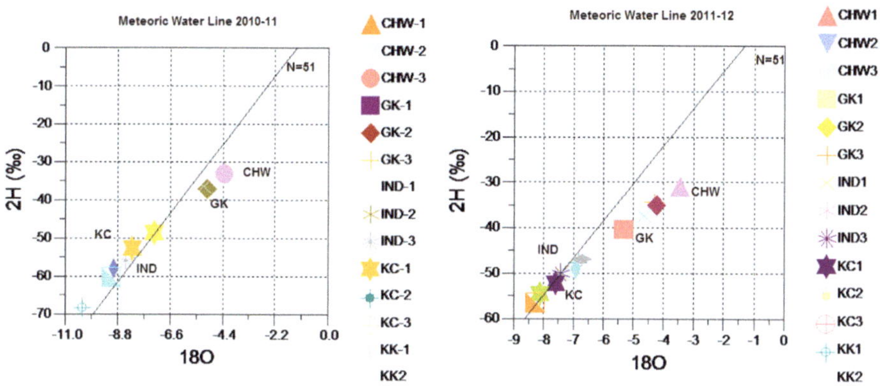

Fig. 4 Meteoric Water Line (MWL) plot: (**a**) 2010-2011 and (**b**) 2011-2012

Isotope Hydrochemistry

The results of present study reveal (Fig. 4) that the variations in δD were found ranging from (min.) –68.325‰ to –33.219‰ (max.) and the variation in $\delta\ ^{18}O$ from –10.266‰ to –4.337‰ in the year 2010-11 (pre--monsoon) and the variation in δD were found ranging from –61.45‰ to –14.978‰ and the variation in $\delta^{18}O$ from –9.405‰ to –1.248‰ in 2010-11 (post-monsoon). In the year 2011-12 variations in δD were found ranging from –56.16‰ to –34.39‰ and variation in $\delta^{18}O$ from –8.243‰ to –4.30‰ (pre-monsoon), and variation in δD were found ranging from –58.44 to –26.133 and variation in $\delta^{18}O$ from –8.587‰ to –2.735‰ (post-monsoon). These results show the depletion in isotopic signature of ground water from North to South-west area.

In this diagram (Fig. 4) water samples plotted close to the world Meteoric Water Line (MWL) suggest a meteoric origin. The plot shows ^{18}O on the X-axis and ^{2}H on the Y-axis. The global meteoric water line is defined by the following equation: $^{2}H = 8.2 \times {}^{18}O + 10.8$. The difference in the stable isotope composition between samples from the northern and those from the southern part of the basin is probably due to regional and/or seasonal variations in the stable isotope composition of ground water. Figure 5 reveals the integrated study of stable isotopes i.e. oxygen-18 relationship with the major anions (nitrate, chloride, fluoride and sulphate). Chemical information integrated with isotopic information can give us the exact status of chemical distribution at a stretch of the study area.

Major Ion Hydrochemistry

In piper plot (Fig. 6) major ions (Datta et al., 1996) were plotted as cation and anion percentage of milliequivalents in two base triangles. The projection reveals certain useful properties of total ion relationships. The total cations in meq/l, and total

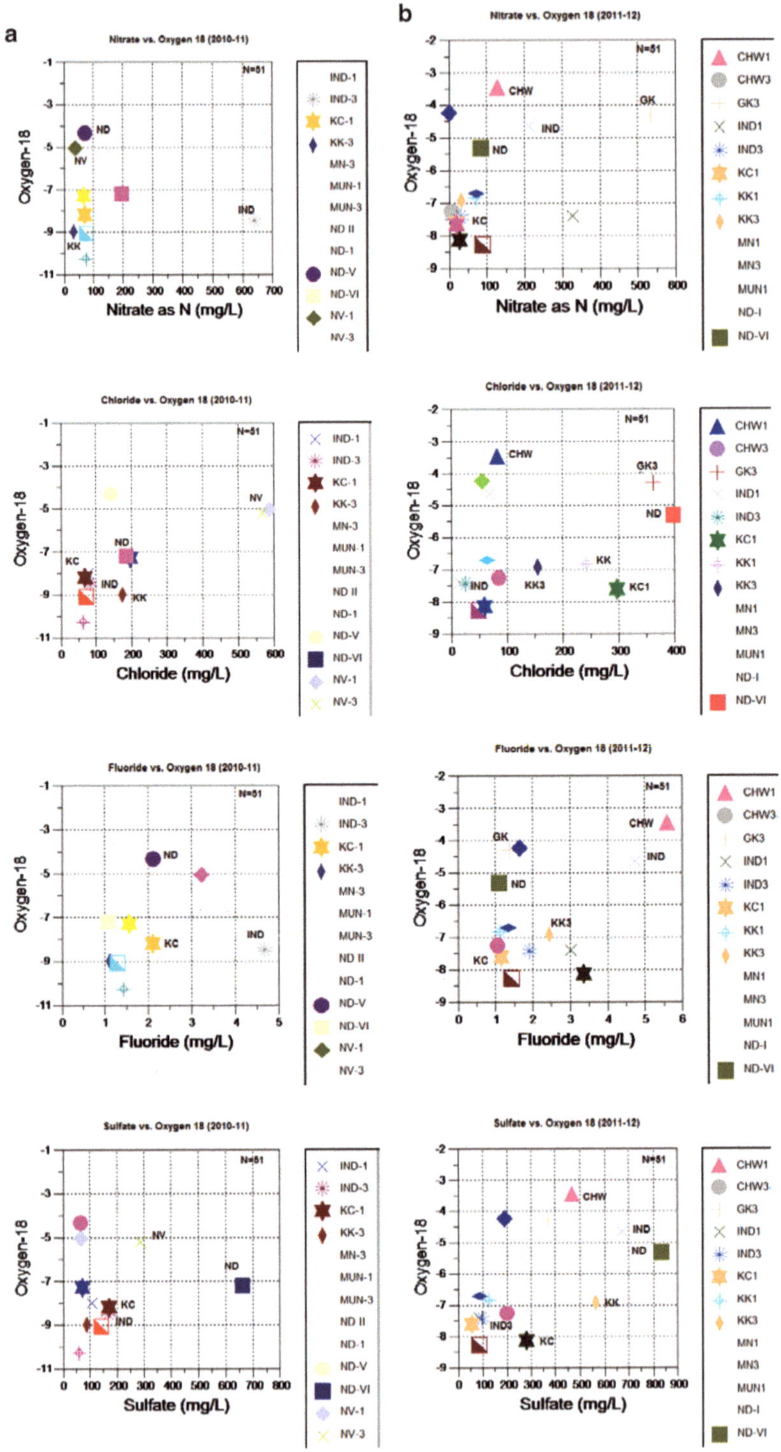

Fig. 5 Relationship of oxygen-18 vs. nitrate, chloride, fluoride and sulphate: (**a**) 2010-11 and (**b**) 2011-12

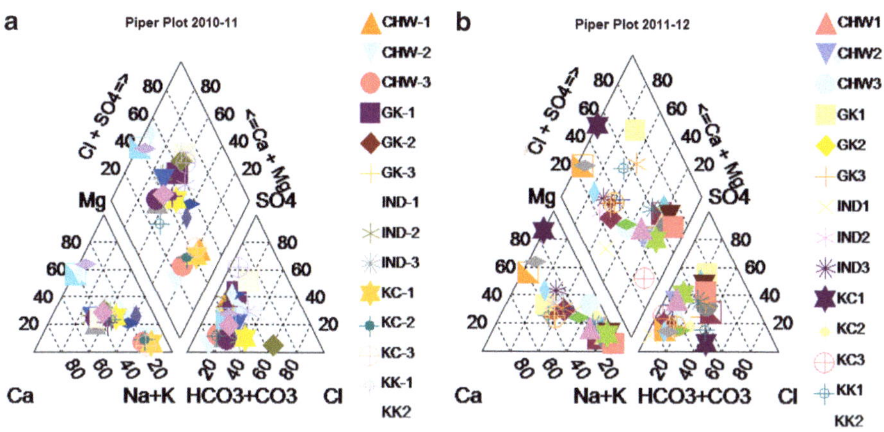

Fig. 6 Piper plot: (**a**) 2010-11 and (**b**) 2011-12

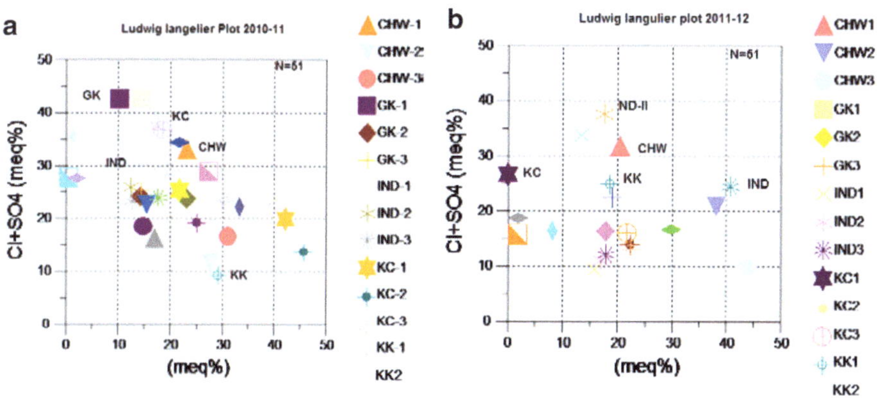

Fig. 7 Ludwig Langelier plot: (**a**) 2010-11 and (**b**) 2011-12

anions in meq/l were set equal to 100%. Every sample was represented by three data points; one in each triangle and one in the projection diamond grid. This plot shows the clustering of samples of the period's viz., pre-monsoon and post-monsoon of the years 2010-11 and 2011-12.

With the help of Ludwig-Langelier plot (Fig. 7) we have observed patterns and correlations between the major cations and anions for multiple samples. By convention, the sum of selected cations is plotted on the X-axis and the sum of selected anions is plotted on Y-axis. Each axis ranges from 0 to 50 meq%. Sample points are calculated as follows:

$$\sum \text{anions} = \text{Cl} + \text{SO}_4 + \text{HCO}_3$$
$$\sum \text{cations} = \text{Ca} + \text{Mg} + \text{Na} + \text{K}$$
$$\%\ \text{Na} = 50\ \text{Na}/\sum \text{cat}$$
$$\%\ \text{Cl} = 50\ \text{Cl}/\sum \text{an}$$

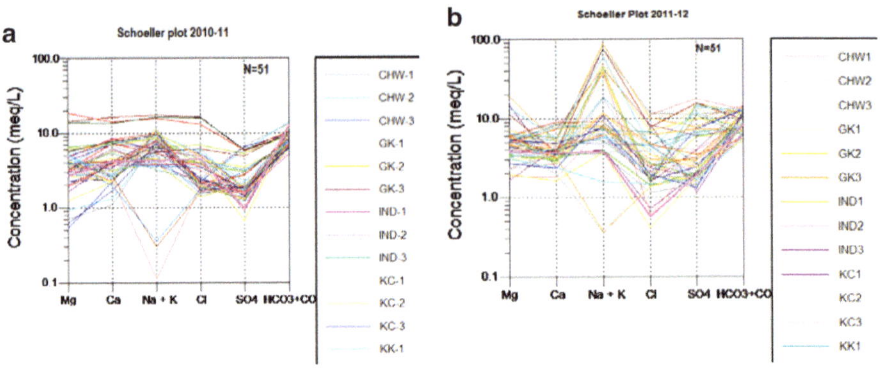

Fig. 8 Schoeller plot: (**a**) 2010-11 and (**b**) 2011-12

Suitable groupings of cations and anions are selected and plotted as percentages:

$$\%Ca + \%Mg = 50 - (\%Na + \%K),$$
$$\%Cl = 50 - (\%HCO_3 + \%SO_4).$$

These plots display the pattern and correlation between the major cations and anions for multiple samples of ground water of two periods viz., pre-monsoon and post-monsoon. The results of this study display that the most of cations and anions were found in between 10-40% and very few found in between 0-10% and 40-50% for the year 2010-11 and most of the cations and anions were found in between 10-50% and very few found in between 0-10% for the year of 2011-12.

The Schoeller plot (Fig. 8) indicates intermixing of ground water that have different chemical compositions and water type, viz., Ca-Mg-HCO_3, Na-Ca-HCO_3-Cl, Mg-Na-HCO_3, Na-HCO_3, Na-Cl-HCO_3, Na-Mg-Cl, Na-Cl and Ca-Mg-Cl. These semi-algorithmic diagrams were developed to represent major ion analyses in meq/l and to demonstrate different hydro chemical water types on the same diagram (Datta et al., 1996).

Figure 9 shows scatter diagram of sodium vs. chloride in mg/l and the relation between sodium and chloride for the total 102 samples from years 2010-11 and 2011-12. We have found good correlation in between 'sum of anions and sum of total ions' ($r = 0.772$), but it was not so significant correlation in between the 'sum of cations and sum of anions' ($r = 0.478$). There were also found very good correlation between "total ions vs. sum of cations" which was highly significant ($r = 0.907$).

Conclusion

Chemical quality of ground water in NCT Delhi varies with depth and space. Brackish ground water mainly exists at shallow depths in northwest, west and Southwest districts with minor patches in north and central districts. In alluvial

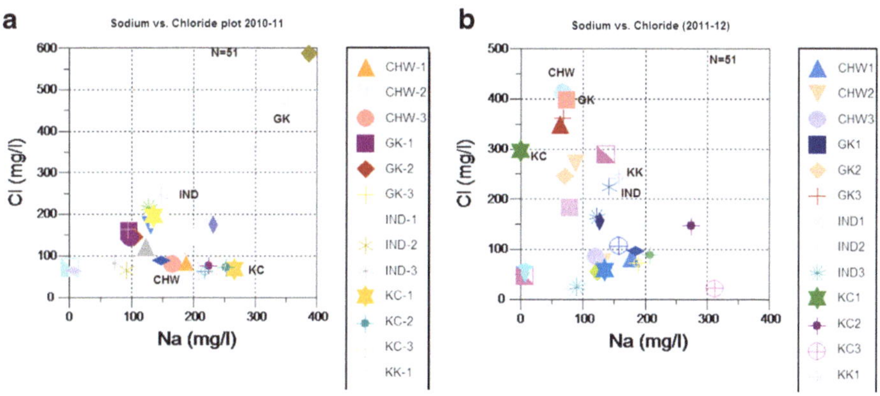

Fig. 9 Sodium vs. chloride: (**a**) 2010-11 and (**b**) 2011-12

formations, the quality of ground water deteriorates with depth, which is variable in different areas. The ground water is fresh at all depths in areas around the ridge falling in central, New Delhi, south and southwest districts and also Chattarpur basin. In the areas west of the ridge, in general, the thickness of fresh water aquifers decreases towards northwest; the thickness of fresh water zones is limited in most parts of west and southwest. In the flood plains of Yamuna, in general, fresh water aquifers exist down to 30-70 m. The fluoride contamination in ground water has been noticed in samples of groundwater collected from southwest and west districts. The high fluoride levels are mostly found in areas where ground water is brackish to saline in nature. The nitrate concentration in ground water has been reported mostly from areas where domestic effluent is discharged into open unlined drains. The high nitrate concentrations are mainly from point source of contamination.

Isotopic data reveals that the aquifer in the studied area does not constitute a homogeneous system in its lateral extent. The groundwater recharge variation from location to location and pumping induced groundwater intermixing (Datta et al., 1996) through different flow pathways results in wide range of spatial variations in stable isotope (^{18}O) signature of ground-water, with $\delta^{18}O$ values as highly enriched +0.59% to as depleted –7.59‰ during 2003-04; from –0.60‰ to –7.70‰ in March 2005, and –1.80‰ to –8.10‰ in October 2005 (Tyagi et al., 2008). The study establishes that ground water has become more vulnerable to contamination and isotope studies should be conducted as part of comprehensive hydro chemical investigation of groundwater vulnerability.

There are indications of pollutants transport from the western, northwestern and southwestern areas to the urbanized and overexploited parts. In the northwestern and western parts, there is evidence of increasing groundwater pollution and leachate transport to ground water through surface drainage. The scope of this study has broader significance, since urbanization can place similar pressures on the water resources worldwide. Hydrogen and isotope signature reveals that the aquifer in the studied area does not constitute a homogeneous system in its lateral extent.

The main sources of groundwater contamination in the Najafgarh drain basin area are as follows: Southwest part of basin at Chawala, Nangli Dairy, Qutub Vihar, Goyala Khurd, Kakrola and Najafgarh area; Agricultural part, cattle farm activities and residential area all along the drain basin area except at southwest parts; Residential (organized and unorganized colonies) and commercial activities, in central urban area of the drain basin at Moti Nagar, Inderlok and Wazirpur; Major industrial units like insecticide, caustic soda, vanaspati, electroplating etc.

References

Adelana, S.M.A., Oalsehinde, P.I. and Vrbka, P. (2003). Isotope and Geochemical characterization of surface and subsurface waters in the semi-arid Sokoto Basin, Nigeria. *African Journal of Science & Technology* (AJST), **4(2):** 80-89.

APHA-AWWA-WEF (2005). Standard Methods for the Examination of Water and Waste Water, 21st Edition. Washington DC.

Aravena, R., Evans, M.L. and Cherry, J.A. (1993). Stable isotopes of oxygen and nitrogen in the source identification of nitrate from septic systems. *Groundwater*, **31:** 180-186.

Bhattacharya, P., Jacks, G., Gustatsson, J.P., Von Bromssen, M., Matin, K., Ahmed, A.A.K., Chatterjee, D. and Bhattacharya, S.K. (2002). Groundwater Arsenic contamination in Bengal Delta Plains in India and Bangladesh.

Bhattacharya, S.K., Gupta, S.K. and Krishnamurthy, R.V. (1985). Oxygen and Hydrogen isotopic Ratios in Groundwater and river water from India. *Proc. Indian Acad. Sciu. (Earth Planet, Sci.)*, **94:** 203-294.

Bohlke, J.K., Smith, R.l. and Hannon, J.E. (2007). Analysis of N and O isotopes in nitrite and nitrate by sequential selective bacterial reduction to N_2O. *Analytcal Chemistry*, **79(15):** 5888-5895.

Brown, J.G. (1996). Movement of metal contaminants in ground water in the Pinal Creek basin, Arizona—Model assessment and simulation of reactive transport. University of Arizona, Tucson. M.S. thesis.

Busenberg, E., Plummer, L.N., Doughten, M.W., Widman, P.K. and Bartholomay, R.C. (2000). Chemical-Isotopic compositions and gas concentrations of ground water at and near the Idaho National Engineering and Environmental Laboratory, Idaho, 1994-97. U.S. *Geological Survey Open-File Report*, 00-81: 55.

CGWB (2006). Groundwater resources of National Capital Territory, Delhi. Central Ground Water Board, New Delhi.

CPCB (1995). Groundwater Quality in problem Area—A status report (part -1V).

Curtis, G.P., Davis, J.A. and Naftz, D.L. (2006). Simulation of reactive transport of uranium (VI) in groundwater with variable chemical conditions. *Water Resources Research*, **42** (doi: 10.1029/2005WR003979).

Datta, P.S., Bhattacharya, S.K., Mookerjee, P. and Tyagi, S.K. (1994). Study of Groundwater occurrence and Mixing in Pushkar (Ajmer) valley, Rajasthan with $\delta^{18}O$ and hydro-chemical Data. *Jour. Geological Society*, **43:** 449-456.

Datta, P.S. and Tyagi, S.K. (1995). Isotopic Investigation on ground water recharge conditions and flow regime in Delhi region—A review. Proceeding International Conference 'Water & Energy 2001'. Vol. II. Oxford & IBH, New Delhi.

Datta, P.S. and Tyagi, S.K. (1996). Major ion chemistry of ground water in Delhi Area. Chemical weathering processes and ground water flow region. *Journal Geological Society of India*, **47:** 179-188.

Datta, P.S., Bhattacharya, S.K. and Tyagi, S.K. (1996). ^{18}O studies of recharge of phereatic Aquifers and ground water flow-paths of mixing in the Delhi area. *Journal of Hydrology*, **176:** 25-36.

Datta, P.S., Tyagi, S.K. and Mookerjee, P. (1997). Estimating contributions to ground water from rainfall and lateral flow in urbanized area based on ^{18}O-Cl relationship. Proceeding international conference on Isotopes in the solar system, Ahmedabad.

Datta, P.S., Deb, D.L. and Tyagi, S.K. (1997). Assessment of ground water contamination from fertilizers in Delhi Area based on ^{18}O, NO_3- and K-composition. *Journal of Contaminant Hydrology*, **27(3-4)**.

Deshpande, R.D., Bhattacharya, S.K., Jain, R.A. and Gupta, S.K. (2003). Distribution of Oxygen and Hydrogen Isotopes in Shallow Groundwater from Southern India: Influence of Dual Monsoon System. *Journal of Hydrology*, **271:** 226-239.

Emanuel, M. (1997). Chemical and Isotopic groundwater hydrology—Applied approach. Marcel Dekker, Inc. 270. Madison Avenue, N Y – 10016 USA.

Fritz, P. and Fontes, J.C. (eds) (1980). Handbook of Environmental Isotope Geochemistry, Vol. I, Vol. II. Elsevier, New York.

Gangal, R.K. and Zutshi, R. (1990). Groundwater pollution and the effects of rain near Khetri copper complex. *Indian J. Environ. Health*, **34(4):** 334-339.

Garg, D.K., Pant, A.B., Agarwal, M.B. and Goyal, R.N. (1990). Seasonal Variations in groundwater quality in Roorkee city. *IJEP*, **10 (9):** 673-676.

Government of India, Greening Delhi Action Plan (2005-2006). Department of Forest and Wildlife, Delhi.

Glynn, P.D. and Plummer, L.N. (2005). Geochemistry and the understanding of ground-water systems. *Hydrogeology Journal*, **13(1):** 263-287.

Kalin, R.M and Long, A. (1993). Application of hydro-geochemical modeling for validation of hydrologic flow, modeling in the Tucson Basin Aquifer, Arizona, USA. IAEA-TECDOC-0777, Vienna, pp. 209-254.

Kalin, R.M. (1996). Basic concept and information for Isotopic Geochemical modeling of ground water systems. IAEA-TECDOC-910, Vienna, pp. 155-206.

Krueger, C.J., Barber, L.B. II, Metge, D.W. and Field, J.A. (1998). Fate and transport of linear alkylbenzene sulfonate in a sewage-contaminated aquifer—A comparison of natural-gradient pulsed tracer test. *Environmental Science and Technology*, **32(8):** 1134-1142.

Leenheer, J.A., Hsu, J. and Barber, L.B. (2001). Transport and fate of organic wastes in ground water at the Stringfellow hazardous waste disposal site, southern California. *Journal of Contaminant Hydrology*, **51(3-4):** 163-178.

Margrita, R. and Vitart, X. (1987-1992). Methodologies and approaches in experimental studies using radioisotope tracers, of pollutant transport in hydrological environment. IAEA-TECDOC-0713, Vienna, pp. 181-208.

Mclarin, W.R. (1996). The groundwater system and nitrate contamination of the unconfined aquifer, Manakau, Horowhenua. Unpublished MSc thesis, Victoria University, Wellington.

Plummer, L.N., Bexfield, L.M., Anderholm, S.K., Sanford, W.E. and Busenberg, E. (2004). Geochemical characterization of ground-water flow in the Santa Fe Group aquifer system, Middle Rio Grande Basin, New Mexico. *U.S. Geological Survey Water-Resources Investigations Report*, **03-4131:** 395.

Revesz, K. and Casciotto, K. (2007). Determination of the delta ($^{15}N/^{14}N$) and delta ($^{18}O/^{16}O$) of Nitrate in solids. RSIL lab code 2897 U.S. Geological survey, Techniques and Methods, book 10, chap. C14.

Roy, T.N. (2000). Impact of Sewage Irrigation on Groundwater Regime of Roorkee Area. M.E. Dissertation (Unpublished). Department of Hydrology, University of Roorkee, Roorkee (Uttaranchal), India.

Shivanna, K., Navada, S.V., Kulkarni, K.M., Sinha, U.K. and Sharma, S. (1998). Application of isotopes techniques to investigate groundwater pollution in India. IAEA-TECDOC-1046, Vienna, pp. 166-184.

Singh, R.P., Chauhan, B.S. and Khan, B.Z. (1997). Groundwater and its pollution. *Asian J. of Chemistry*, **9(4):** 841-844.

Singh, R.P., Chauhan, B.S., Khan, B.Z. and Karri, A. (1998). Characteristics and groundwater pollution potential of solid waste leachate. *Asian J. of Chemistry*, **10(4):** 824-827.

Singhal, D.C., Roy, T.N., Joshi, H. and Seth, A.K. (2003). Evaluation of ground water pollution potential in Roorkee Town, Uttaranchal. *Journal Geological Society of India*, **62**.

Sinha, K. (2003). Scanning the earth to help predict the movement of contaminants in groundwater. Stanford University, Stanford Earth Sciences Report, Fall 2003.

Tyagi, S.K., Datta, P.S. and Bhattaccharya, S.K. (1997). ^{18}O-isotope imaging of ground water in Najafgarh basin, Delhi to assess availability under changing recharge conditions. Proceeding of International Symposium on Emerging trend in Hydrology, Sept 25-27, Roorkee.

Tyagi, S.K., Datta, P.S., Kulshreshtha, S. and Sharma, R.K. (2008). Isotopic and hydrochemical signatures in characterizing pollutants movement in overexploited groundwater aquifers of Delhi State. Proceeding of 3rd WEPA International Forum on Water Environmental Governance in Asia, Malaysia.

Unsal, N. and Afsin, M. (1999). Hydrochemical and isotopic properties of the Mahmutlu and Bagdatoglu mineralized thermal springs, Kirsehir, Turkey. *Hydrogeology Jour.*, **7:** 540-545.

Voss, C.I. and Wood, W.W. (1994). Synthesis of geochemical, isotopic and ground water modeling analysis to explain regional flow in coastal aquifers of Southern Oahu, Hawaii. *In:* Mathematical models and their applications to isotope studies in ground water hydrology. IAEA-TECDOC-777, Vienna, pp. 147-178.

Vrba, J. (1985). Impact of domestic and industrial wastes and agricultural activities on groundwater quality. *In:* Memories of the 18th Congress. *Hydrogeology in the service of man.* **Xviii(i):** 91-117.

Vrba, J. (2003). The impact of aquifer intensive use on ground water quality. *In:* Intensive use of groundwater, Llamas, R. and Custodio, E. (eds). A.A. Balkema, Lisse, The Netherlands.

Helium and Natural Gas Anomalies in Tubewells around Southern Fringes of Bundelkhand Region, Sagar-Damoh District, Madhya Pradesh, India

Arun K. Shandilya

Introduction

Near-surface geochemical prospecting studies have shown that the light hydrocarbon gases and helium can escape from the subsurface reservoirs/source rocks and migrate to the surface. Processes such as effusion and diffusion along with the secondary porosities in the rocks, sediments and soils can cause the gases to diffuse or migrate out in the atmosphere or get mixed with the ground water. Added selective references of near-surface geochemical prospecting occurrences of the surface geochemical anomalies pertaining to the seepage of natural gases indicate the possibility of hydrocarbon rich zones in the subsurface. The surface geochemical studies prove to be a cheaper and faster method to delineate potential hydrocarbon zones where detailed studies can be carried out and also these methods act as direct indicators of the micro-seepage of hydrocarbons.

The seepage history in the Sagar-Damoh areas of Vindhyan basin dates back to 1980's when the first leakage of petroleum gas was reported from the Meerkheri village located on the Sagar-Vidisha road, at about 15 km from the Rahatgarh town. In 1993, another tube well at Rahatgarh was found to have the leakages of natural gas. Shandilya (2007) has reported the occurrence of the petroleum gas in the Sagar district and suggested that the petroleum gas is thermogenic in origin. Shandilya (2008) observed the leakage of natural gas from the bore wells located in the rocks of the Vindhyan Super group in Sagar and Damoh Districts of Madhya Pradesh and suggested subsurface reserves of natural gas in Sagar which may be responsible for these seepages.

Shandilya (2009) observed that this gas is of thermogenic origin. It also suggested the possibility of Petroleum Gas reserve in Southern Bundelkhand

A.K. Shandilya (✉)
Department of Applied Geology, Dr. Hari Singh Gour University, Sagar, M.P., India
e-mail: akshandilya_u@rediffmail.com

N.J. Raju et al. (eds.), *Management of Water, Energy and Bio-resources in the Era of Climate Change: Emerging Issues and Challenges*,
DOI 10.1007/978-3-319-05969-3_13

Region of M.P. Shandilya and Gajbhiye (2010) made the detection of Rare Helium Gas in Sagar District, M.P. Prasanna et al. (2010) carried out detailed investigations of the light gaseous hydrocarbons in the surface soil around Sagar using the geochemical techniques. The authors focused on the detection of absorbed petroleum gases in Vindhyan basin by gas chromatographic method using compositional ratio of the light hydrocarbons such as methane, ethane, propane and butane.

Seepage of natural gas and helium was observed in the agricultural field of Sagar and Damoh districts of Bundlekhand Region. Near-surface geochemical hydrocarbon prospecting techniques were applied to the seepage areas. A useful geochemical method to correlate gases with their source rocks is dC^{13} determination. Coleman et al. (1977), in a study on the leakage of gas from the underground storage reservoir, showed that isotopic analysis is a reliable technique for differentiating between methane from thermogenic and biogenic source. The carbon isotope studies of the methane and other gases was done to know whether these gases are of biogenic or thermogenic origin. The soil, water and gas samples were collected from the tubewells of agricultural field located in various villages namely Meerkheri, Rahatgarh (tahsil), Pipariya-Bhutoli in Garhakota tahsils of Sagar and Mahalwara in Sukha block of Patharia tahsil of Damoh district in the southern fringes of Bundelkhand region.

Geology

The helium and natural gas leakages were observed in the dried tube wells dug onto the rocks of sandstone, shale and limestone of the Rewa and Bhander group of the Vindhyan Super group. These rocks are overlain by the Lameta bed (late Cretaceous) in the western part of the Pipariya and Bhutoli area in Garhakota tahsil. On the western part of present studies area rocks of upper Cretaceous Deccan Trap Basaltic flows exist which are intercalated with intertrappean limestones. Alluvial soil cover forms the topmost layer (Table 1). The Deccan Trap Basaltic rocks are exposed west of the Chinnoua village in Garhakota tahsil. The dips of the sandstone and shale rocks are 10°-15° toward SE direction. The topography is more or less flat with some low lying areas near the stream. The geological formation of Sagar area consists of nine flows of Deccan Traps.

Along with several intertrappean limestone beds, the whole area being studied is underlain to the east and south by the Lameta limestone which forms the plains and few hills to the south. The western fringe of the Rahatgarh and Meerkheri areas are covered mostly by the Basaltic flows of Deccan Traps except few inliers of the shale and sandstone rocks of Rewa and Bhander group of the Vindhyan Super group. The Deccan Trap basalt rocks are mostly covered by the black soil. The Deccan volcanism during late Cretaceous must have generated the proper thermal conditions and acted as a catalyst in triggering Mesozoic hydrocarbon generation processes in the Vindhyan sedimentary basin in Central India (Rangrajan 1978).

Table 1 The stratigraphic succession in Sagar district, M.P.

Age	*Formation*
Recent-Subrecent/Quaternaries	Alluvium soil
	Black soil
Upper Cretaceous To Eocene	Deccan Trap Basalt
	(with Intertrapean limestone)
Lower Cretaceous	Lameta Formation
Late Proterozoic to Middle	Bhander Group
Proterozoic	Rewa Group
	Kaimur Group
Vindhyan Super Group	Semri Group
	Phosphastic dolomite
	Dolomite
Middle Proterozoic	Ferruginous shale
Bijawar Super Group	Ferruginous sandstone
	Iron formation/BHQ
	Quartzite
	Conglomerate bed.
	Quartz reef intrusion
	Pegmatitic intrusion
Early Proterozoic	Ultramafic intrusive
Bundelkhand Granitic Complex	Granite intrusive rocks
	Granite gneiss
	Biotite schist
	Intrusive body
	Dolomitic marble
	Slate
Archean	Ferruginous formation
Mehroni Super Group	Quartzite
	Schist

Modified after Rangrajan, 1978

Observations

On the request of author, the Director Exploration, KDM IPE, ONGC Dehradun and Director, NGRI Hyderbad had sent a team of scientists for the detailed investigation. The team visited the Piparia-Bhutoli and Rahatgarh and Meerkheri area and collected the samples of soil, water and gas. The details of the sampling locality are as follows:

1. **(a) Bhutoli:** The tube wells from where the samples were collected fall under the panchayat and P.O. Chanauaa, tahsil Garhakota, dist. Sagar, M.P. (part of survey of India toposheet no. 55 M/1). It is accessible by a 3 km village road from town Garhakota around 45 km east of Sagar on way to Damoh. The leakage of petroleum gas was reported from eight tubewells in the month

of March 2007. These bore/tube wells vary in the depth of 260 to 400 ft and the top 60 ft is of 8″ diameter with plastic casing. Rest of well is of 6″ dia. depth 340 feet of Shri Bhagwan Singh Yadav (lat: 23°47′59.2″ N, long 79°05′29.6″ E, elevation 448 m). **(b) Piparia:** The tube wells (depth 400 feet) of Sri Asharam Patel S/o Sri Ghappu Patel (lat: 23°48′20″ N, long: 79°50′20.7″ E, elevation 450 m).

2. **Rahatgarh:** Rahatgarh village is located 40 km west of Sagar on way to Bhopal Road. In the bore well of Shri Leeladhar Tiwari (Tiwari Dhawa: lat: 23°57′15.7″ N, long: 79°25′03″E, elevation 484 m), the tube well is situated on the Deccan trap, Vindhyan contact. The leakage of petroleum gas is reported since 1993.
3. **Meerkhedi:** This tube well (lat: 23°45′56″N, Long:78°18′9.6″E, elevation 440 m) from where gas seepage was occurring belongs to Shri Dhan Singh. The area falls in the toposheet of India No. 55 I/5. This tube well is the eastern extremity of village Meerkhedi, 13 km of Rahatgarh on way to Vidisha. This tube well shows the petroleum gas since 1984 as evidenced by enormous bubbling in the water. The tube well is located on the contact of Deccan trap basalt and Vindhyan sand stone (In lier). The quantity of petroleum gas bubbling is increasing day by day.
4. **Mahalwara:** The leakages of the petroleum gas has been reported in 2009 from Mahalwara village in Patharia tahsil of Damoh district. The 378 feet deep tube well is pumping water along with the natural petroleum gas. A 2-3 feet long flame can be seen as sign of natural gas leakage from the well. Leakage can also be seen in bore well at Mahalwara, Damoh distt., which was dug 378 feet in Nov 2008. One-metre high flames can be observed in the presence of burning matchstick.
5. **Batyagarh:** It is located about 25 km from Damoh. In the tube wells the water smells of kerosene/diesel. This has been observed since Nov.-Dec. 2008 onwards. The tube wells are varying in depth from 350 to 400 feet in the Vindhyan limestone rocks.
6. **Patneshwar:** In this village the tube well is situated in the in liers of the Vindhyan rocks among the Deccan trap cover. The tube well was dug in 2000, and depth is about 380 feet. It is also pumping the natural petroleum gas and was reported in November 2011.
7. **Sukha:** In the agricultural field near the limestone quarry of Narsinghgarh Cement (Heidelberg Cement Plant) there is a leakage of petroleum gas along with water. Reported in Dec. 2009.
8. **Mandi Bamora:** A tube well situated in the Deccan trap rocks in Mandi Bamora, depth 390 feet, dug in May 2012 also showed the occurrence of natural gas along with water. The flame of 8-10 feet was observed at the well site in the Public Health Centre at Mandi Bamora. The thickness of the Deccan trap is very low at the site, resting over the rocks of Vindhyan Super group, in the northern-most extend of the Deccan cover.

9. **Konda:** At this village the tube wells have been pumping the gas since the month of Nov. and Dec. and the water is giving the smell of the kerosene in it. It is situated on the rocks of the Vindhyan Super group.
10. **Jhilla:** The tube well is situated on the in lier of the Vindhan rocks. The tube well was dug in Deccan trap rocks, which is overlying the rocks of Vindhyan Super group. The depth of tube wells are varying from 300 to 400 feet.
11. **Khurai:** In Khurai village the leakage of gas was reported by the villagers, which was studied in detail and found that the tube wells are containing the methane gas in it with a kerosene smell.
12. **Bannad 1:** The leakage of gas was reported from this village in Jan. 2012, in a farmer's well, which is also containing the petroleum gas. The well was burning for more than 24 hours.
13. **Bannad 2:** In this village the leakage of the natural petroleum gas has been reported in March 2012, from a tube well which is about 360 feet deep. The gas was coming forcefully in the afternoon and the bubbling sound could be heard from the tube well.
14. **Ghoghra:** The Ghoghra village falls in the Banda tahsil of Sagar district which is about 25 km from Banda, on the way to Patharia. In this village there are six tube wells pumping natural gas along with the water. Two hand pump are also showing the leakages. The gas coming out from these has been burning day and night. With the help of the local administration, the fire was brought under control on 7th May 2012. Continuous gas leakage has been reported from this village. In the agriculture field of the villagers the tube wells were dug in different time, but all of them are pouring the natural gas, which is burning 6 to 8 feet long flame.

The experiment of the gas leakages were done in some of the old tube wells, in which the natural gas is coming along with water and it burns when ignited. Local villagers say that there water is burning ('Paani me Aag'). In this village the hand pumps are also pumping the petroleum gas of indefinite quantity and are burning with flames day and night.

Geochemical Analysis

At ONGC, Dehradun

The samples were analysed in the geochemical laboratory of KDM IPE Kaulagarh Road, ONGC, Dehradun. The finding of the geochemical analysis of the natural gas, water and soil are shown in Table 2.

The Oil and Natural Gas Commission Dehradun has concluded that the seeping gases of Pipariya Bhutoli and Rahatgarh are predominantly methane (72.14 %-84 % in Pipariya Bhutoli and 99 % in Tiwari Dhaba, Rahatgarh Bore well) and are devoid of higher hydrocarbons. The hydrocarbon gases seem to have predominance of

Table 2 Geochemical composition of natural gas and stable isotope values

		Chemical composition % (v/v)						*Isotopic values*
S. No.	*Bore well*	*He*	O_2	N_2	CO_2	C_1	C_{2+}	(δC^{13})
1	Pipariya Bhutoli	0.34	1.6	24.87	0.93	72.14	0.01	−61.5
2	Rahatgarh	0.72	0.65	14.37	0.28	84.00	0.02	−54.0

After Shandilya, 2007

Table 3 Stable isotopic value of hydrocarbon gas in Sagar district, M.P. (India)

Location	*Name of petroleum gas*	*Stable isotopic value* δC^{13}
Piparia-Bhutoli	Methane	−43.6 per mil w.r.t. PDB
	Ethane	−24.6 per mil w.r.t. PDB
Rahatgarh-Mirkheri	Methane	- 54.9 per mil w.r.t. PDB
	Ethane	- 26.4 per mil w.r.t. PDB

After Shandilya, 2008

Table 4 The adsorbed soil hydrocarbon gas concentration in Sagar distt., M.P. (India)

	C1	*C2*	*C3*	*i C4*	*n C4*	*∑C3+*
Minimum	1	1	1	1	1	1
Maximum	104	14	10	09	08	41
Mean	19	2	1	0.5	0.3	4
Standard Deviation	22	4	2	2	1	8

After Prasanna et al., 2010

bacterial methane. The pressure of both the seepages is extremely low. As per the owners, the quantity of gas is diminishing day-by-day. The results of methane from the samples collected in this survey are genetically different from thermogenic gases encountered in exploratory well jabera-1, drilled by ONGC in distt. Damoh (M.P.).

In January 2008 in Bhutoli village, the land owner dug the well up to 400 ft. deep and encountered huge quantity of natural gas which was reported to governmental agencies. In Feb. 2008 two villagers at Pipariya-Bhutoli drilled two more tube wells up to the depth of 300-350 feet about 600-700 m away from the earlier wells in the shales and sandstone rocks of lower Bhander. They could not get the ground water, but were leakages of natural gas, giving a blue flame.

At NGRI, Hyderabad

The samples of petroleum gas, soil and water were also collected by the scientists of NGRI Hyderabad along with the author on 4-5 May 2008. The detail geochemical and stable isotopic studies of the natural petroleum gas, soil and water sample have been done in the laboratories of NGRI (Tables 3 and 4).

Table 5 Adsorbed hydrocarbon ratio in soil samples in Sagar distt., M.P. (India)

Ratio	*Threshold value*	*% of at incidence in survey at Sagar*	*Reference*
C1/C2 + C3	<10	79	16
C1/C2	<16	96	18
C2 /∑C3 +	<1.3	50	17

After Prasanna et al., 2010

The presence of the ethane gas and δ $\mathbf{C^{13}}$ value (in the range of –24.6 per mil. w.r.t. P.D.B. and –26.4 per mil. w.r.t. P.D.B.) indicates a possible thermogenic source for these gases in both the localities. Geochemical signatures of the natural gas, desorbed soil gas and water samples show that the samples of the Piparia-Bhutoli and Rahatgarh-Meerkheri of Sagar district contain 72 % to 99 % of methane and 0.34 %-0.72 % of helium along with the oxygen, carbon dioxide and nitrogen. The stable isotopic δC^{13} value is in the range of –43.6 per mil. w.r.t VPDB for methane and –24.6 per mil. w.r.t PDB for ethane at Piparia-Bhutoli to –54.9 per mil. w.r.t VPDB for methane and –26.4 per mil. w.r.t PDB for ethane at Rahatgarh are indicative of the thermogenic origin and also methane is associated with oil. Bernard (1978) suggested a genetic diagram by correlating (C1/ C2 + C3) ratio with δC^{13} 1 concentration of methane to classify natural gas types. Molecular ratio (C1/C2 + C3) less than 50 are typical for the thermogenic hydrocarbon gases with δC^{13} value ranging between –30 part per mil and –55 part per mil (VPDB) (Table 5). This suggests that most of the samples collected in this study fall in the thermogenic range.

The adsorbed soil gas result indicates the presence of methane, ethane, propane and butane in Sagar district. The studies indicate that these migrated hydrocarbons are of thermogenic origin and the area is warm for hydrocarbon exploration.

Conclusion

The natural gas and helium seeping out in the borewells of the Sagar-Damoh area contain the high amount of methane (72-99 %), and helium (0.34 % to 0.72 %) along with minor amounts of oxygen, nitrogen and carbon dioxide. This indicates that it must have been formed at higher temperature condition at deeper horizons in the Proterozoic Vindhyan sedimentary basin. The reservoir must be lying below the ground at least 500 m or more deep level. The leakage of natural gas through many fine cracks/fracture and feather joints in the sandstone, shales and limestone rocks of the Rewa and Bhander group rocks of the Vindhyan Super group can be the reason for the visible surface anomalies.

The geochemical and stable isotopic studies of gas samples have shown the presence of the methane and ethane gas. The presence of ethane gas collected from the above mentioned localities and the δC^{13} stable isotopic value in the range of –24.6 per mil. w.r.t. to –26.4 per mil w.r.t. PDB is indicative of the thermogenic sources. The study suggests that this area is likely to contain a potential gas reserve.

However detailed studies in the area should be carried out to know the commercial viability of the reserves.

Acknowledgement The author is thankful to the Director, NGRI Uppal Road Hyderabad, and the Director, KDMIPE, ONGC, Dehradun for the detailed geochemical and stable isotopic analyses of the various samples of gas, water and soil. The author is highly thankful for the critical comments and valuable suggestions on this manuscript by Prof. N.S. Gajbhiye, Vice Chancellor, and Prof. S.P. Banerjee (Rtd Professor), Deptt. of Chemistry, Dr. Hari Singh Gour University, Sagar, M.P. (India).

References

Allen, M.R., Stott, P.A., Mitchell, J.F.B., Schnur, R. and Delworth, T.L. (2000). Quantifying the uncertainty in forecasts of anthropogenic climate change. *Nature*, **407:** 617-620.

Bernard, B.B. (1978). Light Hydrocarbon in Sediments. Ph.D dissertation. Texas A&M University, USA.

Coleman, D.D., Meents, W.F., Liu, C.L. and Keogh, R.A.S. (1977). Isotopic identification of leakage gas from underground storage reservoir – A progress report III. *Petrology*, **111:** 1-10.

Fuex, A.N. (1977). The use of stable carbon isotopes in Hydrocarbon exploration. *Jour. Geochem. Explor.*, **7:**155-188.

Lones, V.T. and Drozd, R.J. (1983). Prediction of oil and gas potential by near surface geochemistry. *Ameri. Asso. Petr. Geol. Bull.*, **983:** 67,932-67,952.

Horvitz, L. (1979). Near surface evidence of hydrocarbon movement from depth problems of petroleum migration. *Am. Asso. Pet. Geology.*, **10:** 241-269.

Prasanna, M.V., Rasheed, M.A., Madhavi, V.T., Kalpana, G., Patil, D.J. and Dayal, A.M. (2010). Light gaseous hydrocarbon anomalies in the near surface soils of Sagar District, Vindhyan Basin, India. *Current Science*, **99(11):** 1586-1590.

Rangrajan, K. (1978). Geology of Sagar district and western part of Damoh distt. Memoirs, GSI, Vol.109.

Raina and Tiwari (2007). Unpublished report of ONGC 2007.

Shandilya, Arun K. (2007). Natural Petroleum Gas found in Piparia-Bhutoli- Rahatgarh in Sagar District, Bundelkhand Region M.P. *University Journal- MadhyaBharati*, **53:** 122-124.

Shandilya, Arun K. (2008a). Discovery of Natural Petroleum gas in the rocks of the Vindhyan Super Group in Sagar Distt., Bundelkhand Region, M.P. *Qtrly Jour. of GARC*, **16(1):** 24-27.

Shandilya, Arun K. (2008b). Fuel Ho – Leakages of Gas in Tube wells spew natural gas. The News Magazine, March 2008, 31-32.

Shandilya, Arun K. (2008c). Natural Gas leakages from Bore wells in the rocks of the Vidhyan Super group in Sagar and Damoh Distts., M.P. *Journal Himalayan Geology*, **29(2):** 174-175.

Shandilya, Arun K. (2008d). Natural Gas Reserves in Sagar District., *"Srujun" University News Letter*. **9:** 19.

Shandilya, Arun K. (2009a). Gas in the Backyard. *Business India Magazine*, 144.

Shandilya, Arun K. (2009b). Thermogenic Petroleum Gas found in tube wells of Sagar District, M.P. *Madhya Bharati, Diamond Jubilee Volume*, 314-319.

Shandilya, Arun K. (2009c). The possibility of Petroleum Gas reserve in Southern Bundelkhand Region, M.P. *Vishwavidyalaya Samvad*, **2:** 3.

Shandilya, Arun K. and Gajbhiye, N.S. (2010). Discovery of Rare Helium Gas in Sagar District, M.P., India. Proc. Vth International Conference, South Valley University QENA, Egypt. Jan 2-7, 2010.

Shandilya, Arun K. (2011). The Discovery of ural petroleum and Helium gas in Tube wells of Sagar-Damoh Distt. in Vindhyan Rocks, M.P. Proc. Sedimentary Basins of India, Gondwana Geological Society, Nagpur 2011.

Shandilya, Arun K. (2012). Seepage of natural gas from tubewells in Bundelkhand regioin in Sagar-Damoh district, M.P. *Int. Jour. ONGC Bulletin*, **47(2):** 155-159.

Shandilya, Arun K. (2012). Gaseous Anamolies of Natural Gas around Sagar Distt., M.P. *Int. Jour. of Chemical and Life Science*, **5:** 1054-1060.

Shandilya, Arun K. (2013). The Stable Isotopic studies of Natural Gas in Sagar Area, M.P. *Madhyabharati Journal*, Dr. HSGVV, **51**.

Section II
Energy and Bio-resources Management

Production of Renewable Energy and Waste Water Management from Vetiver Grass

Ashutosh Kumar and Ram Prasad

Introduction

Environmental issues such as climate change, global warming and energy supplies need to augment existing power generation facilities and have got the attention to develop gasification of biomass source, solar energy, wind energy etc. Biomass corresponds to various kinds of recently dead and living plant cell, anatomy, morphology and composition of which is distinct for different parts and species of plant (Shafizadeh, 1982).

Municipal residues, wastes, forestry, agricultural residues and dedicated energy crops are the major raw materials for renewable energy generation. Wastes are now typically less expensive to supply to end point users, and will like to play an important role in early development of renewable energy demand. However, analyses of future demand for renewable energy indicate that these are not able to meet the required production in future. Thus dedicated energy crops will be required to meet the demands of a growing renewable energy market. At present scenario, biomass globally supplies about 50 EJ (1 exajoule = 10^{18} J) that is 10 % of annual primary energy consumption at global level (IEA Report, 2006).

Worldwide biomass ranks fourth as an energy resource, providing approximately 14 % of the world's energy needs; biomass is the most important source of energy in developing nations, providing 35 % of their energy (Sudha and Ravindranath, 1999).

Biomass is carbon lean fuel as the carbon released during combustion in the form of CO_2 is fixed during plant growth by photosynthesis. Wood is a highly efficient biomass fuel but it is too valuable to be used for power generation and in timber

A. Kumar (✉) • R. Prasad
Department of Chemical Engineering and Technology, Indian Institute of Technology (BHU), Varanasi 221005, India
e-mail: akumar.rs.apc@iitbhu.ac.in

N.J. Raju et al. (eds.), *Management of Water, Energy and Bio-resources in the Era of Climate Change: Emerging Issues and Challenges*,
DOI 10.1007/978-3-319-05969-3_14

industry it is better used as a construction material. So the plant residue bagasse, animal dung, stacks of maize, husks of rice, coconut and renewable perennial biomass of herbaceous plants are major feedstock for energy production.

Vetiver is perennial grass of family gramineae commonly known as Kasa. It is hydrophytes, often dominant in fresh-water swamps, flood plains and on stream banks. However, it can also grow well under alternating very wet and very dry conditions, at average temperatures of 25 °C-35 °C (Seaforth and Tikasingh, 2002). Indian Railways has 114,500 kilometres (IRYB, 2009–10) of total track over a route of 65,000 kilometres and to prevent soil erosion from the elevated track with respect to ground level, on both sides of it vetiver is grown. Major quantity of vetiver grass leaves are burnt by rural people for cooking and oil extraction industry; it harms the atmosphere releasing carbon dioxide as a major green house gas. Combustion of fossil fuel produces substantial greenhouse gases (GHG) and toxic gases such as CO_2, CH_4, SO_2, NO_x and other pollutants, causing global warming and acid rain (Meng et al., 2006a). So production of power and energy resource from biomass via sustainable method lower the emission of environmentally prejudicious gases like SO_2 and NO_x and mitigates environmental issue and challenges such as global acid formation, global warming etc. (IEA Report, 2006; Jenkins et al., 1998; Malkina et al., 2002; Rezaiyan et al., 2005).

Chemical composition of biomass constitutes carbon as a major element together with a mixture of organic molecules containing hydrogen, oxygen, nitrogen and also small amount of alkaline earth, alkali and heavy metals. In nature hemicelluloses comprise roughly one-fourth to one-third of the most plant materials and some important applications of these polysaccharides have been demonstrated during the past two decades. Hemicelluloses show a potential fermentation feedstock for the production of sugars, furfural and used as an additive in paper-making, pharmaceutical, cosmetic and food areas (Fenden et al., 2002; Sun and Tomkinson, 2002). A novel product like hydrogel has been also exploited since hemicelluloses are, in their native form, responsible for the interaction with water (Gabrielii et al., 2000). In addition, with a chemical modification, a hydrophobic film from maize bran hemicelluloses has been created as a bioplastic (Fenden et al., 2002).

Hemicellulose samples from the vetiver grass is an arabinoxylan mainly consisting of a backbone of 1, 4-linked xylopyranosyl units. The $(1 \rightarrow 4)$-xylopyranose backbone is substituted on O-2 and/or O-3 by single residue or short chains. These branches may be single arabinose residue, single glucuronic acid residue or a short chain of sugar residues containing arabinose, xylose and galactose. So it shows a great potential to use as energy production (Oraphin et al., 2004). There is absolute contingency to augment utilization of biomass by tapping the wastes, unused plants and its residue. Conventional crops production used for energy can also be accelerated, with considerate account of land availability and food demand. Ligninocellulosic crops could be produced on coastal, devalued and ancillary agricultural lands and furnish the bulk of biomass resources.

Fig. 1 Vetiver grass

Experiment

Plants of vetiver grass (Fig. 1) were collected from Banaras Hindu University campus and separated into three categories: stems, leaves and total plants. These were dried at 110 °C in the oven for 24 hrs and then ground in a mixer-grinder and sieved (mesh no. 18–100).

Experimental Setup

The schematic diagram of experimental set up of the pyrolysis of biomass is represented in Fig. 2 and the role of different components of experimental set up is mentioned in Table 1.

Pyrolysis of Biomass

At first 10 g powder of vetiver grass was taken in the fabricated quartz reactor and it was flushed out by nitrogen for 30 minutes to replace the air present in reactor. Now the temperature of reactor was increased at a rate of 5°C/min from room temperature to 500 °C. Gas flow rate was measured with the help of digital gas flow meter until production of gas completely stopped. The pyrolysed gas produced

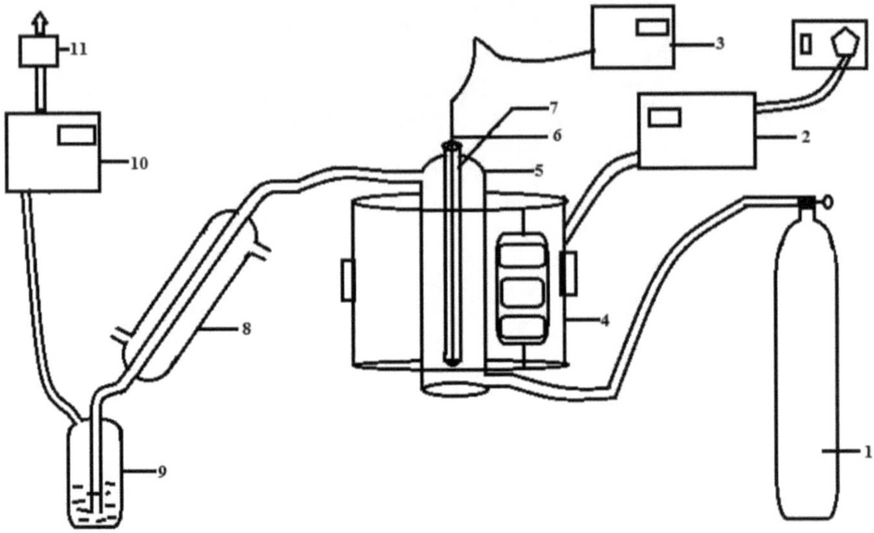

Fig. 2 Schematic diagram of the pyrolysis experimental set up (1. Nitrogen gas; 2. Microprocessor-based temperature controller; 3. Temperature indicator; 4. Split-open furnace; 5. Reactor; 6. Thermocouple; 7. Thermowell; 8. Condenser cum cooler; 9. Gas bubbler; 10. Digital gas flow meter; and 11. Gas sampler)

Table 1 Constituent components and their functions of experimental set up

Components of experimental set up	*Functions*
Vertical quartz cylindrical reactor of diameter 4.5 cm with a cone and socket joints	For the pyrolysis of biomass (vetiver grass)
Microprocessor-based temperature controller	To maintain the constant heating rate
Thermocouple attached to temperature indicator	To monitor the temperature of the biomass bed
Split-open furnace	To heat the reactor at desired rate
Condenser cum cooler	To condense liquid products of gas
Gas bubbler	To retain the condensed part during pyrolysis such as liquid products and water soluble components of gas
Digital gas flow meter	To measure the gas flow rate
Syringe pump	For the injection of water whenever required
Nitrogen cylinder	To flush out oxygen from the reactor in the beginning and to maintain inert atmosphere at the end of the experiment

was collected in a gas sampler and it was analyzed with the help of a Nucon gas chromatograph (model no. 5765) equipped with pora pack Q column, methanizer and FID detector under the following conditions: Carrier gas – Nitrogen, Fuel gas – Hydrogen, Oven temp – 60 °C, Detector and injector temperatures – 80 °C each, Methanizer temperature – 300 °C and Gas sample size – 500 μl.

In the second experimental series, catalytic effect of Na_2CO_3 on the gas flow rate and amount of gas released on different separated parts of vetiver grass was analyzed. 10 g powder of vetiver grass was thoroughly admixed with 10 ml of 1 % Na_2CO_3 solution and dried in oven at 110 °C for overnight. Ten gram of dried material was taken into the reactor and pyrolysis was done as described above.

In third experimental series, the optimum concentration of Na_2CO_3 for the pyrolysis of stem part of vetiver grass was determined.

In fourth experimental series, water gas was produced and measured by passing steam over char residue obtained by pyrolysis of the stem of vetiver grass at 426 °C. Steam was generated from water introduced at the rate of 0.33 ml/min with the help of a syringe pump.

In fifth experimental series, activated char was prepared using char residue obtained by pyrolysis of stem treating with 10 % H_2SO_4 for six hours, followed by washing with distill water to remove H_2SO_4 and drying at 110 °C. Further, activation of acid treated and dried char was carried out by passing CO_2 gas at 400 °C.

In sixth experimental series, activated char was used to remove Cr^{+6} from the prepared standard solution of $K_2Cr_2O_7$ (40 ppm). One gram of the activated char was taken in conical flask having 100 ml of solution. The content of the flask was shaken on electric shaker and at the interval of 1 hr Cr^{+6} content was analyzed with the help of atomic spectrophotometer.

Gasification of Char by Production of Water Gas

After pyrolysis of stem of vetiver grass at 500 °C when gaseous product stopped coming out, the gasification of the char was carried out by passing steam over char residue at 426 °C. Steam was generated from water introduced at the rate of 0.33 ml/min with the help of a syringe pump. The flow rate of the generated water gas was measured with the help of a digital gas flow meter.

Determination of Calorific Value of Biomass

Calorific values of powdered material of stem, leaves and total plant were determined separately with the help of a bomb calorimeter.

Result and Discussion

Analysis of Gas Collected on Pyrolysis of Vetiver Grass

Composition analysis of the gas sample collected on pyrolysis of the vetiver grass was carried out with the help of a gas chromatograph. Methane (CH_4) and carbon

Table 2 Composition of pyrolyzed gas

Gas component	*Composition (%)*
CH_4	50
CO	2
CO_2	48

Table 3 Comparison of calorific value of biomass fuels

S. No.	*Biomass*	*Heating value, kcal/kg*
1	Wood	3500
2	Cattle dung	3700
3	Bagasse	4400
4	Wheat and rice straw	2500
5	Cane trash, rice husk and vegetable wastes	3000
6	Coconut husks, dry grass and crop residue	3500
7	Ground nut shells	4000
8	Coffee and oil palm husks	4200
9	Cotton husks	4400
	Experimental data of present study	
10	Vetiver grass (Total plant)	4900
11	Vetiver grass (Only stem)	6000
12	Vetiver grass (Only leaves)	3900

www.indiasolar.com 2012

dioxide (CO_2) were found as major amounts, whereas carbon monoxide (CO) was found in trace amount. Composition of different gaseous products with their weight % is given in Table 2.

Calorific Value of Vetiver Grass

Calorific values of different parts of vetiver grass were determined in a bomb calorimeter. The experimental data obtained in the present study is compared with the values reported in literature for different biomass in Table 3. It can be seen that the present data is comparable to reported data. It is very clear from the table that the calorific value of stem part of vetiver grass is very high in comparison to total plant and leaves.

Pyrolysis of Powdered Material of Various Parts of Vetiver Grass

Figure 3 shows the flow rate of gas generated on pyrolysis of various parts of the vetiver grass at different temperatures. It can be seen from the figure that the

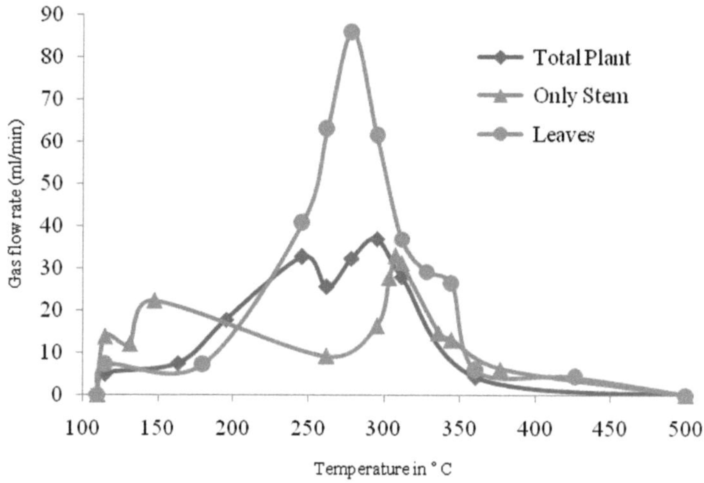

Fig. 3 Gas flow rate vs. temperature graph of total plant (♦), only stem (▲) and leaves (●)

pyrolysis of various parts of vetiver grass started generating gases at 114.7 °C and stopped around 400 °C. In case of total plant and stem of the vetiver grass initially slower rate of gas generation occurred with increase in temperature and then it increased rapidly resulting in two maxima in both the cases at different temperatures and a minimum at 262.3 °C. For total plant the maxima occurred at 245 °C and 295 °C on the other hand for stem the maxima occurred at 147.5 °C and 307.4 °C. Whereas, the pyrolysis of leaves start generating gases at 180.3 °C and stopped at 426.3 °C resulting in a single maximum at 278.7 °C.

During pyrolysis of biomass highest flow rate was observed at 295 °C, 311 °C and 278 °C for vetiver plant, stem and leaves respectively. Gas flow rate in case of leaves shows less variation. It may have occurred due to presence of less amount of lignin-cellulosic material in cell wall in leaves compared to stem. Hemicellulose, cellulose and lignin decompose typically in the temperature ranges of 200–400, 275–400 and 200–500 °C, respectively (Hirunpraditkoon and Garcia, 2008). Pyrolysis of lignin started at lower temperature than that of cellulose, which covers a rather wide temperature interval (Liu et al., 2008). The principal component of vetiver grass was reported as cellulose (ca. 30–35 %), hemicelluloses (ca. 40 %) and lignin (ca. 10 %) (Pawadee et al., 2003). Amount of lignin, cellulose and hemicellulose varies with different parts of the plant. Stem of plant is stiffer than leaves due to presence of lignin and leaves mainly composed of hemicellulose and cellulose. In the studies of alpha cellulose pyrolysis, significant weight loss was observed at 247-280 °C due to dehydration (Chang et al., 2011). In case of pyrolysis of stem and total plant of vetiver grass, gas flow rate goes down at 262.3 °C which may be due to endothermic dehydration of constituent components (cellulose).

Yield of char obtained for plant, stem and leaves were 0.4182 g/g, 0.3411 g/g and 0.4061 g/g of biomass, respectively. Assuming ideal gas law, the weight of gas produced was 0.1707 g/g, 0.2086 g/g and 0.1722 g/g of biomass of total plant, stem and leaves, respectively. Thus the amount of tar, water and other liquid products was calculated by the difference, equal to 0.4111 g/g, 0.4503 g/g and 0.4217 g/g of total plant, stem and leaves, respectively.

Effect of Catalyst (Na_2CO_3) on Pyrolysis of Powdered Material of Various Parts of Vetiver Grass

Few inorganic salts like carbonate, chromates and chlorides show propitious effect on the rate of pyrolysis reaction (Rabah and Eldighidy, 1989). Among alkali catalyst Na_2CO_3 has better catalytic effect compared to other catalyst for biomass pyrolysis (Meng et al., 2006b). Catalysts reduce the tar content and also improve the gas product quality. Figure 4 shows the flow rate of gas generated on pyrolysis of varhious parts of the vetiver grass admixed with Na_2CO_3 catalyst at different temperature. It can be seen from Fig. 4 that the pyrolysis of various parts of vetiver grass started generating gases at 114.7 °C and stopped around 400 °C. In case of total plant and leaves of the vetiver grass, initially slower rate of gas generation occurred with increase in temperature and then it increased rapidly resulting in two maxima in both the cases at different temperatures and a common minimum at 311.5 °C. For total plant the maxima occurred at 270.5 °C and 327.9 °C on the other hand for leaves the maxima occurred at 278.7 °C and 344.3 °C. It should be noted that contrary to without catalyst, with catalyst leaves show two peaks, whereas for stem single maximum occurred at 245.9 °C. Variation in the maxima occurred due to its constituent composition.

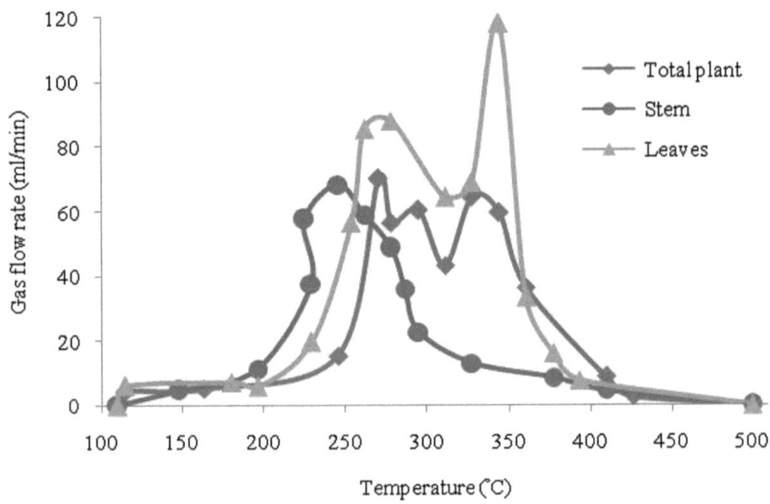

Fig. 4 Gas flow rate vs. temperature graph of 1 % Na_2CO_3 treated total plant (♦), only stem (●) and leaves (▲)

Alvarez and Vazquez (2004) reported the peak of pyrolysis temperature at 300 °C for the hemicellulose and 360 °C for the pure cellulose without catalyst. The present experimental data with catalyst show that the maxima shift toward low temperature compared to the reported values by Alvarez and Vazquez (2004).

Yields of char obtained with 1 % Na_2CO_3 treated total plant, stem and leaves were 0.3842 g/g, 0.3754 g/g, and 0.4332 g/g of biomass respectively. Assuming ideal gas law the weight of gas produced was 0.1942 g/g, 0.2170 g/g and 0.1863 g/g for 1 % Na_2CO_3 treated total plant, stem and leaves, respectively. Thus the amount of tar, water and other liquid products was calculated by the difference equal to 0.4216 g/g, 0.4076 g/g and 0.3805 g/g of 1 % Na_2CO_3 treated total plant, stem and leaves, respectively. The data show that on treatment of biomass with Na_2CO_3 produced more amount of gas and less amount of tar in comparison to without catalyst. The temperature range for maximum gas production with catalyst is very narrow (200-300 °C) as compared to without catalyst (262-377 °C) for stem. Therefore, stem was selected for further studies.

Optimization of Catalyst Concentration for Pyrolysis of Stem of Vetiver Grass

Optimum concentration of catalyst is necessary to minimize the cost of biomass pyrolysis process. Figure 5 shows the effect of catalyst concentration on flow rate of gas generation on pyrolysis of stem of vetiver grass at different temperatures. It can be seen from the figure that the pyrolysis of the stem started generating gases at 114 °C and stopped around 450 °C. In case of 0.5 % and 1.5 % Na_2CO_3 treated stem

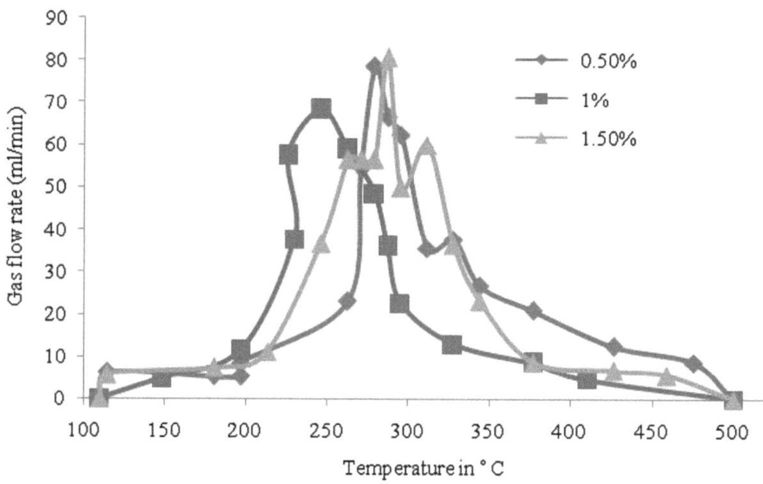

Fig. 5 Gas flow rate vs. temperature graph of 0.5 % (♦), 1 % (■), and 1.5 % (▲) Na_2CO_3 treated stem of vetiver

of the vetiver grass initially slower rate of gas generated and with increase in temperature it increased rapidly resulting in two maxima in both the cases at different temperatures and a minimum at 311.5 °C and 295.1 °C, respectively. For 0.5 % Na_2CO_3 treated stem of the vetiver grass the maxima occurred at 278.7 °C and 327.9 °C, on the other hand for 1.5 % Na_2CO_3 treated stem the maxima occurred at 286.9 °C and 311.5 °C. Whereas, the pyrolysis of 1.0 % Na_2CO_3 treated stem of the vetiver grass started generating gases at 147.5 °C and stopped at 409.8 °C resulting in a single maximum at 245.9 °C.

Yields of char obtained for 0.5 %, 1 % and 1.5 % Na_2CO_3 treated stem were 0.3488 g/g, 0.3754 g/g and 0.3578 g/g of biomass, respectively. Assuming ideal gas law, the weight of gas produced was 0.2296 g/g, 0.2170 g/g and 0.2184 g/g for 0.5 %, 1 % and 1.5 % Na_2CO_3 treated stem, respectively. Thus the amount of tar, water and other liquid products was calculated by the difference equal to 0.4216 g/g, 0.4076 g/g and 0.4238 g/g of 0.5 %, 1 % and 1.5 % Na_2CO_3 treated stem, respectively.

It is to be noted that in all the cases of Na_2CO_3 treated samples, comparable total amount of gases were produced. However, maximum gas flow rate occurred at the lowest temperature of 245.9 °C for 1 % Na_2CO_3 treated stem of the vetiver grass in comparison to other samples.

Water Gas Production

When steam was passed over red hot char left after pyrolysis (4.127 g) at 426 °C, water gas is produced as per following equation:

$$C + H_2O(\text{steam}) \rightarrow CO + H_2 \quad (1)$$

The flow rate of water gas produced with time is plotted in Fig. 6. A maximum flow rate of 9.44 ml/min of water gas resulted at 21 minutes of the start of the experiment. The gas production stopped after 110 minutes of continuous experiment when reducible carbon in the char is exhausted. A total of 390 ml of water gas was produced at ambient temperature and pressure.

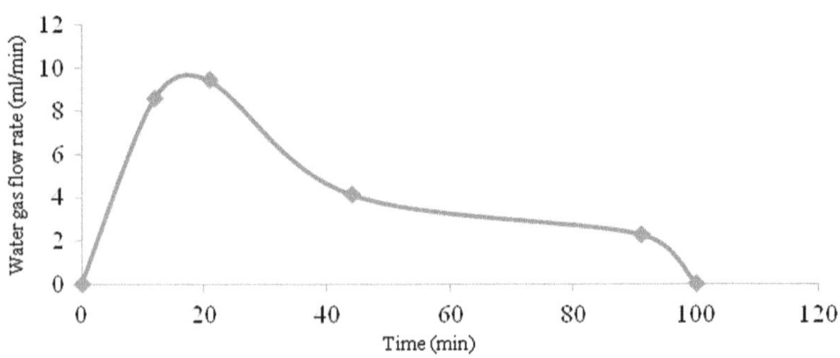

Fig. 6 Water gas flow rate vs. time graph of char prepared from stem of vetiver grass

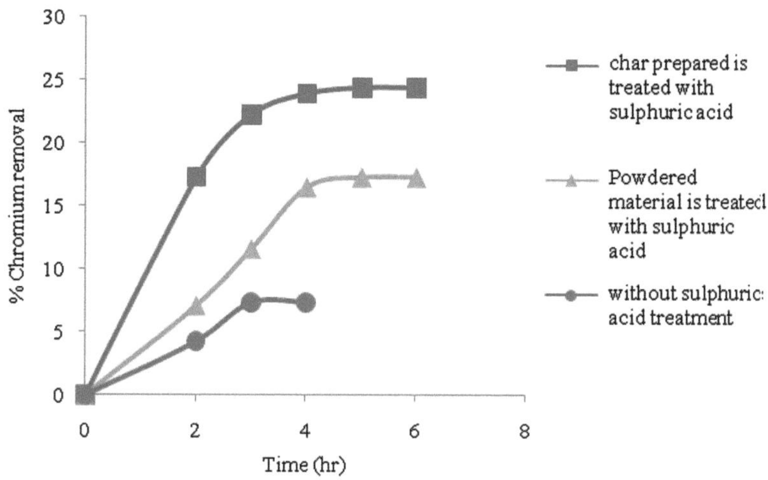

Fig. 7 % Chromium removal at various time

Removal of Chromium

Activated carbon prepared from char remained after pyrolysis was utilized for waste water treatment for Cr^{+6} removal. The results of chromium removal is shown in Fig. 7. Adsorption of chromium is initially increased rapidly and then it becomes constant with increase of time. The optimum time for adsorption was observed 240 min. It can be seen from Fig. 7 that removal efficiency of chromium is highest for activated carbon which was treated with H_2SO_4 after char was prepared from stem powdered material. The powdered materials treated with H_2SO_4 and activated by carbon dioxide show relatively higher efficiency to remove chromium with respect to the activated carbon prepared without treatment with H_2SO_4 and CO_2. Carbon dioxide molecules can diffuse into the narrower micro pores where the nitrogen molecules cannot diffuse because of the increase in the diffusion rate of the gas molecules and the absence of diffusion barrier.

Since nitrogen has a pore penetrator function, carbon dioxide deepens and widens pores, especially at high temperatures. That is it increases the ratio of mesopore volume. Because carbon dioxide gas reacts with elementary graphite micro crystals in the carbon structure and it removes carbon atoms burning at the pore sides during activation, it leaves an activated carbon having a more defective structure that contained different types of pores (Guzel and Tez, 1993).

Conclusion

Vetiveria zizanoides is grown deliberately to prevent soil erosion, typically at the borders of agricultural lands and sides of the railway track, since this plant does not hinder the growth and nutrient uptake of other plants. At the end of summer season stem of vetiver is burnt as it has negligible market value at present.

In this study, it is found that vetiver grass stem has high calorific value compared to other biomass feedstock used for energy production; therefore there is much potential of it to be used as energy crop for bio-energy. Due to the seasonal burning of vetiver grass stem, carbon dioxide is released which is a harmful greenhouse gas. It can be reduced by using vetiver grass as bio-energy. The activated carbon prepared from char obtained after pyrolysis can be used to control waste water pollution and fuel cell. As India is developing agricultural nation, finding productive uses for waste biomass is a great asset for the country, especially when it contributes to the area of energy production.

References

Chang, A.C.C., Chang, H.F., Lin, F.J., Lin, K.H. and Chen, C.H. (2011). Biomass gasification for hydrogen production. *International Journal of Hydrogen Energy*, **36:** 14252-14260.

Chaikumpollertb, O., Methacanona, P. and Suchiva, K. (2004). Structural elucidation of hemicelluloses from vetiver grass. *Carbohydrate Polymers*, **57:** 191-196.

Fenden, E., Granet, R., Zerrouki, R., Krausz, P., Saulinier, L., Thibault, J.F., Rosier, J. and Petit, C. (2002). Hydrophobic films from maize bran hemicelluloses. *Carbohydrate Polymers*, **49:** 1-12.

Gabrielii, I., Gatenholm, P., Glasser, W.G. and Kenne, L. (2000). Separation, characterization and hydrogel-formation of hemicellulose from aspen wood. *Carbohydrate Polymers*, **43:** 367-374.

Guzel, F. and Tez, Z. (1993). The characterization of the micropore structures of some activated carbons of plant origin by N_2 and CO_2 adsorptions. *Separation Science and Technology*, **28:** 1609-1627.

Hirunpraditkoon, S. and Garcia, A.N. (2009). Kinetic study of vetiver grass powder filled polypropylene composites. *Thermochimica Acta*, **482:** 30-38.

IEA (International Energy Agency). IEA bioenergy annual report (2006). http://www.energytech.at/pdf/iea_bereport06.pdf.. Accessed 5 January 2013.

Indian Railways Year Book (2009-2010). Ministry of Railways, Government of India. 2011. Retrieved 26 December 2012.

Indiasolar Web. http://www.indiasolar.com/cal-value.htm. Accessed 5 January 2013.

Jenkins, B.M., Baxter, L.L., Miles Jr, T.R. and Miles, T.R. (1998). Combustion properties of biomass. *Fuel Process Technol*, **54:** 17-46.

Liu, Q., Wang, S., Zheng, Y., Luo, Z. and Cen, K. (2008). Mechanism study of wood lignin pyrolysis by using TG–FTIR analysis. *Journal of Analytical and Applied Pyrolysis*, **82:** 170-177.

Malkina, I.G. and Pykh, Y.A. (2002). Sustainable energy resources, technology and planning. WIT Press, Southampton (UK).

Meng, N., Leung, D.Y.C., Leung, M.K.H. and Sumathy, K. (2006a). An overview of hydrogen production from biomass. *Fuel Process Technology*, **87:** 461-472.

Meng, N., Leung, M.K.H., Sumathy, K. and Leung, D.Y.C. (2006b). Potential of renewable hydrogen production for energy supply in Hong Kong. *International Journal of Hydrogen Energy*, **31:** 1401-1412.

Methacanona, P., Chaikumpollertb, O., Thavornitia, P. and Suchiva, K. (2003). Hemicellulosic polymer from vetiver grass and its physicochemical properties. *Carbohydrate Polymers*, **54:** 335-342.

Rabah, M.A. and Eldighidy, S.M. (1989). Low cost hydrogen production from waste. *International Journal of Hydrogen Energy*, **14:** 221.

Rezaiyan, J. and Cheremisinoff, N.P. (2005). Gasification technologies – A primer for engineers and scientists. CRC Press (Taylor & Francis Groups), Boca Raton (FL).
Seaforth, C. and Tikasingh, T. (2002). Final report: A study for the development of a handbook of selected Caribbean herbs for industry.
Shafizadeh, F. (1982). Introduction to pyrolysis of biomass. *Journal of Analysis and Applied Pyrolysis*, **3:** 283-305.
Sudha, P. and Ravindranath, N.H. (1999). Land availability and biomass production potential in India. *Biomass and Bioenergy*, **16:** 207-221.

Replacing Conventional Fuels through Biogas for Mitigating the Threats related to Climate Change in India: A State-wise Assessment for Emission Reduction

Mrinalini Goswami, Sunil Nautiyal, S. Manasi, Prasanta Bez, K. Bhaskar, and Y.D. Imran Khan

Introduction

Energy plays a crucial role in the socioeconomic development of the country as elsewhere in the world. The past five-year period from 2004–05 to 2008–09 witnessed the economy grow at an average rate of 8.5%, despite the worldwide financial crisis affecting the second half of the five-year period. For the expected economic growth trend (8-9%) to continue during the Twelfth Five Year Plan (2012-2017), the nation requires an annual growth of 6.5% per year in energy supply (GOI, 2011). In India, nearly half of the commercial, primary energy demand is met by coal, but the share of coal in India's total primary energy constitutes about 38%, while the contribution of non-commercial energy such as firewood, cattle dung and agricultural waste forms about 28% of the total primary energy consumed (GOI, 2006a, b).

It is evident that biomass is an important source of fuel for cooking purpose; it contributes 18% of the total primary energy use. Overall trends in total primary energy consumption show a decoupling of energy consumption with increasing GDP. Over the period 1990 to 2005, GDP multiplied by 2.3 times while energy consumption by 1.8 (Stephane et al., 2009). India is the fifth largest emitter of greenhouse gases (GHGs) in the world, accounting for approximately 4.7% of the total global emissions, while China is the largest emitter with 23% of total global GHG emissions (GOI, 2009). Various studies have reported that the use of wood for fuel is one of the major factors responsible for the degradation of forested landscapes, thereby reducing carbon sinks in the country (Osei, 1993; Nautiyal

M. Goswami • S. Nautiyal (✉) • S. Manasi • P. Bez • K. Bhaskar • Y.D.I. Khan
Centre for Ecological Economics and Natural Resources (CEENR), Institute for Social and Economic Change (ISEC), Dr. V.K.R.V. Rao Road, Nagarabhavi, Bangalore 560 072, India
e-mail: sunil@isec.ac.in

N.J. Raju et al. (eds.), *Management of Water, Energy and Bio-resources in the Era of Climate Change: Emerging Issues and Challenges*,
DOI 10.1007/978-3-319-05969-3_15

and Kaechele, 2008). Our study focuses on the replacement possibilities of polluting fuels with renewable energy sources and ensuring of emission abatement through manure management.[1]

Worldwide, about 2.4 billion people lack a ready access to safe and reliable energy while about 1.6 billion people do not have access to electricity. The problem of energy deficit is found to be acute in developing countries such as India, where about 89% of the rural and about 28% of urban households depend on polluting sources of energy such as firewood, chips and dung cakes for meeting their cooking needs (TERI, 2010). Further, in terms of expenditure per household, fuel for cooking and lighting purposes accounts for about 8% of the total expenses (NSSO, 2009-10).

In India, about 600 million people do not have access to electricity and about 700 million people are using biomass as their primary energy resource for cooking (PHD Chambers of Commerce, 2011). The integrated Energy Policy (Ministry of Renewable Energy) states that the total amount of LPG required for providing cooking energy to 1.5 billion persons amounts to around 55 million tonnes of oil equivalent (Mtoe). About 42% of people had access to clean LPG for cooking in 2005. With respect to the rural–urban divide, in 2005, it was only 9% of the rural households that had access to LPG, while 57% of urban households had access to LPG. About 28 kg of firewood and chips (13.03 kilograms of oil equivalent 'kgoe') and 1.82 kg of LPG (2.01 kgoe) are used by Indian households for cooking purpose on a per capita per month basis (NSSO, 2007). This order of the magnitude of energy requirement, if it were to be met by fuelwood *perse*, could raise CO_2 emissions from the current level of one billion tonne (Bt) to 5.9 Bt per year by 2031–2032 (Ravindranath and Balachandra, 2009). An average household dependent on fuel wood consumes 1800 kg of fuel in a year; 159 kg of kerosene per year per household; and 120 kg of LPG per year per household in the case of kerosene and LPG dependent households respectively. The annual fuelwood requirement of the country is estimated to be about 250 m tonnes (Mt), more than 80% of which is for domestic consumption. The availability of fuelwood from forests on a sustainable basis is said to be about 17 Mt (GOI, 2001).

Domestic energy sources like LPG and kerosene are subsidized by the government. Moreover, subsidy is extended to all domestic LPG users, irrespective of their economic status. Apart from contributing to GHG emissions the burning of inferior fuels has a tremendous impact on health, education and overall empowerment of girls and women who spend large part of their everyday life in collecting fuelwood.[2] It is estimated that the inhalation of indoor smoke is responsible for over 400,000 deaths annually, mostly among women and children (SEI, 2009).

---

[1] The term "biogas" in this paper refers to biogas produced out of livestock manure with a generalized 60% of methane composition. GHG emission is taken to be CO_2 equivalent, manure management for methane. The term "states" includes all the 35 states and union territories of India.

[2] 85 million households spend 30 billion hours annually on fuelwood gathering (A study in North India as stated in the Integrated Energy Policy).

There has been an increasing loss of sequestered carbon in view of an unsustainable fuelwood harvesting over a period of time in the absence of an environment-friendly alternative. An earlier study had pointed out that by 2006, an estimated 139 m tonnes of fuelwood would be harvested above the sustainable supply from regulated sources (Bahuguna et al., 2004). An estimate by TERI puts India's fuelwood requirement at 228 m tonnes, out of which 128 m tonnes is met through a sustainable supply and the remaining 100 m tonnes through an unsustainable harvest (Aggarwal et al., 2009). The annual estimated harvest of wood from forests amounts to 1.23 m tonnes while the same from trees outside forested areas to 19.25 m tonnes. The total carbon stock of the country's forests estimated to be 6663 m tonnes (GOI, 2011), is getting affected because of unsustainable fuelwood collection. In the Indian context, statistics regarding the volume of fuelwood harvested (legally) are available but there is no estimate of areas coming under selective harvesting. Even the volume of timber harvested for timber and fuelwood is highly debated as the estimated consumption exceeds the recorded production (Haripriya, 2001). Thus the actual assessment of the loss of sequestered carbon seems very difficult.

Bio-energy as an Alternative Domestic Fuel; Climate Change Mitigation Dimension

Livestock population in India during 2007-08 stood at 530 million. With about 28% of the world's total cattle population, India ranks first in the world (Arora et al., 2010). The share of total dung, used as domestic fuel and farm yard manure, in the total value of livestock sector output amounts to 9% (GOI, 2011); the rest is dumped unused and releases methane into atmosphere. Methane which traps 21 times (IPCC, 2007) more heat than carbon dioxide accounts for 80% of global warming. Livestock contributes about 37% of anthropogenic methane, mostly through enteric fermentation and manure management. It is reported that livestock production accounts for 18% of the global GHG emissions including methane (CH_4) through enteric fermentation and manure management; nitrous oxide (N_2O) through animal manure; and carbon dioxide (CO_2) through land-use change caused by demand for food grains, grazing land and agricultural energy use. Livestock manure management is also a significant source of CH_4 emission (Swamy and Bhattacharya, 2006). India's livestock contributes 1.09 Tg (9%) methane through manure management (as of 2003). Dairy buffaloes and indigenous dairy cattle together constitute 60% of the total methane emission. The three high methane emitter states are Uttar Pradesh (14.9%), Rajasthan (9.1%) and Madhya Pradesh (8.5%) (Chabra et al., 2007).

The promotion of renewable energy generation through these unused resources in India has been supported by various legislations and policies such as Energy Conservation Act (2001), New and Renewable Energy Policy (2005) and Integrated Energy Policy (2009). National Biogas and Manure Management Programme

(NBMMP) a major initiative of the Government of India, was introduced in the early 80s to promote more than 12 million family type biogas plants. These biogas plants cost around Rs 16,000 each and is of two cubic metre (cum) size[3] fed by manure of 4-6 cattle and estimated to suffice cooking energy requirements of four member families. A family size biogas plant substitutes for 316 litres of kerosene, 5535 kgs of firewood and 4400 kgs of cattle dung cake used as fuels and reduces emissions of NO_x, SO_2, CO and volatile organic compounds into the atmosphere by 16.4, 11.3, 987.0 and 69.7 kg per year, respectively (Pathak et al., 2009). As of 2006, the total community-level biogas plants numbered 3902, a small number compared to India's 600,000 villages. Another approach towards supporting biogas plants – financing/subsidy mechanisms – under Clean Development Mechanisms of Renewable Energy projects as envisaged by the Kyoto Protocol, has become an important contributor in the last 4-5 years. Depending on the size of domestic digesters, their efficiency and productive period in a given year, each can save an annual combustion of 1.5-4.0 tonnes of fuelwood, equivalent to an emission reduction of 0.75-2.0 tonne C per digester (Pretty et al., 2002). It is estimated that 79 MT fuelwood can be saved by utilizing the available biogas potential in India. The corresponding carbon emissions avoided would amount to 15.8 MT annually, assuming that 40% of fuelwood requirements are from non-sustainable sources and 0.5 T C of dry wood (Ravindranath and Balachandra, 2009). Electricity is also produced by biogas plants installed under the Biogas Distributed/Grid Power Generation Programme (BGPG) launched by MNRE in 2006. Under this programme, projects are developed at the village level by community organizations, institutions, or private entrepreneurs, and the electricity produced is sold to individuals or communities or to the grid. The unit capacity ranges from 3 KW to 250 KW. MNRE reports 73 projects installed with total capacity of 461 kW (Arora et al., 2010). There are a few research studies (Sirohi and Michaelowa, 2004; Ravindranath et al., 2005) related to the estimation of biogas potential in India, which serve as a basis for the development of policies and schemes with a view to implementing bioenergy generation from livestock manure. The absence of a systematic research examining the actual emission reduction through manure management and conventional fuel replacement with state specific criteria has hampered planning for this energy alternative. In this backdrop, the aim of the paper is to estimate the potential of methane based bio-energy generation from livestock manure and to formulate strategies towards utilizing that potential in mitigating climate change related threats besides offering possible alternatives towards fulfilling the domestic energy needs.

[3] Estimated for central parts of the country with material costs as of 2010-11.

Methodology

A comprehensive data regarding methane emission from various sources is not readily available so far. Through this present research study, authentic data was collected from different government departments' data banks and past research works. With respect to some of the data, details which vary across different sources have been crosschecked with standard sets of available data to arrive at the most reliable numbers. The data on livestock population and energy consumption for domestic purpose has been obtained from the data repository of websites of the Department of Animal Husbandry, Dairying and Fisheries (GOI, 2012), Ministry of Statistics and Programme Implementation (Govt. of India) and www.indiastat.com and synthesized following appropriate procedures. To understand, compare and contrast the relevance of the data and our observations across the sectors and states, a holistic approach in the context of a chosen objective was developed.

Biogas Yield Capacity

Methane emissions of different types of livestock through manure management are calculated on the basis of Livestock Census 2007, Department of Animal Husbandry, Dairying and Fisheries. The amount of excreta produced per day by different animals depends upon the breed, age and weight of animals and the total quantities of fodder fed to them. Large difference prevails in biogas generation potential and emissions between rural and urban areas. The proportion of rural livestock to the total livestock population works out to about 95% in India (Dholakia and Pandya, 2011). As for equivalence in size, animals belonging to the same family and comparable dung generation as stated in different data sources (Khendelwal and Mahdi, 1986; Ravindranath et al., 2000; Mang, 2005; Mahimairaja, 2008; Kalbande et al., 2011) are grouped into Cattle, Buffaloes, Sheep, Goats, Horses and Pigs. Largely, lower values of excreta generation per animal have been considered. The dung yield values per day per animal used in the calculations are 7.5, 10, 0.1, 10 and 2.5 for cattle, buffaloes, sheep and goats, horses and pigs respectively. The biogas yield capacity of one kg of animal excreta tending to 36 L for cattle, sheep, goats and horses and 30 L for buffaloes and 60 L for pigs, have been derived from other research findings where production rates are found as high as 66 L/kg for pig manure and as low as 20 L/kg for cattle manure depending on varied climatic regions. The procurable livestock excreta for different animals varies, viz. cattle 60%, buffaloes and pigs 80%, and others 30% of the total excreta generated per day.

Methane Emission Factors

There are several studies related to methane emission estimations based on IPCC Tier I and Tier II approaches; the factors adopted for this study are 2.75, 4.15, 0.21, 0.22, 2.18 and 6 per animal per annum for cattle, buffaloes, sheep, goats, horses and pigs respectively. As the methane emission factors for cattle vary from 2 to 3.5, the average is taken. The emission factors are chosen from literature survey and the selected factors derived across different climatic regions (cool: $<15°C$, temperate: 15-25°C and warm: $>25°C$) and methane-producing potential of manure (Swamy and Bhattacharyya, 2006). These factors have been developed following country-specific guidelines under IPCC Tier II approach for cattle and buffaloes, and Tier I approach for other livestock. This method is based on animal population of the States and Union Territories grouped into cool, temperate and warm regions according to meteorological data from 391 stations. These factors are considered as this is the recent updated research data available based on the IPCC guidelines. The total methane emissions per state are converted into CO_2 equivalents taking Global Warming Potential (GWP) of methane as 21 times that of CO_2 over a period of 100 years (IPCC, 2007).

Conventional Fuels and Energy Content

Data on fuelwood, kerosene and LPG consumption for each state with rural-urban divide for 2009 have been extracted from NSS reports (66th round survey). The emission factors in terms of CO_2 for per kg of fuelwood, per litre of kerosene and per kg of LPG amounts to 1.83 kg, 2.54 kg and 2.9 kg respectively (IPCC, 2006b). The energy content of different fuels referred to in this study has been sourced from O'Sullivan and Barnes (2006) (Table 1). The energy content of different fuels varies with fuel quality and combustion efficiency as explained in different studies. One of the widely cited LPG energy contents relates to 46.1 MJ/kg (Hargreaves, 2003). The extent to which households are able to extract useful energy from a particular fuel depends on the technology used.

Table 1 Energy content and efficiency of different fuels

Fuel type	*Energy content, MJ/Kg*	*Useful energy, MJ/kg*	*Approx. quantity of fuel required to provide 5 GJ of useful energy for cooking in kg*
LPG	45.5	27.3	180
Kerosene (pressure)	43	23.6	210
Kerosene (wick)	43	15.1	330
Firewood (Efficient stove)	16	4	1250
Firewood (Traditional stove)	16	2.4	2000
Biogas (60% methane)	22.8 MJ/cum	–	365 cum
Dung	14.5	1.7	2900

Adapted from O'Sullivan and Barnes, 2006.

Loss of Carbon Sequestration

The loss of carbon sequestration has been calculated by using two different approaches: volume of biomass used as fuel and forest cover lost on account of fuelwood harvesting. According to MoEF (2007), the total sustainable availability of wood from all sources, public and private amounts to about 127 million cum per year. Of the total estimated fuelwood consumption for the year 2009 (397 cum), the unsustainable collection accounted for 270 cum that can be considered as the primary cause for the degradation of forests. This 68% is considered for estimating the loss of carbon sequestration following the calculations mentioned below.

The Good Practice Guidance for LULUCF (IPCC, 2006a) explains the methods for the loss of carbon sequestration due to fuelwood gathering. The calculation for the same involves the following.

$$L_{\text{fuelwood}} = FG \times D \times BEF2 \times CF$$

where L_{fuelwood} is annual carbon loss due to fuelwood gathering, tonnes C. yr^{-1}, FG = annual volume of fuelwood gathering, $m^3\ yr^{-1}$, D = basic wood density, tonnes d.m. M^{-3}, BEF2 = biomass expansion factor for converting volumes of extracted roundwood into the total aboveground, and CF = carbon fraction of dry matter (default = 0.5), tonnes C (tonne d.m.)$^{-1}$.

The biomass expansion factor varies from 1.3 to 3.4 for different types of forests. Some estimation shows biomass expansion factors (total above ground biomass/volume-based biomass to 7 cm in diameter) of 1.50 for stands with volumes <90 m^3/ha and 1.35 for stands with volumes 290 m^3/ha for Indian forests (Iverson et al., 1994). The biomass expansion factors have also been calculated as Hardwood = 1, Spruce-fir = 1 and Pine = 0.81-0.95 (Sharma et al., 2011). The value of BEF-2 considered for this study is 1.35.

Another method (Chatterjee, 2004) for the estimation of carbon sequestration loss considers 1 cum of stemwood ~ 1 cum fuelwood ~ 2.3 cum of total biomass loss.[4,5]

$$DM = V_{\text{fuelwood}} \times 2.3 \times 0.43^5$$

$$SC_{\text{loss}} = DM \times 0.50$$

where DM is the dry matter content of the total biomass lost due to the exploitation of fuelwood unsustainably, V_{fuelwood} is the total volume of fuelwood acquired unsustainably and SC_{loss} is the loss of sequestered carbon.

---

[4] The ratio of total biomass to usable stem biomass was assumed by the German Bundestag to be 1.6 for closed forest and 3 for open forest. One cum of stem wood is, therefore, taken to be equivalent to 2.3 m^3 of the total biomass.

[5] This value varies in relation to factors like plant species, geoclimatic conditions etc. Dry matter content with solid wood amounts to 0.45 tonnes per m^3 (IPCC, 2006).

Canada's Model Forest Programme (2000) is one of the highly discussed programmes in this respect besides being cited as one of the UNFCCC-REDD methods. Conversion factors from vegetation cover to carbon are:

1. Total wood volume = vegetation cover × 1.454 × 0.396 (in m^3)
2. Total dry matter biomass = wood volume × 0.43 (in tonnes)
3. Total carbon = dry matter biomass × 0.5 (in tonnes)

Estimation of Emission Reduction

The loss of methane during production across digesters varies. In the IPCC Guidelines for National Greenhouse Gases Inventories, it is given as 0-8 g of CH_4 lost during per kg of waste treated on wet basis (following Tier I approach). Emission reduction by utilizing potential livestock excreta for the production of biogas is a factor of methane emission through manure management, methane leakage in the production process, emission from inefficient combustion of biogas and mitigation of emission due to replacement of other polluting fuels with biogas. CH_4 lost to the atmosphere during the biogas production process is taken as 10% (UNFCCC, 2008) default leakage rate of 10% of the total production and the emission from inefficient burning as 5 kg of CH_4/TJ energy production (IPCC, 2006b).

Thus CH_4 emission due to leakage can be calculated as:

$$ME_{leakage} = BP \times 0.1 \times 0.6$$

where $ME_{leakage}$ is methane emission due to leakage in the production process, BP is biogas potential, 0.1 is assumed fraction loss and 0.6 is assumed percentage of methane in the biogas produced.

$$ME_{burning} = \left(5 \times E_{biogas} \text{ in TJ}\right) \text{kg}$$

where $ME_{burning}$ is the total methane emitted during inefficient burning and E_{biogas} is the total energy generated from biogas.

Keeping the current situation in view, the total emission (in terms of CH_4) reduction can be calculated for four hypothetical scenarios for the country:

Scenario 1: Replacement of fraction of fuelwood with biogas
Scenario 2: Replacement of LPG and fraction of fuelwood
Scenario 3: Replacement of kerosene and fraction of fuelwood
Scenario 4: Replacement of LPG, kerosene and fraction of fuelwood

$$ER_m = ER_{mm} - \left[ME_{leakage} + ME_{burning}\right] + ER_{[fuel/fuels\ in\ scenario\ 1-4]}$$

where ER_m is methane emission reduction, ER_{mm} is emission reduction through manure management and $ER_{[fuel/fuels\ in\ scenario\ 1\text{-}4]}$ is reduction by replacing fuel/fuels under different scenarios.

Mapping of Distribution of Bio-energy Potential

The state-wise biogas potential is calculated under different scenarios such as firewood, kerosene and LPG, while their replacement percentage is calculated on the basis of livestock population across states; later average percentage of biogas potential of each state is calculated using these three values. Following of which, average percentage of biogas potential is created in the GIS database for analysing the data set in the spatial domain using ArcGIS 9.2 software. The whole range of high and low value percentage of replacement potential is reclassified into four classes and four classified values are categorized into four sensitivity zones according to their values i.e. very high, high, medium and low. The sensitivity zones are prepared by dividing the whole state potential replacement capacity with four group ranges as presented in Table 2, which is colour coded in the GIS domain.

The biogas potential density of rural India has been generated using ArcGIS 9.2 software by attaching the attribute data with the spatial data and converted to the raster by classifying the biogas potential in cum per capita per year and reclassified using the spatial analyst tool in the GIS software into five different classes of biogas density. The methane emission reduction potential of CO_2 through manure management is calculated and the data set created with spatial variability in the Geo database of the spatial domain using ArcGIS 9.2; further the data set is united with the GIS layer and converted into four different groups for analysis and these groups are weighted to four spatial zones.

Table 2 Classification of bio-energy potential (for replacing conventional fuels) for mapping

Classified % replacement capacity	*Sensitivity zone*
<16%	Low
16-32%	Medium
32-48%	High
48-64%	Very high

Results

According to the 58th round of the National Sample Survey Organisation (NSSO), more than 51% of the urban population has switched over to LPG as cooking fuel (PISCES, 2011). The energy supplied by three different sources varies largely across states. The contribution of fuelwood is as high as 95% in Nagaland and 94% in Arunachal Pradesh and as low as 4% in Delhi. It is evident from Table 3 that 80.8% of the household energy consumption flows from fuelwood and the rest from kerosene and LPG. The biogas potential in all the states and union territories can replace 29.52% of the total energy supplied by a combination of the three aforementioned fuels. Out of 206.8 million tonnes of fuelwood consumed annually, rural areas accounts for 91%. Nearly 38% of the rural fuel wood potential of biogas can be derived from (Dholakia and Pandya, 2011) 188.9 million tonnes of fuelwood used. The potential of biogas in rural areas is about 45,809 million cum annually, approximately 95% of India's total procurable biogas from livestock manure. Our calculation (data not presented here) based on various data source shows that rural India can generate about 61 cum per capita annually biogas for the rural population and reduction in methane emission up to 0.016 tonnes per capita annually.

The total quantity of livestock excreta generated across states is 4.98 million tonnes per day; 3.49 million tonnes of that is collectible. If that entire quantity is utilized, 151,057 MLD of biogas could be generated. Among all states, Uttar Pradesh has the highest biogas potential of 19,410 MLD. The states with more than 10,000 MLD biogas potentials are Andhra Pradesh, Madhya Pradesh, Maharashtra, Rajasthan and West Bengal. The states with a much higher production potential of biogas are Mizoram, Sikkim and Goa. Cattle and buffaloes together account for 57% of the total manure-based biogas generation in India, whereas pig population for 42% of the same.

With 95% of the livestock population in rural areas, the rural population has the potential to harness 48,220 million cum of biogas annually. Uttar Pradesh and Andhra Pradesh with a potential of producing 6462 million cum and 4821 million cum annually, rank top among states. In terms of biogas density, rural Uttar Pradesh

Table 3 Consumption of different fuels and their energy terms in India in the year 2009 and biogas potential as per livestock population 2007

Fuel type and energy	*Consumption*	*Energy consumed in terms of the given fuel in TJ*	*% contribution to the total* energy consumed*
Fuelwood	206.8 million tonnes	33,08,861	80.8
Kerosene	6867.76 million litres	295,314	7.2
LPG	10.83 million tonnes	492,869	12
Biogas potential	53,047.37 million cum	12,09,480	29.52**

* This total is for fuelwood, kerosene and LPG.

** Biogas has the potential to replace 29.52% consumption of other three fuels.

accounts for the highest density of 114,797 cum per square kilometre (sq km), followed by West Bengal and Bihar among 28 states. Most of the Union Territories show comparatively good biogas density, mainly because of their relatively low rural population and discrepancy in livestock population. Although Uttar Pradesh is bestowed with the highest biogas share in India, its per capita availability of biogas is quite low (47 cum/year), in view of its population-livestock ratio. Rajasthan (95 cum/year), Madhya Pradesh (89 cum/year) and Chhattishgarh (85 cum/year) accounts for the highest per capita biogas potential with respect to their respective rural population among the 28 states. Kerala has the lowest per capita availability of biogas potential per year with a value 13. With an optimum production and utilization of biogas in rural areas, rural populace can reduce methane emission by 0.74 kg per capita annually through manure management (equivalent to 15.54 kg CO_2e per capita annually). The biogas potential density of rural India shown in Fig. 1 helps predict the zones with a higher density of biogas potential. The map shows clearly that the central and western parts of India and two states in the northeastern region can harness a reasonably good quantity (>72 cum/capita/year) of biogas.

The contribution of GHGs constitutes 88%, 4% and 8% of fuelwood, kerosene and LPG respectively while their respective energy shares accounts for 81%, 7% and 12%. Besides, direct combustion emissions, there is a significant emission factor associated with LPG and kerosene through production and transportation. The production of LPG signifies higher energy consumption and release of more GHGs into the atmosphere. The production and transport emissions for LPG are 8.5 and 0.6 g-c in CO_2 equivalents per mj and 5.7 and 0.7 for kerosene respectively (Bailis et al., 2004). The total methane emission from manure of all the six categories of livestock studied across states are 3.9 million tonnes (Livestock Census, 2007). The emission of methane from the livestock categories are: Cattle 0.83 million tonnes, Buffaloes 0.5 million tonnes, Sheep 0.01 million tonnes, Goats 0.03 million tonnes, Horse 0.003 million tonnes and Pigs 2.5 million tonnes. Uttar Pradesh contributes 0.43 million tonnes of methane from livestock manure management and highest across states. The methane emission reduction potential in terms of CO_2 from manure management across states is grouped into four classes and converted according to their values (Fig. 2). As seen, Arunachal Pradesh and Andhra Pradesh have the highest emission reduction potential of CO_2 comparatively.

GHG emission abatement potential of biogas generation is a function of procurable manure, loss of methane during biogas production and emission from combustion of biogas as fuel. Total of 0.58 million tonnes of methane can be trapped from the entire cycle of manure management in India. Two states show negative reduction in methane emission levels as they belong to temperate zone where emission from manure is minimal and Andhra Pradesh has the highest methane emission reduction potential from biogas production amounting to 0.015 million tonnes. The yield of biogas varies with temperature in this study as the yield values are generalized for the country, also indicating Andhra Pradesh with

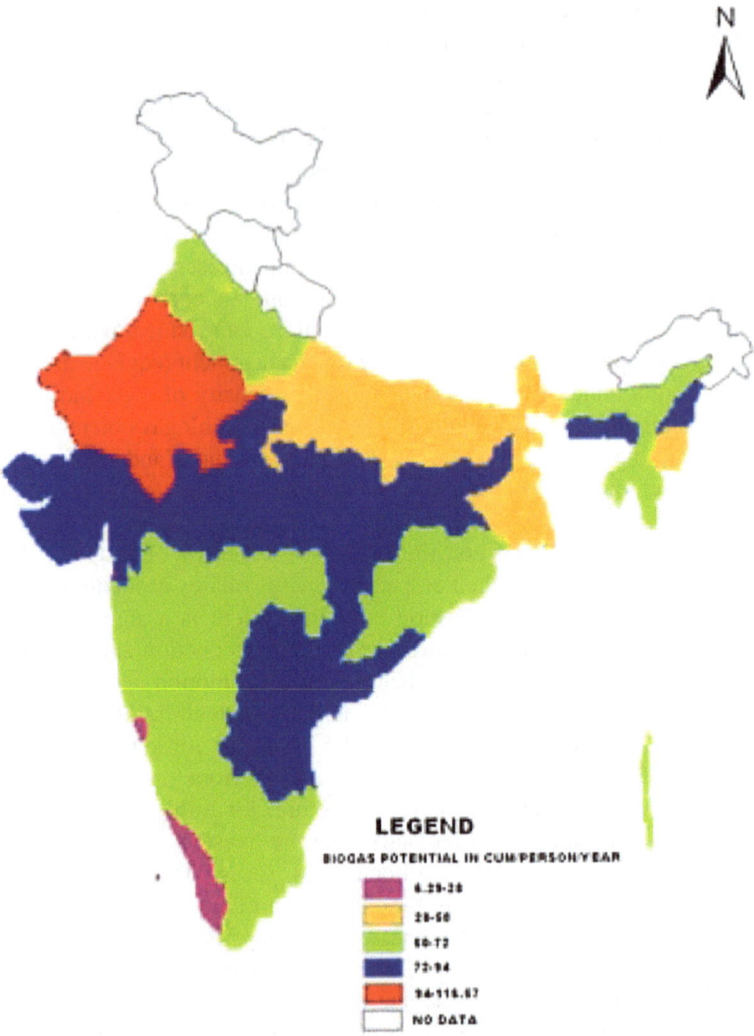

Fig. 1 Map of India showing density of potential biogas density across the states in cum/capita/year

the highest (0.15 million tonnes/year) emission reduction potential from manure management from biogas generation.

In Fig. 3, the states consuming more than 10 million tonnes fuelwood annually are shown with a fuelwood consumption replacement potential of biogas and a consequent reduction in emission in terms of CO_2e. The total emission in terms of CO_2 from fuelwood and kerosene amounts to 378 million tonnes and 17 million tonnes per year respectively; clean fuel LPG shows 32 million tonnes of emission in 2008-09. Figure 3 indicates Chhattisgarh in the highest sensitivity zone in biogas

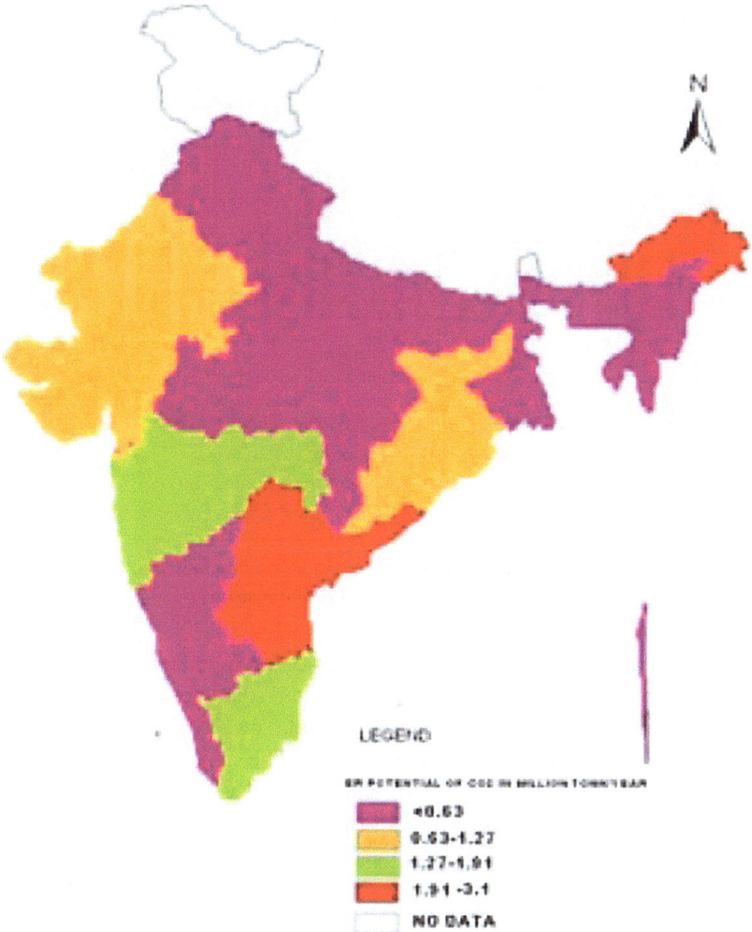

Fig. 2 Map of India showing the emission reduction potential across the country from manure management by utilizing the tappable manure for biogas generation

replacement potential as the livestocks' potential is relatively high due to consumption demand; similarly, the states of Kerala, Goa, Tripura, Mizoram and Delhi show a low potential zone for harnessing biogas because of low livestock population.

The replacement biogas potential (conventional fuel) of different states has been analyzed. Biogas has the potential to replace about 37% of fuelwood consumption in the country and about 30% of the total energy supplied by three conventional fuels viz. fuelwood, kerosene and LPG. The replacement of fuelwood can be achieved with varied percentage shares for different states i.e. as high as 82% for Delhi and as low as 4% for Kerala. The replacement of fuel wood with LPG is low in Chandigarh (6%) and Delhi (4%), where the consumption of domestic cooking energy is higher and livestock population is low.

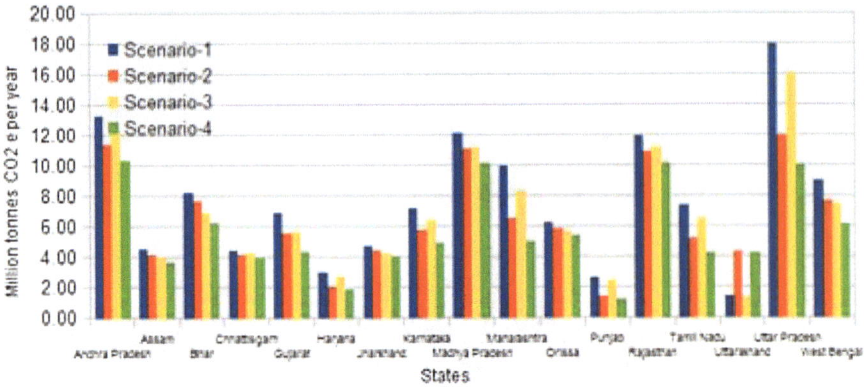

- Scenario 1: Replacement of fraction of fuelwood with biogas
- Scenario 2: Replacement of LPG and fraction of fuelwood
- Scenario 3: Replacement of Kerosene and fraction of fuelwood
- Scenario 4: Replacement of LPG, Kerosene and fraction of fuelwood

Fig. 3 Potential of biogas to replace fuelwood in states consuming fuelwood more than 10 million tonnes per year and the corresponding reduction in GHG emissions

An evaluation of the most efficient scenario for replacing conventional fuels with biogas to reduce GHG emissions is calculated (data not presented here) and if we consider biogas to replace only fuelwood, a total of 138.33 million tonnes GHGs per year can be reduced in fuelwood combustion. The same is achieved by LPG and fraction of fuelwood by 114.24 million tonnes annually. Given the higher potential of emission reduction, the replacement of fuelwood with biogas may be considered as the most appropriate scenario (Fig. 3), followed by replacement of fuelwood with kerosene and fraction of fuelwood (i.e. 122 million tonnes of GHG per year). The least efficient scenario is the replacement of fuelwood with all LPG and kerosene and fraction of fuelwood with emission reduction potential of 97.91 million tonnes of GHGs annually.

The loss of sequestered carbon is the result of an unsustainable exploitation of forest resources for fuelwood; it has been calculated using the IPCC recommended method as mentioned in the methodology. The unsustainable harvesting of fuelwood accounts for 68% of the total fuelwood consumption. The calculations based on biomass volume reveal a total loss of 82 million tonnes of sequestered carbon due to fuelwood use for the year 2008-09. The sequestered carbon lost in rural and urban areas amounts 75 million tonnes and 7 million tonnes respectively. The shares of rural and urban areas in the sequestrated carbon loss constitute 91% and 9% respectively, corresponding to the use of fuel wood. Uttar Pradesh, the largest consumer of fuelwood, has squandered 20.5 million tonnes and 0.80 million tonnes of carbon from its forests in rural and urban areas respectively. Karnataka and Rajasthan occupy 2nd and 3rd position in losing their sequestered carbon due to an unsustainable fuelwood gathering.

The computation of the total volume of carbon lost includes: (a) carbon transferred to forest products (in the form of biomass), (b) carbon released from forest biomass into the atmosphere due to clearing of forests or forest fires; and (c) carbon released to soil pools. The unsustainable harvesting of fuelwood implies a partial clearing of forests and the consequent degradation. When the forests are subjected to partial clearing, 85% of the stem biomass gets transferred to wood products, 10% remains on the stumps, and 5% is transferred to the soils (Haripriya et al., 2005).

Discussion

With a 13% of livestock population of the world, India contributes about 17% of GHG emission to the atmosphere from livestock sector (FAO, 2006). Although a major part of that emission is accounted for by enteric fermentation which cannot be managed. The manure, being an invaluable underused resource, possesses the potential to reduce the GHG contribution to global warming by replacing polluting fuels with renewable energy sources and to reduce emission through manure management.

Energy consumption patterns are extremely complex besides being one of the determinants of a nation's economy; household energy consumption for cooking and lighting is an issue associated with per capita energy sufficiency and overall development of a population. In 2005, India's total GHG emissions amounted to 1866 million tonnes CO_2e, while India's per capita GHG emissions to 1.7 metric tonnes of CO_2e. The results of five different independent studies show that in absolute terms India's annual GHG emissions in 2031 could reach between 4.0 billion tonnes of CO_2e and 7.3 billion tonnes under the BAU (Business-as-usual) scenarios (Climate Modeling Forum, 2009). An expert committee on integrated energy policy believes that it is possible to reduce India's energy intensity by upto 25% from the current level and GHG intensity of the economy by as much as one-third. This excludes mostly the non-commercial domestic fuels as energy intensity is based on commercial energy sources. Thus energy sources in this sector demand a special attention in the context of climate change uncertainties.

The energy potential of cow and buffalo dung comes to 562.2 PJ and 336 PJ respectively as estimated by Ravindranath and Balachandra (2009). This potential can be used by 17 million biogas plants and 150,000 community biogas plants for reducing GHG emissions by 5 TC/year and 10.8 TC/year respectively (Ravindranath et al. 2000). The climate change benefits in terms of carbon emission reductions are to the tune of 110 TC per year provided the available potential of bio-energy technologies are utilized (Ravindranath and Balachandra, 2009). The present study reveals that an optimum biogas production has the potential of reducing GHG emissions by 138.33 million tonnes of CO_2e by way of replacing fuelwood and 0.58 (only methane) million tonnes of CO_2e from manure management. Under CDM activity by the mid 2012 India has 2231 approved projects with a carbon credit of 730.8 million tonnes worth Rs 20,1730 million. The present price of carbon credit is worth Rs 276 per tonne while the entire spectrum of bio-energy envisages a reasonable earning through carbon credit if considered under CDM and would be helpful for self-sustenance. A further analysis of per capita biogas availability at the village level for actual project implementation would help develop a village-based model for biogas generation, distribution and utilization for ensuring self-sufficient domestic energy production under CDM initiatives.

Rural areas with more than 70% of the population and 95% of livestock population of the country consume 91% of the total fuelwood harvested. The replacement of fuelwood has been evaluated as the best suitable scenario for promoting and implementing biogas projects. Given these facts, it is apparent that the adoption of biogas as a major domestic energy source by fuelwood consuming households in the rural areas would be the most appropriate option, economically and environmentally. The total manure generated per day by all the six categories of studied animals works out to approximately 5 million tonnes, while about 3.5 million tonnes are collectible for generating 151,057 MLD biogas. In view of the states having varied ownership of livestock population, geo-climatic conditions, household energy needs and conventional fuel use patterns, it is difficult to generalize the statement regarding the biogas potential and the replacement prospect of fuelwood with kerosene and LPG. It requires a detailed plan at the state, district and much smaller regional levels. This paper has investigated the potential at the state level; a further assessment of demand and availability of resources (procurable animal excreta) is required in a decentralized manner for strategy development so as to implement biogas generation plants.

The regions with a reasonably high annual temperature and per capita biogas potential in rural areas can be considered as priority areas for implementing biogas plant projects for replacing conventional fuels and also for the consequent abatement of GHG emissions. The estimated potential of biogas plants was 1.23 crore family type plants; under NBMMP there was a cumulative achievement of 35% by the year 2011. This target had a global warming mitigation potential of 120 Mt CO_2 equiv. $year^{-1}$ and US $1197 million as carbon credit under the CDM (Pathak et al., 2009). During the 11th Five Year Plan, there was an outlay of Rs. 562.00 crore for the promotion and implementation of biogas plants. On the other hand, the GOI is paying Rs 1973.6 crore as subsidy for the year 2010-11 for LPG cylinders for the entire country, which was to meet 12% of the energy consumed through fuelwood, kerosene and LPG, whereas biogas with a potential to supply about 30% of that total energy received comparatively little financial support. The expenditure on LPG is a recurring one besides being subject to a continuous inflation with the increasing cost of production, transportation and import. In the 12th Five Year Plan, the convergence of NBMMP with Indira Awas Yojna has been recommended. The cost of biogas plants varies with structural specifications of different models, for instance it is 30% and 50% more in hilly states and northeastern states respectively than in other parts of the country.

Although a good number of biogas plants have been installed, a significant slip-back has been noticed. Surveys conducted in various regions of India have found the proportion of functional plants to be from 40% to 81% (Dutta et al., 1997;

Bhat et al., 2001). To overcome this, an on-the-ground assessment based on conventional fuel availability and willingness to shift should be done along with a mass awareness agenda on climate change mitigation. Domestic biogas digesters have numerous challenges to overcome for a continued proliferation in the 21st century. Designs delivering lower cost, improved robustness, functionality, ease of construction, operation and maintenance would help market penetration of biogas plants (Bond and Templeton, 2011). However, in the absence of a proper technological up-gradation, operation and management, this valuable technology may become a source of environmental problems both at the local and global levels (Khoiyangbam, 2010). Therefore, it is necessary to rethink policies in evolving a flexible subsidy structure rather than the existing fixed subsidy structure. In view of the ineffective implementation of government-aided programmes, the recommendation for the convergence is not an effective way to promote; rather needs to be treated as a criterion for availing support under IAY. A detailed cost benefit analysis is of utmost importance for financial reshuffling under household energy sources considering both environmental internalities and externalities.

References

Aggarwal, A, Paul, V. and Das, S. (2009). Is India ready to implement REDD-PLUS: A Preliminary Assessment. Disscussion Paper presented in COP-15.

Agriculture Research Data Book (2001). Indian Agricultural Statistics Research Institute, Pusa, New Delhi.

Annual Report: 2011-12 (2012). Department of Animal Husbandry, Dairying and Fisheries, Ministry of Agriculture, Government of India.

Arora, D.S., Busche, S., Cowlin, S., Engelmeier, T., Jaritz, J., Milbrandt, A. and Wang, S. (2010). Indian Renewable Energy Status Report. Background Report for DIREC, NREL; US Department of Energy.

Bahuguna, V.K., Mitra, K., Capitrano, D. and Sargal, S. (2004). Root to Canopy: Regenerating Forests through Community-State Partnerships. Delhi: Winrock International and Commonwealth Forestry Association. CIFOR (Centre for International Forestry Research) news online, Fuelwood and the poor in the last decade.

Bailis, R., Pennise, D., Ezzati, M., Kammen, D.M. and Kituyi, E. (2004). Impacts of Greenhouse Gas and Particulate Emissions from Woodfuel Production and End-use in Sub-Saharan Africa. 2nd World Conference and Technology Exhibition on Biomass for Energy and Industry, Rome, Italy, 10-15 May, 2004.

Basic Animal Husbandry Statistics (2010). Department of Animal Husbandry, Dairying and Fisheries, Ministry of Agriculture, Government of India.

Bhat, P.R., Chanakya, H.N. and Ravindranath, N.H. (2001). Biogas plant dissemination: Success story of Sirsi, India. *Energy for Sustainable Development*, **5(1):** 39-46.

Bond, T. and Templeton, M.R. (2011). History and future of domestic biogas plants in the developing world. *J Energ for Sust Dev*, **15:** 347-354.

Chabra, A., Manjunath, K.R. and Panigrahy, S. (2007). Assessing the role of Indian Livestock in Climate Change. Workshop Proceedings; Impact of Climate Change on Agriculture, ISRO, ISPRS. http://www.isprs.org/proceedings/XXXVIII/8-W3/B6/5-14.pdf

Chatterjee, K. (2004). CDM in the Forestry Sector in India. Development Alternatives, New Delhi, India. www.devalt.org. Last accessed on 25th June 2012.

Climate Modeling Forum (2009). India's GHG Emissions Profile: Results of Five Climate Modelling Studies. http://moef.nic.in/downloads/home/GHG-report.pdf.

Dholakia, RH. and Pandya, M.B. (2011). Estimating Urban and Rural Incomes in Gujarat 1993-94 to 2004-05. W.P. No. 2011-09-02, 2011, http://www.iimahd.ernet.in.

Dutta, S., Rehman, I.H. and Malhotra, P.V.R.P. (1997). Biogas: The Indian NGO experience. Tata Energy Research Institute, New Delhi, India.

Emissions from Livestock and Manure Management (2006). IPCC Guidelines for National Greenhouse Gas Inventories. IPCC.

FAO (2006). Livestock's long shadow: Environmental issues and options. Steinfeld, H., Gerber, P., Wassenaar, T., Castel, V., Rosales, M. and de Haan, C. (eds). http://fao.org/docrep.

GOI (2001). Report of the task force on greening India for livelihood security and sustainable development. Planning Commission of India, Government of India.

GOI (2006). Report of the Working Group on Animal Husbandry & Dairying. Eleventh Five-year Plan (2007-12). Planning Commission of India, Government of India.

GOI (2006). The Eleventh Five-year Plan (2007-12), Energy-Chapter. Planning Commission of India, Government of India.

GOI (2009). India's GHG Emissions Profile. Results of Five Climate Modeling Studies. Climate Modelling Forum; Supported by Ministry of Environment and Forest, Government of India.

GOI (2011). Report of the Working Group on Animal Husbandry & Dairying. Twelfth Five-year Plan (2012-17). Planning Commission of India. Government of India.

GOI (2012). Annual Report: 2011-12. Department of Animal Husbandry, Dairying and Fisheries, Ministry of Agriculture, Government of India.

Hargreaves, R. (2003). Mitigating the Impacts of Climate Change on the Built Environment. Study Report No. 118. BRANZ.

Haripriya, G.S. (2001). Accounting for the forest resources in the national accounts in India. *Environ Res Econ*, **19:** 73-95.

Haripriya, G.S., Sanyal, S., Sinha, R. and Sukhdev, P. (2005). The Value of Timber, Carbon, Fuelwood, and Non-Timber Forest Products in India's Forests. Monograph 1 of the Green Accounting for Indian States Project (GAISP). TERI Press.

India's GHG Emissions Profile (2009). Results of Five Climate Modeling Studies. Climate Modelling Forum. Supported by Ministry of Environment and Forest, Government of India.

IPCC (1996). Good Practice Guidance and Uncertainty Management in National GHG Inventories.

IPCC (2006). Good Practice Guidance for LULUCF. www.ipcc.nggip.iges.or.jp/public/gpglulucf/gpglulucf_files/Chp3/Anx_3A_1_Data_Tables.pdf.

IPCC (2006). Guidelines for National Greenhouse Gas Inventories: Energy.

IPCC (2007). Fourth Assessment Report: Climate Change. http://www.ipcc.ch/publications_and_data/publications_and_data_reports.shtml#1.

Iverson, L.R., Brown, S., Prasad, A., Mitasova, H., Gillespi, A.J.R. and Lugo, A.E. (1994). Use of GIS for Estimating Potential and Actual Forest: Biomass for Continental South and Southeast Asia. Reprinted from: Virginia H. Dale. Springer-Verlag.

Jain, G. (2010). Preparation Energy Poverty and Income Poverty Linkages: Indian Experience. PEA-AIT International Conference on Energy and Sustainable Development: Issues and Strategies. Asian Institute of Technology, Thailand.

Kalbande, S.R., Kamble, A.K and Gangde, C.N. (2011). Bioenergy assessment and its integration for self sufficient renewable energy village, Karnataka. *J. Agric. Sci.*, **24(2):** 207-210.

Khendelwal, K.C. and Mahdi, S.S. (1986). In biogas technology: A practical technology. Tata McGraw-Hill, New Delhi.

Khoiyangbam, R.S. (2010). Environmental Implications of Biomethanation in Conventional Biogas Plants. *Iranica Journal of Energy & Environment*, **2(2):** 181-187.

Livestock Census (2011). All India Summary Report. Department of Animal Husbandry, Dairying and Fisheries, Ministry of Agriculture, Government of India. www.dahd.nic.in.

Mahimairaja, S., Dooraisamy, P., Lakshmanan, A., Rajannan, G., Udayasoorian, C. and Natarajan, S. (2008). Composting technology and organic waste utilization in Agriculture. A.E. Publications. Coimbatore. www.agritech.tnau.ac.in.

Mang, H.P. (2005). Biogas Sanitation Systems. Chinese Academy of Agricultural Engineering, Beijing.

National Sample Survey Reports (2007). National Sample Survey Organization, Ministry of Statistics and Programme Implementation, Government of India, 2004-05 to 2009-10.

Nautiyal, S. and Kaechele, H. (2008). Fuel switching from wood to LPG can benefit the environment. *Env Imp Ass Rev*, **28:** 523-532.

NSSO (2010). Level and pattern of consumer expenditure. National Sample Survey Organization, Ministry of Statistics and Programme Implementation, Government of India, 2009-10.

O'Sullivan, K. and Barnes, D.F. (2006). Energy Policies and Multitopic Household Surveys: Guideline for Questionnaire Design in Living Standard Measurement Surveys. Energy and Mining Sector Board, Discussion Paper No. 7. World Bank.

Osei, W.Y. (1993).Woodfuel and deforestation – Answers for a sustainable environment. *J Environ Manage*, **37:** 51-62.

Pathak, H., Jain, N., Bhatia, A., Mohanty, S. and Gupta, N. (2009). Global warming mitigation potential of biogas plants in India. *J Environ Monit Mon Assess*, **157:** 407-418.

PHD Research Bureau (2011). Comparative study on power situation in Northern and Central states of India. PHD Chambers of Commerce, India.

PISCES (2011). Fuelwood Report: Working Paper. Prepared for DFID by the PISCES RPC Consortium. Policy Innovation System for Clean Energy Security, Chennai, India.

Planning Commission (2001). Report of the task force on greening India for livelihood security and sustainable development. Planning Commission of India, Government of India.

Planning Commission (2006). Report of the Working Group on Animal Husbandry & Dairying. Eleventh Five Year Plan (2007-12). Planning Commission of India, Government of India.

Planning Commission (2006). The Eleventh Five Year Plan (2007-12), Energy-Chapter, Planning Commission of India, Government of India.

Planning Commission (2011). Faster, Sustainable and More Inclusive Growth: An Approach to the Twelfth Five Year Plan. Planning Commission of India, Government of India.

Planning Commission (2011). Report of the Working Group on Animal Husbandry & Dairying. Twelfth Five Year Plan (2012-17). Planning Commission of India. Government of India.

Pretty, J.N., Bal, A.S., Xiaoyun, L. and Ravindranath, N.H. (2002). The role of sustainable agriculture and renewable-resource management in reducing greenhouse-gas emissions and increasing sinks in China and India. *Phil. Trans. R. Soc. Lond. A*, **360:** 1741-1761.

Ravindranath, N.H. and Balachandra, P. (2009). Sustainable bioenergy for India: Technical, economic and policy analysis. *Energy Ener*, **34:** 1003-1013.

Ravindranath, N.H., Somashekara, H.I., Nagarajaa, M.S., Sudhaa, P., Sangeethaa, G., Bhattacharya, S.C. and Salam, P.A. (2005). Assessment of sustainable non-plantation biomass resources potential for energy in India. *Biomass and Bioenergy*, **29(3)**.

Ravindranath, N.H., Usha Rao, K., Natarajan, B. and Monga, P. (2000). Renewable energy and environment – A policy analysis for India. Tata McGraw-Hill, New Delhi.

Reducing Greenhouse Gas Emissions in India: Financial mechanisms and opportunities for EU-India collaboration. Stockholm Environment Institute. 2009.

Sharma, C.M., Gairola, S., Baduni, N.P., Ghildiyal, S.K, and Suyal, S. (2011). Variation in carbon stocks on different slope aspects in seven major forest types of temperate region of Garhwal Himalaya, India. *J. Biosci.*, **36(4):** 701-708.

Sirohi, S. and Michaelowa, A. (2004). CDM potential of Dairy Sector in India. Discussion Paper, Hamburg Institute of International Economics. ISSN 1616-4814.

State of Forest Report, India (2011). Forest Survey of India, Ministry of Environment and Forest, Government of India. http://www.fsi.org.in/sfr_2011.htm.

Stephane, R.C., McNeil, M. and Sathaye, J. (2009). India Energy Outlook: End Use Demand in India to 2020. Ernest Orlando Lawrence, Berkeley National Laboratory.

Swamy, M. and Bhattacharya, S. (2006). Budgeting anthropogenic GHG emission from Indian livestock using country-specific emission coefficients. *Curr. Sci.*, **91:** 1340-1353.

TERI (2010). Cooking with cleaner fuels in India: A strategic analysis and assessment. Policy Brief 2, The Energy Research Institute (TERI).

UNFCCC (2008). The United Nations Framework Convention on Climate Change. http://unfccc.int/meetings/unfccc_calendar/items/2655.php?year=2008.

Chronic Arsenicosis Induced Oxidative Stress in Cattle: Role of Zn and Se

Jeevan Ranjan Dash, Bakul Kumar Datta, Samar Sarkar, and Tapan Kumar Mandal

Introduction

High levels of arsenic in the ground water are a matter of concern, especially in the Indo-Bangladesh region where over a million people are reported to be suffering from arsenic poisoning. This kind of slow, low level, inevitable poisoning has caused serious concerns about the health of all living species in these areas (Majumder et al., 2012).

Arsenite (As^{III}) and arsenate (As^{v}) are the two major forms of the inorganic arsenic. While arsenite has a tendency to react readily with the sulphydryl groups of proteins and thus inhibit biochemical pathways, arsenate acts as a phosphate analogue and interferes with phosphorylation reactions (Valko et al., 2005). Most of the arsenate absorbed is reduced to arsenite in blood. The toxic effects manifested by both the molecules are very similar. However, the trivalent species (arsenite) is the biologically active form and the major source to arsenic toxicity represents a potential threat to the environment, human health and animal health due to their carcinogenic and other effects (Singh et al., 2004). Arsenic can result in acute and chronic toxicity. Zinc is one of the essential trace metals which have been studied for its protective value against arsenic. Selenium is known to promote the biliary excretion of exogenous selenium and selenite also augments the excretion of arsenic into bile (Flora et al., 1999; Gregus et al., 1998). Therefore in this experiment it has been tried to use zinc oxide-sodium selenite mixture as a

J.R. Dash • B.K. Datta • T.K. Mandal (✉)
Dept. of Pharmacology and Toxicology, West Bengal University of Animal and Fishery Sciences, 37 K.B Sarani, Kolkata 37, India
e-mail: drtkm48@gmail.com

S. Sarkar
Dept. of Veterinary Medicine, Ethics and Jurisprudence, West Bengal University of Animal and Fishery Sciences, Kolkata 37, India

N.J. Raju et al. (eds.), *Management of Water, Energy and Bio-resources in the Era of Climate Change: Emerging Issues and Challenges*,
DOI 10.1007/978-3-319-05969-3_16

mitigation compound to reverse arsenicosis in cattle of some arsenic affected areas of West Bengal so as to curb the arsenic load in food chain which may have ultimate benefit to human being.

Material and Methods

Design of Experiment

Cattle from the six experimental villages namely Mitrapur, Nonaghata, Mandal Hat, Goetra, Ghetugachi and Dakhin Panchpota from Haringhata and Chakdah blocks of Nadia district, reported with high arsenic contamination, were covered under this study. The arsenicosis affected animals were selected as per the criteria depicted by Dash et al. (2013). Total 48 stressed cows affected with arsenicosis were selected for this experiment of which 22 animals were kept as untreated control and 26 animals were treated with Zn-Se mixture. The farmers were supplied with zinc oxide and sodium selenite mixture and advised to feed the selected animals at the dose rate of 10 mg/kg (Zinc oxide) and 0.1 mg/kg (sodium Selenite) orally once daily for six months and the samples were collected in monthly intervals for analysis.

Chemicals

All chemicals of analytical grade were purchased from Rankem Pvt. Ltd., E-Merck (India), and Sigma Aldrich (USA). The kits used was from Cogent, India.

Enzymatic Assays

5% RBC haemocylate was prepared in chilled phosphate buffer (pH 7.0). Superoxide dismutase (SOD), catalase (CAT) and malonaldehyde (MDA) were estimated by standard procedure of Misera and Fridivich (1972); Aebi (1974) and Nair and Turner (1984), respectively.

Hormone Analysis

Plasma FSH (Follicle stimulating hormone), LH (Leutinising Hormone) and PrL (Prolactin) levels were analyzed using the AccuBind ELISA Microwells kit (Monobind Inc. USA).

Total Arsenic Analysis

The milk, faeces and hair samples were digested with triacid mixture containing nitric acid, perchloric acid and sulphuric acid in the ratio of 10:4:1 according to the modified method of Datta et al. (2010). Total arsenic estimation was performed by atomic absorption spectrophotometer (AAS) with vapour generation accessories (VGA 77) (Varian AA 240, Russia).

Statistical Analysis

The data were analyzed by paired 2-tailed 't'-test using SPSS 17.0 software. $P < 0.05$ were considered significant.

Results

Stress Enzymes Status

There was a significant increase ($p < 0.05$) in the SOD (Fig. 1) and CAT (Fig. 2) level from 30th day and MDA (Fig. 3) level decreased significantly ($p < 0.05$) from 90th day onwards in the group treated with Zn-Se mixture.

Hormonal Status

FSH (Fig. 4a) and LH (Fig. 4b) levels were increased from 60th day onwards and Prl (Fig. 4c) was found to increase significantly ($p < 0.05$) from 30th day of initiation of the treatment.

Discussion

Oxidative stress is an imbalance between free radical generation and the antioxidant defense system. Many reports evidenced a decrease in the levels of antioxidants after arsenic exposure (Son et al., 2001). Decreased antioxidant levels due to arsenic toxicity is reported both in human (Wu et al., 2006) and animals (Das et al., 2012) and this effect is due to generation of reactive oxygen species (ROS) (Shi et al., 2004) leading to cell damage and death.

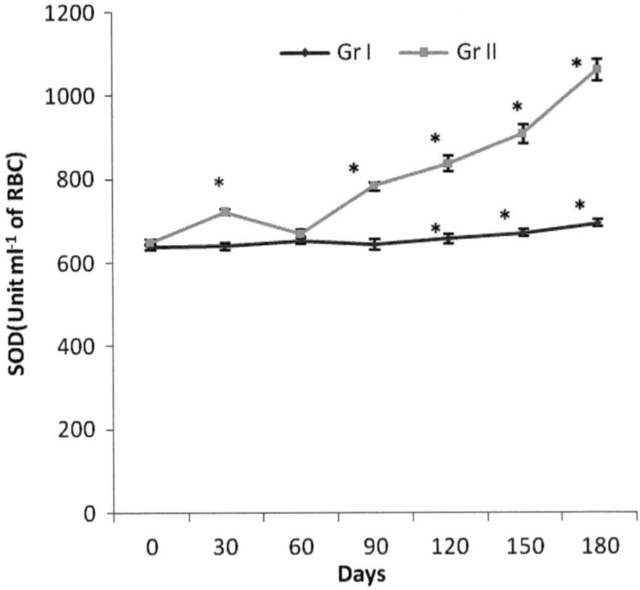

Fig. 1 SOD (unit ml^{-1} of RBC) status of cattle from experimental villages before and after oral administration of zinc and selenium combination for six months. Gr I: Untreated control group ($n = 22$), Gr II: Treated with zinc oxide and sodium selenite @ 10 and 0.1 mg/kg for 180 days ($n = 26$). * $p < 0.05$, compared to respective 0 day

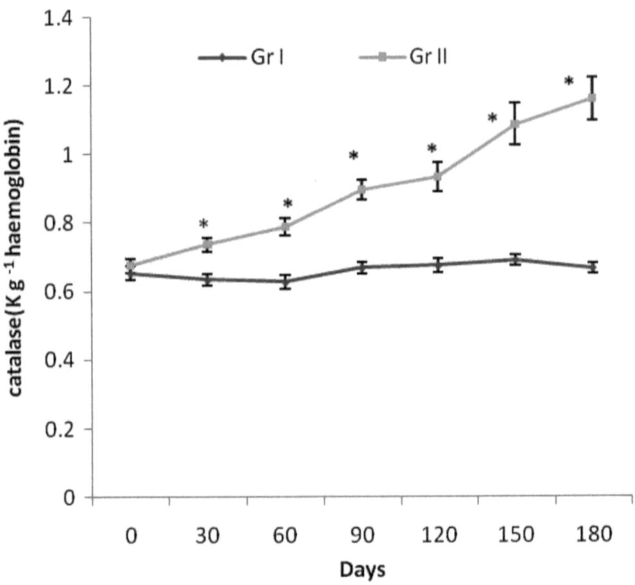

Fig. 2 Catalase (K g^{-1} haemoglobin) status of cattle from experimental villages before and after oral administration of zinc and selenium combination for six months. Gr I: Untreated control group ($n = 22$), Gr II: Treated with zinc oxide and sodium selenite @ 10 and 0.1 mg/kg for 180 days ($n = 26$). * $p < 0.05$, compared to respective 0 day

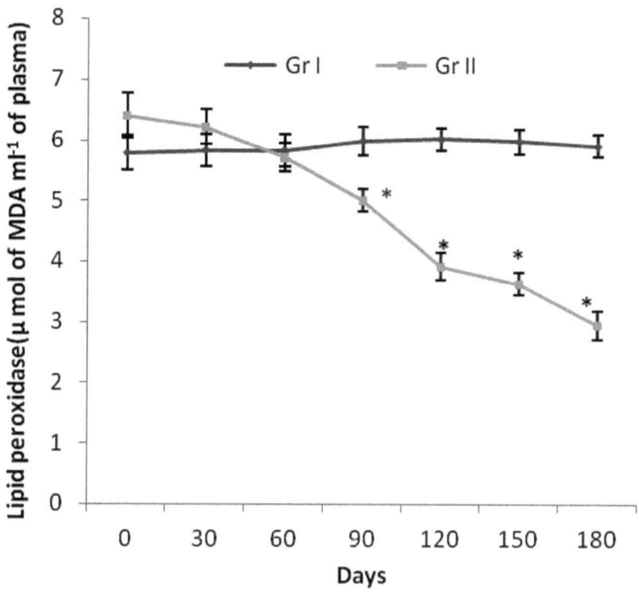

Fig. 3 Lipid peroxidase (μmol of MDA ml^{-1} of plasma) status of cattle from experimental villages before and after oral administration of zinc and selenium combination for six months. Gr I: Untreated control group ($n = 22$), Gr II: Treated with zinc oxide and sodium selenite @ 10 and 0.1 mg/kg for 180 days ($n = 26$). * $p < 0.05$, compared to respective 0 day

Lipid peroxidation is one of the important consequence of oxidative stress (Kumaraguruparan et al., 2002). Lipids, the most susceptible for peroxidative damage due to low energy necessary for the initiation of the process as well as the presence of unsaturated bonds (Balasinska, 2004). In the present study MDA production was decreased by 53.82% by the end of 180th day of treatment period compared to the 0th day in the group treated with the Zn-Se mixture, may be due to enhancement of cells natural protective system and increase in GSH production (Liu et al., 2003).

SOD (Superoxide dismutase) catalyzes the dismutation of superoxide anions and prevents the subsequent formation of hydroxyl radicals (Imlay and Linn, 1988). CAT (Catalase) catalyzes the removal of H_2O_2 formed during the reaction catalyzed by SOD (Kono and Fridovich, 1982). In this study, plasma SOD and CAT level was gradually increased from 30th day in the treated group than the control group and recorded a hike of 63.58% and 71.59% respectively at the end of 180th day of treatment which suggested that decreased accumulation of superoxide anion radical might be responsible for decreased lipid peroxidation following treatment with Zn-Se. This effect may be due to pronounced elimination of arsenic by Zn supplementation. This finding is in support of our previous findings that total arsenic concentration in milk decrease non-significantly though significant effect was observed from 150th day, faecal arsenic increase significantly ($p < 0.05$) from

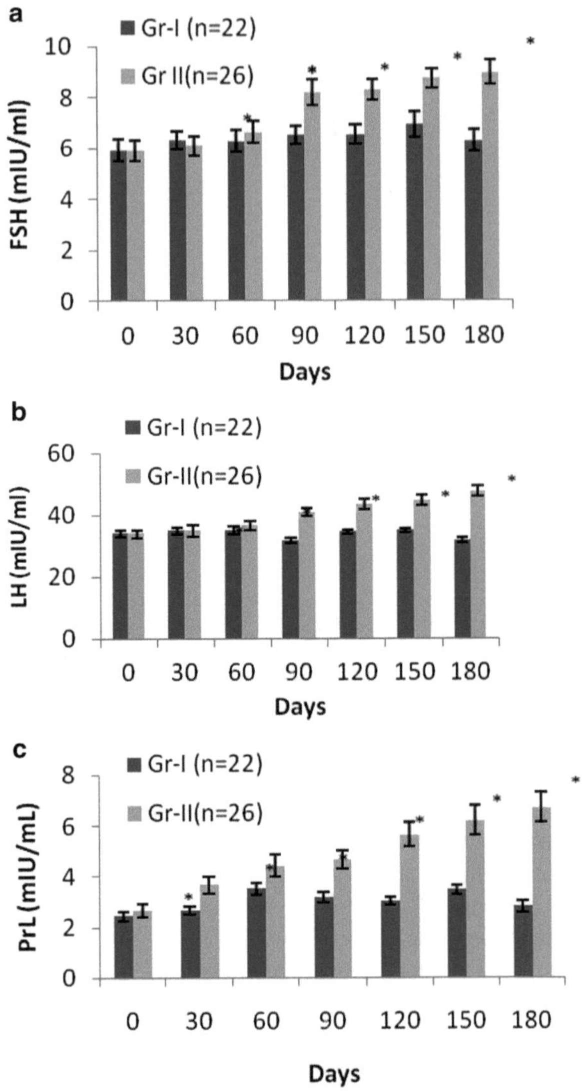

Fig. 4 (**a**) FSH, (**b**) LH and (**c**) PrL levels in plasma of cattle from experimental villages before and after oral administration of zinc and selenium combination for six months. $* = p < 0.05$, n = number of animals

90th day onwards and accumulation of total arsenic in hair decrease gradually but not significantly in cattle treated with Zn and Se mixture (Dash et al., 2013). It is also in corroboration with the findings of Modi et al. (2006) who have reported that iron or zinc either alone or in combination with monoisoamyl di mercapto succinic acid (DMSA) produce more pronounced elimination of arsenic in male mice. Selenium may be enhancing the protective effect against oxidative damage by augmenting the excretion of arsenic into bile (Flora et al., 1999) and it was in support of our previous finding that concentration of faecal arsenic increase gradually from 90th day onwards in cattle treated with Zn-Se.

Dairy cows undergo massive metabolic adaptations during the onset of lactation, and it was postulated that some of these physiological events may negatively impact the health of the dairy cows (Sordillo et al., 2009). During this experimental period we observed a significant increase in the levels of FSH and LH from 60 days onwards and PrL level from 30 days onwards in the treated group (Figs. 4 a, b, c) which can be due to decreased oxidative stress by rapid elimination of arsenic from the body of the animals receiving Zn-Se treatment.

Conclusion

From the above study it can be concluded that zinc oxide-sodium selenite mixture can potently reduce the oxidative stress induced by arsenicosis in cattle by enhancing the dismutation of superoxide radicals and reducing the lipid peroxidation by enhancing the excretion of arsenic from the body and thereby may enhance the physiological performance of the animal and can curb the entry of arsenic into the human food chain.

Acknowledgement We acknowledge West Bengal University of Animal and Fishery Sciences and National Agriculture Innovation Project (NAIP), ICAR, New Delhi in collaboration with World Bank for financial support.

References

Aebi, H. (1974). Catalase: Methods of enzymatic analysis. Ed. Bergmeyer HV **2:** 674-677.

Balasinska, B. (2004). Evaluation of antioxidant status in living organisms. *Med Weter*, **60:** 579-583.

Das, T.K., Mani, V., Kaur, H., Kewalramani, N., De, S., Hossain, A., Banerjee, D. and Datta, B.K. (2012). Effect of vitamin E supplementation on arsenic induced oxidative stress in goats. *Bulletin of Environmental Contamination and Toxicology*, **89(1):** 61-66.

Dash, J.R., Datta, B.K., Sarkar, S. and Mandal, T.K. (2013). Chronic arsenicosis in cattle:Possible mitigation with zinc and selenium. *Ecotoxicology and Environmental Safety,* **92**:119-122.

Datta, B.K., Mishra, A., Singh, A., Sar, T.K., Sarkar, S., Bhatacharya, A., Chakraborty, A.K. and Mandal, T.K. (2010). Chronic arsenicosis in cattle with special reference to its metabolism in arsenic endemic village of Nadia district, West Bengal, India. *Science of the Total Environment*, **409:** 284-288.

Flora, S.J.S., Kannan, G.M. and Kumar, P. (1999). Selenium effects on gallium arsenide induced biochemical and immunotoxicological changes in rats. *Chemico-Biological Interactions*, **122:** 1-13.

Gregus, A., Gyurasics, A. and Koszorus, L. (1998). Interactions between selenium and group Va-metalloids (arsenic, antimony and bismuth) in the biliary excretion. *Environmental Toxicology and Pharmacology*, **5:** 89-99.

Imlay, J.A. and Linn, S. (1988). DNA damage and oxygen radical toxicity. *Science*, **240:** 1302-1309.

Kono, Y. and Fridovich, I. (1982). Superoxide radicals inhibit catalase. *Journal of Biological Chemistry*, **257:** 5751-5754.

Kumaraguruparan, R., Subapriya, R., Kabalimoorthy, J. and Nagini, S. (2002). Antioxidant profile in the circulation of patients with fibroadenoma and adenocarcinoma of the breast. *Clinical Biochemistry*, **35:** 275-279.

Liu, L., Trimarchi, J.R., Navarro, P., Blasco, M.A. and Keefe, D.L. (2003). Oxidative stress contributes to arsenic induced telomere attrition, chromosomal instability and apoptosis. *Journal of Biological Chemistry*, **278:** 31998-32004.

Majumder, A., Biswas, T. and Kole, S.C. (2012). Geobacillus Stearothermophilus – An Arsenic Oxidizing Bacterium Isolated from Arsenic Polluted Field of West Bengal, India. National Conference on Biotechnology and the Environment, NIT, Durgapur.

Misera, H.P. and Fridivich, I. (1972). The role of superoxide dismutase ion in the auto-oxidation of epinephrine and a single assay for super-oxide dismutase. *Journal of Chemotherapy*, **247:** 3170-3185.

Modi, M., Kaul, R.K., Kannan, G.M. and Flora, S.J.S. (2006). Preventive Effects of Zinc and N-acetylcysteine Administration against Arsenic Intoxication in Rats. *Journal of Trace Elements in Medicine and Biology*, **20:** 179-204.

Nair, V. and Turner, G.A. (1984). The thiobarbituric acid test for lipid-peroxidation, structure of the adduct with malonaldehyde. *Lipids*, **19:** 804.

Shi, H., Shi, X. and Liu, K.J. (2004). Oxidative mechanism of arsenic toxicity and carcinogenesis. *Molecular and Cellular Biochemistry*, **255:** 67-78.

Singh, T.S. and Pant, K.K. (2004). Equilibrium, kinetics and thermodynamic studies for adsorption of $As^{(III)}$ on activated alumina. *Separation and Purification Technology*, **36:** 139-147.

Son, M.H., Kang, K.W., Lee, C.H. and Kim, S.G. (2001). Potentiation of arsenic-induced cytotoxicity by sulfur amino acid deprivation (SAAD) through activation of ERK1/2, p38 kinase and JNK1: The distinct role of JNK1 in SAAD-potentiated mercury toxicity. *Toxicology Letters*, **121:** 45-55.

Sordillo, L.M. and Aitken, S.L. (2009). Impact of oxidative stress on the health and immune function of dairy cattle. *Veterinary Immunology and Immunopathology*, **128:** 104-109.

Valko, M., Morris, H. and Cronin, M.T. (2005). Metals, toxicity and oxidative stress. *Current Medicinal Chemistry*, **12:** 1161-1208.

Wu, M.M., Chiou, H.Y., Hsueh, Y.M., Hong, C.T., Su, C.L. and Chang, S.F. (2006). Effect of plasma homocysteine level and urinary monomethylarsonic acid on the risk of arsenic-associated carotid atherosclerosis. *Toxicology and Applied Pharmacology*, **216:** 168-175.

Macro-benthos Diversity in a Headwater Stream Affected by Tea and Paddy Agricultural Runoff, Sri Lanka

H.L.K. Sanjaya, H.B. Asanthi, and U.A.D. Jayasinghe

Introduction

It is well documented that soil, organic matter, fertilizer and pesticides from agricultural lands are transported into adjacent streams with surface runoff changing the quality of those water ways (Neuman and Dudgeon, 2002). The eroded soil particles transported over the agricultural lands include nutrients, pesticides and their residuals. Therefore, soil erosion can cause both physical and chemical impacts in adjacent aquatic systems (Merrigton et al. 2002). Even a thin film of fine sediments in aquatic systems can eliminate or reduce the sensitive taxa population of macro invertebrates, such as Ephemeroptera and Plecoptera (Evans, 1996; Leeks, 1995). However, pesticides transportation is one of the severe impacts on aquatic ecosystem compared to other chemicals such as artificial fertilizers (Liess and Schulz, 1999; Merrigton et al. 2002) altering the dynamics of macro-benthos communities in streams.

The effects of agricultural runoff have been monitored by several authors through the assessments of stream water quality and its biota (Liess and Schulz, 1999; Neuman and Dudgeon, 2002; Song et al., 2009). The changes of physical and chemical nature of water can create diverse biological effects ranging from severe to subtle effect levels (Bartram and Ballance 1996). Ecological methods which are based on the community structure and the presence or absence of species are more prominent in literature than the physical and chemical methods (Bartram and Ballance 1996).

Particularly, macro-benthos are suitable in ecological approach of monitoring an ecosystem (Bartram and Ballance 1996). The abundance and the distribution of macro-benthos are affected by various physical and chemical conditions of water

H.L.K. Sanjaya (✉) • H.B. Asanthi • U.A.D. Jayasinghe
Department of Limnology, University of Ruhuna, Matara, Sri Lanka
e-mail: hlksanjaya@yahoo.com

N.J. Raju et al. (eds.), *Management of Water, Energy and Bio-resources in the Era of Climate Change: Emerging Issues and Challenges*,
DOI 10.1007/978-3-319-05969-3_17

such as water depth, water velocity, and organic matter content in sediment and toxicant levels in water and sediments. According to Mandawille (2002) macro-benthos diversity is used to predict the health of water bodies in many studies as they are species rich, sedentary and long lived and therefore can indicate long-term effects on their habitat. Bengtsson (1998) has documented that the long-term goal of managing and developing sustainable ecosystems mostly depend on the understanding of the linkages between key species or functional groups rather than focusing only on the species diversity.

The macro-benthos are secondary producers which form important links in the food webs of aquatic ecosystems (Sharma et al., 2009). Furthermore, they play a major role in transferring energy from the first trophic level to second trophic level in fresh water ecosystems. Therefore, secondary production can be used to evaluate the trophic potential within a particular ecosystem (Tumbolo and Downing, 1994). The quality of a particular ecosystem can be predicted by estimating the trophic potential (Wilber and Clarke, 1998). The higher diversity together with the higher secondary production indicates better environmental quality and vice versa (Munari and Mistri, 2007). Also, the secondary production of macro-benthos can be used as an important tool for monitoring the changes in an ecosystem.

This study assessed the impact of agricultural runoff from both tea and paddy cultivations to a small headwater stream at Wathurawa, Sri Lanka. Headwater streams are defined as all the first and second order streams which make 2/3 of the total length of a river network, and large rivers can be fed by hundreds of thousands of headwater streams (Leopold et al., 1964). These stream ecosystems are critical sites for organic matter processing and nutrient cycling (Bilby and Likens, 1980; Wallace et al., 1997). Sri Lanka is blessed with 103 perennial rivers which are fed with numerous headwater streams radiating from hill country region towards Indian Ocean. As for the use of considerable amounts of agrochemicals for tea and paddy cultivations, and more obvious soil erosion in these hill slopes, it is essential to assess the impacts on macro-benthos diversity and their productivity as the second trophic level for explaining the quality of these headwater streams.

Materials and Methods

Study Site

Wathurawa is a mountainous remote village surrounded by a forest reserve which is located in between the margin of Matara and Galle administrative districts (Fig. 1). Nearly 300 families living in this area are engaged in paddy and tea cultivation as their major livelihood. These cultivations are extended upto the reserved forest margin towards the crest. Wathurawa stream, which originates from this hill top and later connects with the Gin River, was selected for this study. The stream valley is used for paddy cultivation due to the ease of getting water and draining back to the

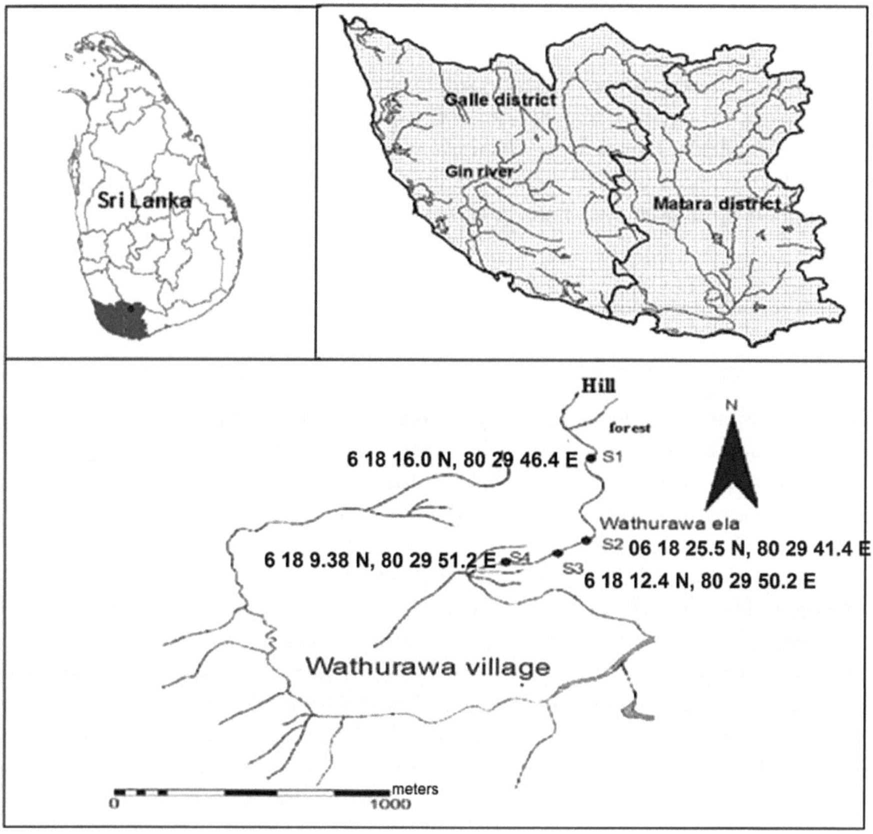

Fig. 1 The sampling sites in Wathurawa stream (S1, S2, S3 and S4)

downstream. Tea plantations are extended throughout the hill slopes facing the stream valley. Four Sampling sites along the stream reach were selected in respect to the ease of access and the points affected by agricultural runoff as follows.

The site 1 (S1) was located at the natural forest margin and it was considered as the reference site. The site 2 (S2) and site 3 (S3) were located at the point of connection of run-off from tea land to the stream while site 4 (S4) was selected as a place where the run-off from both tea and paddy lands are connected to the stream. There was no more variation of the substrate characteristics at all four sampling sites and slightly shaded with stream bank vegetations. Total length of the selected steam reach was about 600 m with sites apart from each other approximately by 100 to 200 m. Samples were taken during the rainy season (average rainfall 579.3 mm) from November 2010 to February 2011 twice a month as both the fungicides application and paddy cultivation are done concurrently throughout this season.

Measurement of Physico-chemical Parameters

Bottom water temperature (YSI 85, Japan), pH (pH meter: AD 111, Hungary), flow rate (Digital flow meter: Kenek VP 1000, Japan), and depth (Digital depth gauge: Hondex, PS-50C, Japan) were determined using relevant instruments and total suspended solids, total dissolved solids, NO_3^--N, and PO_4^{-3} concentrations were determined by standard methods described in APHA (1985).

Macro-benthos Sampling and Analysis

Macro-benthos were collected using a Surber sampler of 0.5 mm mesh size and with a quadrate area of 25 cm^2. The Surber sampler was positioned against the stream flow and vigorously stirred the substrate area at nine random points per each site during each sampling occasion. Altogether 54 replicates were collected during six sampling efforts. The benthos collected at the cod end of the Surber sampler were preserved in 4 % formalin and identified to the Family level following the keys of Benjamin et al. (2003), Bouchard (2004), Fernando (2002), Hartmann (2006), and Pescador et al. (1995). They were quantified as individuals/m^2 under each taxa and Alpha diversity of macro-benthos was estimated by counting the number of individuals in each taxonomic group and calculating Shannon-Weiner diversity index for four sampling sites separately. EPT-Index and taxa richness were also calculated for the total number of families including Orders Ephemeroptera, Plecoptera and Tricoptera at each site (Mandawille, 2002).

The dry weight and the Ash Free Dry Weight (AFDW) of taxonomic Orders were determined by drying (110 °C for 24 h) and incineration (550 °C for 1 h) for benthos of each taxonomic group. Secondary production was esti-mated by using the empirical formula described by Brey (1990), which has been followed by Dara and Duoglas (1998) for miscellaneous macro-benthos groups.

Secondary productivity was estimated for each site using following formula (Brey, 1990):

$$\log P = -0473 + 1.007 \log B - 0274 \log W$$

where P is annual production (g AFDW m^2yr^{-1}), B = annual biomass (g AFDW m^{-2}); and W = individual weight (g AFDW).

Data Analysis

Karl Pearson's correlation coefficient was used to quantify the relationship between hydrological attributes and biological attributes. Mean values of hydrological attributes in each study site was used to calculate the inter relationships (r-values) according to Sharma et al. (2008). One way ANOVA was performed to analyze the variation of hydrological and biological attributes between sampling sites using SPSS 16.0 package. Cluster analysis and correspondent analysis (PCA) were done for the physico-chemical parameters at sampling sites with the help of Primer-5 package. Multi-Dimensional Scaling (MDS) was performed to observe the variation of diversity of macro-benthos at each sampling site during the study period.

Results

Physico-chemical Parameters

The mean values of physico-chemical attributes at four sites are given in Table 1. Wathurawa stream is a shallow stream and the mean depth ranges from 0.15 m to 0.18 m. Water velocity varied within a small range from 0.227 to 0.274 m/s and the highest mean velocity was 0.274 ± 0.017 m/s at the site 2 and the lowst velocity was 0.227 ± 0.009 m/s at the site 3 (Table 1).

There were no significant spatial differences of pH, TSS, TDS, NO_3^- and PO_4^{3-} at the sampling sites. However, the highest TSS (174 mg/L) and PO_4^{3-} (0.033 mg/L) were observed at the downstream site (S4). The highest NO_3^- concentration (1.21 mg/L) was observed at S3 and the value was much lower than the EU standards for drinking water (50 mg/L) (Merrigton et al. 2002).

The cluster analysis based on all physico-chemical attributes shows that the similarity between S1 and S2 is about 22 % and the similarity between S3 and S4 is 10 %. Therefore, Fig. 2 shows that the group separation and the similarity between the two groups was 12 % (Fig. 2).

Macro-benthos Diversity and Other Indices

Immature stages of several insect orders such as Plecoptera, Odonata, Diptera, Ephemeroptera, Tricoptera, Hemiptera, Coleoptera and Lepidoptera and some other taxa of Amphipoda, Class Turballaria, Class Gastropoda and Gordea were observed in the stream.

The orders and families found in the stream are given in Table 2. All together 43 taxa were found from the benthos in Wathurawa stream and the highest taxa richness was 36 at site 1.

Mean EPT index values were 8.67 ± 1.05, 6 ± 0.63, 6.67 ± 0.67 and 5.17 ± 0.85 at the sites of S_1, S_2, S_3, and S_4 respectively. According to paired test of Tukey HSD the EPT index was significantly different between the site 1 and the site 4 ($P < 0.05$). The calculated values of mean Shannon-Wiener diversity index

Table 1 Mean values of physico-chemical attributes at the four sampling sites in Wathurawa stream

Site	*pH*	*Bottom water Temperature (°C)*	*Flow rate (m/s)*	*TSS (mg/L)*	*TDS (mg/L)*	NO_3^- *(mg/L)*	PO_4^{3-} *(mg/L)*	*Depth (m)*	*DO (mg/L)*
1	6.95 ± 0.18	23 ± 0.2	0.241 ± 0.02	2.42 ± 0.55	68.25 ± 18.95	0.56 ± 0.15	0.013 ± 0.003	0.15 ± 0.01	10.84 ± 0.26
2	6.95 ± 0.14	24.1 ± 0.4	0.274 ± 0.02	4.58 ± 1.11	104.12 ± 9.61	0.70 ± 0.24	0.01 ± 0.002	0.17 ± 0.01	11.1 ± 0.15
3	6.81 ± 0.22	24.6 ± 0.5	0.227 ± 0.01	8.12 ± 3.34	90.12 ± 21.24	0.92 ± 0.15	0.01 ± 0.003	0.17 ± 0.01	10.86 ± 0.20
4	6.56 ± 0.27	24.9 ± 0.6	0.267 ± 0.01	36.42 ± 27.60	81.42 ± 19.15	0.85 ± 0.05	0.03 ± 0.017	0.18 ± 0.01	10.97 ± 0.17

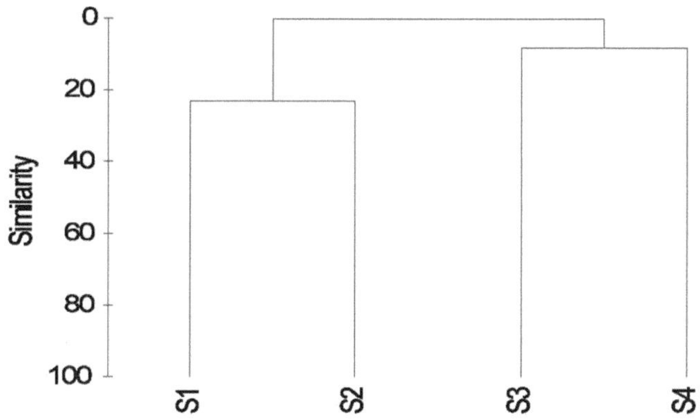

Fig. 2 Cluster analysis for physico-chemical parameters

(S-W index) of the benthic macro-invertebrates are shown in Fig. 3. According to that, the highest index value (3.54 ± 0.14) was at the S1 and the lowest value (2.91 ± 0.33) was at S4. However, there was no significant difference among mean Shannon-Wiener index values at the four stations.

Taxa richness (TR) values were 20.17 ± 1.54, 17 ± 1.13, 16.67 ± 0.88 and 12.33 ± 2.16 at the sites of S1, S2, S3 and S4 respectively (Fig. 4). According to one way ANOVA there was a significant difference of taxa richness at the four sampling sites ($P < 0.05$). Tukeys HSD multiple comparisons test shows a significant difference of TR between S1 and S4 ($P = 0.008$). It was observed as a clear decline of the sensitive taxa (Ephemeroptera, Plecoptera, Odonata, Tricoptera) towards the downstream (Fig. 5).

The highest density of Tricoptera (36.3 % ind/m^2) was found at S1 and the lowest density (22.9 % ind/m^2) was found at S2. However, the highest density of Class Gastropoda was recorded (11.1 % ind/m^2) at S4 and the lowest density (2.66 % ind/m^2) was recorded at S1. The highest density of Odonata was 28.1 % ind/m^2 at the S1 and the density of Plecoptera, which is one of the most sensitive groups, was zero at S4.

The highest value of Shannon-Wiener diversity index of the macro-benthos was 3.54 ± 0.14 at the S1 and the lowest value was 2.91 ± 0.33 at S4. However, there was no significant difference of Shannon-Wiener index at the four sampling sites.

Also, there were some significant correlations between the physico-chemical parameters and density of the macro-benthos. Table 3 gives the Pearson's correlation values and the related parameters. The NO_3^- shows significant negative correlations with the abundance of orders of Ephemeroptera, Plecoptera and Odonata.

MDS (Multi-Dimensional Scaling) shows the variations of the diversity at S1, S2, S3 and S4 (Fig. 6). Arrows are indicating the degree of variations of Shannon-Wiener diversity at S_2, S_3 and S_4 sites compared to the reference site (S1). The highest variation can be observed between S1 and S4.

Table 2 Orders and Families of macro-invertebrates found in Wathurawa stream

		Number of individuals at sampling sites			
Order	*Family*		*S1*	*S2*	*S3*
Plecoptera	Nemouridae	11	0	0	0
	Perlidae	80	47	10	0
Odonata	Clorocyphidae	2	2	0	0
	Calopterygidae	2	2	2	0
	Euphaedae	46	43	15	0
	Lestida	2	0	2	22
	Libellulidae	5	5	0	0
	Platistictidae	311	162	91	0
	Gomphidae	33	29	22	79
	Cordulidae	0	2	0	12
Tricoptera	Helicopsychidae	87	32	140	0
	Hydropsychidae	329	159	64	22
	Polycentropodidae	21	0	2	88
	Leptoceridae	50	8	61	4
	Limnephilidae	8	0	2	95
	Lepidostomatidae	14	2	0	11
	Glossosomatidae	0	9	0	0
	Odontoceridae	9	2	0	2
Ephemeroptera	Leptophlebiidae	117	109	67	12
	Ephemeridae	9	7	14	56
	Caenidae	15	9	4	14
	Baetidae	8	11	15	2
Diptera	Tipulidae	52	40	24	10
	Simulidae	46	4	8	22
	Athericidae	4	0	0	0
	Chironomidae	4	0	0	0
Amphipoda	Atyidae	4	51	38	0
	Potamonidae	0	4	15	4
Coleoptera	Scirtidae	23	20	20	0
	Hydrophilidae	2	11	0	15
	Helophoridae	7	19	0	5
	Psephenidae	20	20	17	20
	Elmidae	34	58	41	45
Class Oligocheata		20	4	8	7
Gordea	Gordiidae	0	0	0	2
Class Gastropoda	Thiaridae	34	18	44	7
	Pilidae	4	15	19	38
	Planorbidae	0	0	2	26
Lepidoptera	Pyralidae	4	0	4	0
	Tortricidae	2	0	2	4
Class Turballaria		0	2	0	0

(continued)

Table 2 (continued)

		Number of individuals at sampling sites			
Order	*Family*		*S1*	*S2*	*S3*
Hemiptera	Naucaridae	7	17	17	2
	Veliidae	0	2	0	6
Total		1426	925	770	639

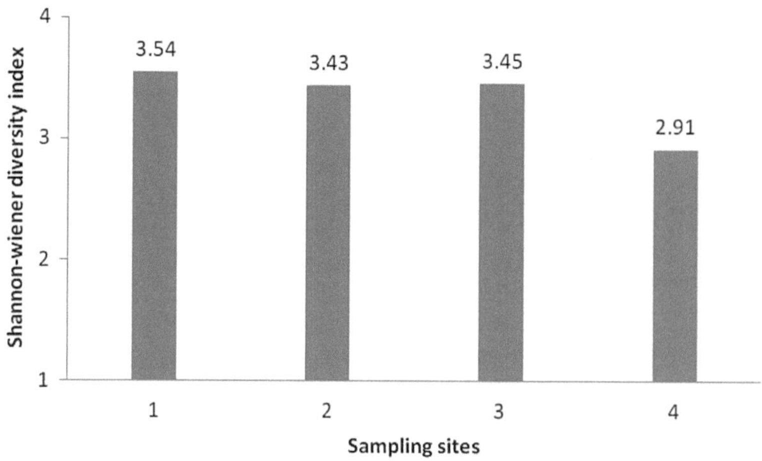

Fig. 3 Mean Shannon-Wiener diversity index values at the four sampling sites in Wathurawa stream

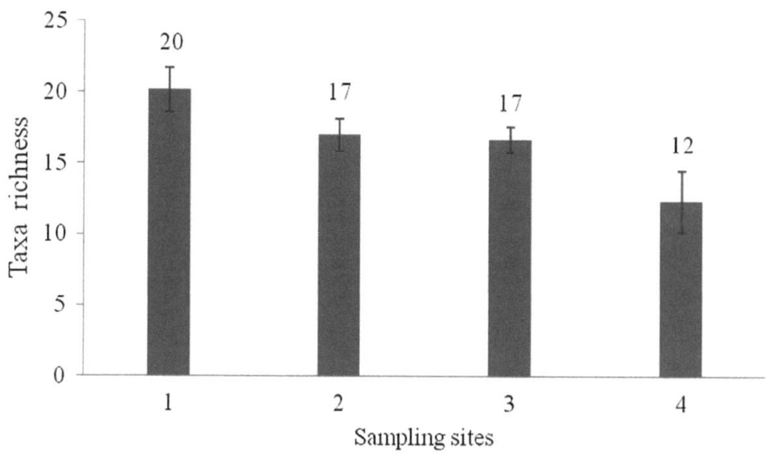

Fig. 4 Taxa richness at the four sampling sites in Wathurawa stream

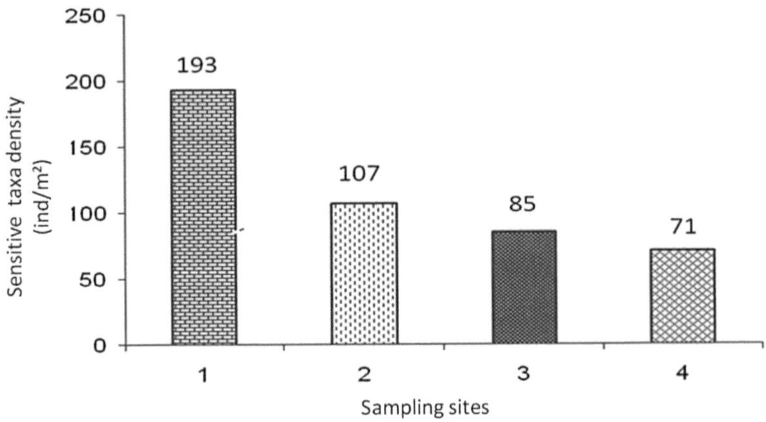

Fig. 5 Sensitive taxa density at the four sampling sites in Wathurawa stream

Table 3 Pearson's correlation values between the physico-chemical parameters and the density of macro-benthos

Order	*TSS*	PO_4^{3-}	*TDS*	NO_3^-
Gordea	0.989	0.989	0.989	
Hemiptera	–	–	0.918	–
Ephemeroptera	–	–	–	–0.96
Plecoptera	–	–	–	–0.96
Odonata	–	–	–	–0.96
Class Gastropoda	–	–	–	0.934

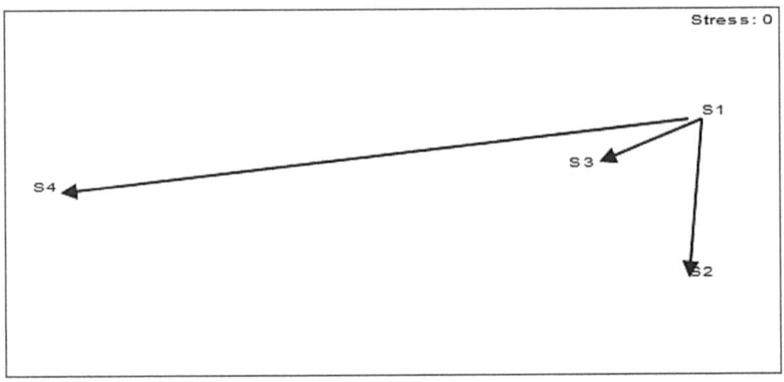

Fig. 6 The variations of Shannon-Wiener diversity index at the sampling sites in Wathurawa stream (MDS stress = 0.00)

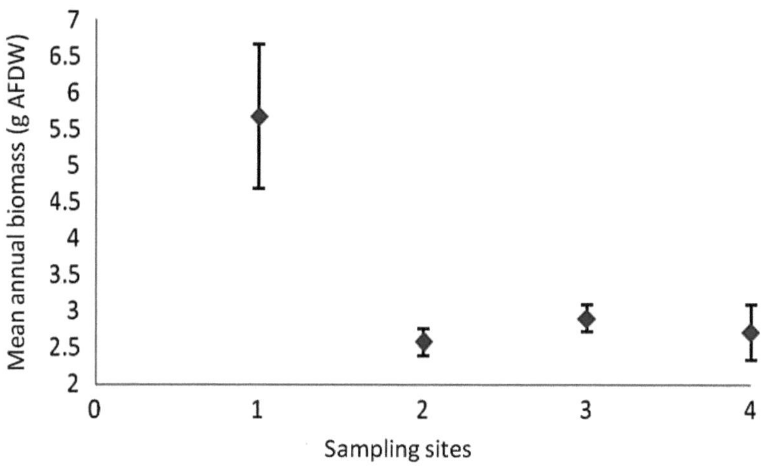

Fig. 7 Mean annual biomass at the four sampling sites

Macro-benthos Secondary Productivity

The highest secondary productivity (8.43 g AFDW m^{-2} yr^{-1}) was obtained at S1 while the lowest secondary productivity (3.95 g AFDW m^{-2} yr^{-1}) was found at S4. The mean annual biomass of macro-benthos at the four sites is shown in Fig. 7.

Discussion and Conclusions

The cluster separation of site 3 and site 4 according to all physico-chemical characteristics explains the effects of agricultural runoff over the tea and paddy cultivations. Even though, the spatial variation of water quality parameters are not significant, the highest values were observed at the site 3 and site 4 with the effects of agricultural runoff. However, long-term environmental changes can be predicted by assessing the community structure of macro-benthos (Sharma et al., 2009). There were well published evidences for correlations between water quality and abundance of macro-benthos such as Thani and Phalaraksh (2008) who observed that the Plecoptera families except Nemouridae and Perlidae were low in tropical countries due to comparatively higher temperature. Further, Sharma et al. (2009) have observed a negative correlation between Plecoptera and mean water temperature in Tons river, Doon valley, India. Similarly, significant negative correlations were observed for Plecoptera, Odonata and Diptera with the mean water temperature in the present study.

Total Suspended Solids (TSS) is another considerable factor for determining the community structure of the aquatic ecosystem. Merrigton et al. (2002) have revealed that the erosion events are short lived but the impact can be very persistent.

The highest TSS value was recorded at the site 4 (36.42 ± 27.59 mg/L) where the runoff from both paddy and tea cultivations join the stream. Also, the significant negative correlation between TSS and Shannon-Wiener diversity index shows the importance of TSS on macro-benthos community structure. According to Leek (1995) and Evans (1996), the Ephemeroptera and Plecoptera are severely affected by thin films of fine sediments when there is higher TSS in water. These evidences are supported by the absence of Plecoptera at the site 4 due to higher TSS in the present study.

Particularly, NO_3^- is the most prevalent soil mineral in well drained and well aerated soil and it has an ability to accumulate in soil with higher concentrations (Merrigton et al. 2002). However, nitrate concentration in Wathurawa Ela stream was lower than the EU standard limits for drinking (50 mg/L). The strong negative correlations between nitrate concentration and the abundance of sensitive orders explain the complete absence of the most sensitive order of Plecoptera (Stonefly) at the site 4.

Mandawille (2002) has revealed that the Gastropods are highly pollutant tolerance species. In fact a significant positive correlation between Gastropod density and nitrate concentration has been observed by Sharma et al. (2009) in Tons river. Meantime, the present study shows a positive correlation between Gastropod density and nitrate concentration. Phosphate also can be identified as a major component that can cause environmental problems even at a low concentration (20 μg/L) (Merrigton et al. 2002). The phosphate concentration ranged from 10 μg/L to 30 μg/L and there was a significant negative correlation between phosphate concentration and the Shannon-Wiener diversity index in the present study.

The highest mean diversity index value (3.54 ± 0.14) was found at the site 1 which is located near the forest margin considered as the reference site and it is the lowest disturbed site by anthropogenic activities. The lowest diversity was recorded at the site 4, due to the runoff from both paddy and tea cultivations. According to Sharma et al. (2009), the higher diversity index indicates good habitat quality while the less diversity indicates degraded habitat quality. However, Shannon-Wiener diversity index does not imply any information about the sensitivity of the organisms to the pollution level. Therefore, no significant differences of S-W index among the four sampling sites were observed in this study. Particularly the Stoneflies (Plecoptera), which is the most sensitive taxonomic group among macro-benthos, were absent at the site 4 and the site has been represented by higher tolerant and medium tolerant group of Ephemeroptera (May fly) Families. EPT index was significantly different between site 1 and site 4 which explains the variations of habitat quality with the effects of agricultural runoff.

Munari and Mistri (2007) have stated that higher annual secondary productivity is a better indicator for good quality environmental condition. Therefore, it has been confirmed the effects of agricultural runoff on the reduction of secondary production at the site 4 in present study. According to Clarke (2008), there is still an argument, whether the headwater streams are taxonomically rich or not. Even though, there were effects of agricultural runoff on degradation of the habitat quality and productivity, the present study on the headwater stream has been able to identify 41 families of macro-benthos at the four sampling sites during the rainy period. However, there should be a systematic approach to mitigate the effects of agricultural runoff not only in headwater streams but also for the entire environment. An interesting finding of Song (2009) was the remaining of a grass or woody vegetation border width of 10–20 m to mitigate effectively the effects of agricultural runoff on sensitive and important habitats.

References

American Public Health Association (1985). Standard Methods for the Examination of Water and Wastewater. 60th ed, Washington, DC 20005.

Bartram, J. and Ballance, R. (1996). Water Quality Monitoring: A Practical Guide to the Design and Implementation of Freshwater Quality Studies and Monitoring Programmes.

Benjamin, K.J., Abby, M., James, B.S., Kerry, L., Neal, D. and Ellen, F. (2003). Family level key to the stream invertebrates of Maryland and surrounding areas. 3rd ed. 580 Taylor Avenue, Maryland.

Bengtsson, J. (1998). Which species, What kind of diversity, Which ecosystem function: Some problems in studies of in relations between biodiversity and ecosystem function. *Applied Soil Ecology*, **10:** 191-199.

Bilby, R.E. and Likens, G.E. (1980). Importance of organic debris dams in the structure and function of stream ecosystems. *Ecology*, **61:** 1107-1113.

Bouchard, R.W. (2004). Guide to aquatic macro invertebrates of the upper Midwest. St. Paul, MN. University of Minnesota, St. Paul, MN.

Brey, T. (1990). Estimating productivity of macrobenthic invertebrates from biomass and mean individual weight. *Meeresforsch*, **32:** 329-343.

Clarke, A., Nally, M.R., Bond, N. and Lake, S.P. (2008). Macroinvertebrate diversity in head water streams: A review. *Freshwater Biology*, **53:** 1707-1721.

Dara, H.W. and Duoglas, G.C. (1998). Estimating secondary production and benthic consumption in monitoring studies: A case study of the impacts of dredged material disposal in Galveston bay, Texas.

Evans, R. (1996). Soil erosion and its impacts in England and Wales. Friends of the earth, London.

Fernando, C.H. and Weerawardhana, S.R. (2002). Sri Lanka freshwater fauna and fisheries. Kitchener, Ontario, Canada N2B 2B9.

Hartmann, A. (2006). Regional capacity building workshop on the macroinvertebrate taxonomy and systematic for evaluating the ecological status of rivers in the Hindu Kush-Himalayan (HKH) region. University of natural resources and applied life science, Vienna.

Leeks, G.J.L. (1995). Effect of agricultural soil erosion on water courses. National Rivers Authority. R & D Note 409.

Liess, M. and Schulz, R. (1999). Linking insecticide contamination and population response in an agricultural stream. *Environmental Toxicology and Chemistry*, **18(9):** 1948-1955.

Leopold, L.B., Wolman, M.G. and Miller, J.P. (1964). Fluvial processes in geomorphology. W.H. Freeman and company, Sanfrancisco.

Mandawille, S.M. (2002). Benthic macroinvertebrates in freshwaters-taxa tolerance values, metrics, and protocols. Soil and water conservation society of metro Halfax.

Merrigton, G., Winder, L. and Redman, M. (2002). Agricultural pollution. Spon press, London.

Munari, C. and Mistri, M. (2007). Structure and secondary production of the macrobenthic community in an aquatic transition environment of the gulf of Olbia Mediterranean sea. *Marine Science*, **36(3):** 216-226.

Neuman, N. and Dudgeon, D. (2002). The impact of agricultural runoff on stream benthos in Hong Kong China. *Water Research*, **36:** 3103-3109.

Pescador, M.L., Rasmussen, A.K. and Harris, S.C. (1995). Identification manual for the caddisfly (Tricoptera) Larvae of Florida. Clarion University of Pennsylvania, Pennsylvania.

Song, Y.M., Leprieur, F., Thomas, A., Lek-ang, S., Chon, T. and Lek, S. (2009). Impact of agricultural land use on aquatic insect assemblages in the Garonn river catchment (SW France). *Aquatic Ecology*, **43:** 999-1009.

Sharma, R.C., Arambam, R. and Sharma, R. (2009). Surveying macro invertebrate diversity in the Tons river, Doon valley, India. *Environmentalist*, **29:** 241-254.

Thani, I. and Phalaraksh, C. (2008). A preliminary study of aquatic insect diversity and water quality of Mekong river, Thailand. *KKU Science Journal*, **36:** 95-106.

Tumbolo, M.L. and Downing, J.A. (1994). An empirical model for the prediction of secondary production in marine benthic invertebrate populations. *Marine Ecology Progress Series*, **114:** 165-174.

Xinzheng, L., Jinbao, W., Hongfa, W. and Baolin, Z. (2005). Secondary production of macro benthos from the East China Sea. *Applied Environmental Biology*, **11(4):** 459-462.

Wallace, J.B., Eggert, S.L., Meyer, J.L. and Webster, J.L. (1997). Multiple trophic levels of a forest stream linked to terrestrial litter inputs. *Science*, **277:** 102-104.

Wilber, D.H. and Clarke, D.G. (1998). Estimating secondary production and benthic consumption in monitoring studies: A case study of the impacts of dredged material disposal in Galveston bay texas. *Estuaries*, **21(2):** 230-245.

Bioremediation and Detoxification of Xenobiotic Organic Compounds in Landfill Leachate by *Pseudomonas* sp. ISTDF1

Pooja Ghosh, Mayank Krishna, Mihir Tanay Das, and Indu Shekhar Thakur

Introduction

Increase in municipal solid waste (MSW) production has been a consequence of rapid population growth and urbanization in the past decade. It is startling to know that presently Delhi itself generates 8000 tonnes/day of MSW and to make the situation worse, it is projected to rise to 17,000-25,000 tonnes/day by the year 2021 (Talyan et al., 2008). The most commonly employed method for MSW management is landfill disposal. Landfill requires a close environmental engineering surveillance in its design and operation, as it is likely to generate leachate which would potentially contaminate nearby ground water and surface water (Mor et al., 2006). In spite of being economically viable, generation of heavily polluted leachate constitutes a major drawback of landfills. In the absence of leachate and landfill gas collection systems, these landfills are a major source of groundwater contamination and air pollution, including the generation of greenhouse gases (Talyan et al., 2008; Gardner et al., 1993).

Leachate from a landfill varies widely in composition depending on the age of the landfill and type of waste that it contains (Slack et al., 2005). It is a highly contaminated liquid containing dissolved organic matter, inorganic macro components, heavy metals and xenobiotic organic compounds such as halogenated organics (PCBs, dioxins etc.). Leaching of organic chemicals and heavy metals from landfills to ground water and surface water is a potentially important environmental problem, with such chemicals being toxic, estrogenic and carcinogenic to both terrestrial and aquatic organisms (Cozzarelli et al., 2011; Huset et al., 2011). So there is a need to develop reliable and sustainable options to manage leachate generation and treat it effectively.

Various physico-chemical methods such as adsorption, membrane filtration, chemical precipitation, ion-exchange and reverse osmosis are used for leachate

P. Ghosh • M. Krishna • M.T. Das • I.S. Thakur (✉)
School of Environmental Sciences, Jawaharlal Nehru University, New Delhi 110067, India
e-mail: isthakur@hotmail.com

N.J. Raju et al. (eds.), *Management of Water, Energy and Bio-resources in the Era of Climate Change: Emerging Issues and Challenges*,
DOI 10.1007/978-3-319-05969-3_18

treatment. However, due to certain limitations of these methods such as high cost and limited versatility, biological methods of treatment are increasingly being investigated (Kapoor and Viraraghavan, 1995). Many bacteria particularly *Burkholderia* sp., *Pseudomonas* sp., *Sphingomonas* sp. and *Bacillus* sp. have been reported to be potential in degrading various persistent organic pollutants (Kim et al., 2003; Resnick and Gibson, 1996). In spite of their great potential, however, leachate treatment processes employing bacteria have not been fully investigated.

Though bioremediation is a promising tool for removal of pollutants from contaminated sites, removal of contaminants may not always correspond to a reduction in health risk (Lemieux et al., 2009). Sometimes the metabolites produced as a result of microbial treatment may even be more toxic than the parent compounds (Gillespie et al., 2007). Hence bioremediation studies need to be coupled with toxicological evaluation of the sample after microbial treatment. In vitro models using human cancer cell lines have become well-established tools for rapid and accurate evaluation of toxicity at acute, chronic and sub-chronic levels with fair reproducibility (Chang et al., 2007; Tai et al., 1994). Hepato-carcinoma cell line like HuH7 express many nuclear receptor proteins that regulate the expression of xenobiotic metabolizing enzymes, including cytochrome P450 1A1 (CYP1A1) responsible for the metabolism of multiple endogenous and exogenous chemicals making these cells ideal in vitro models for toxicological studies (Tai et al., 1994). Methyltetrazolium (MTT) and alkaline comet assays are commonly used for comparative toxicity evaluation. The MTT assay is an overall indicator of cytotoxicity and is based on the ability of living cells to reduce dissolved MTT (yellow) into insoluble formazan (blue) in the presence of mitochondrial succinate dehydrogenase (Mosmann, 1983). Comet assay which is based on the principle of quantification of the amount of denatured DNA fragments migrating out of the nuclei during electrophoresis has been used to evaluate the genotoxic potential (Widziewicz et al., 2012).

The present investigation was designed to evaluate the potential of alkalotolerant dibenzofuran degrading *Pseudomonas* sp. ISTDF1 for bioremediation of landfill leachate and to study its detoxification efficiency using HuH7 cell line.

Materials and Methods

Chemicals

Naphthalene, Phenanthrene, Fluorene, Pyrene, Fluoranthene, Pentachloro-phenol, Phthalate and N, O-bis(trimethyl-silyl)trifluroacetamide (BSTFA) and all cell culture related reagents, i.e. Dulbecco's minimal essential medium (DMEM), fetal bovine serum, and antibiotic antimycotic solution were procured from Sigma–Aldrich (India). All solvents and other chemicals (HPLC grade) were purchased from Merck, India.

Sampling and Sampling Site

Leachate samples were collected from Okhla landfill site (28°30′48″N, 77°17′4″E), Delhi in May 2012 (Pre-monsoon season). It was collected from three sampling points across the landfill and later on mixed to obtain a composite sample. Sample collected was stored at 4 °C until complete analysis.

Extraction of Compounds from Leachate

Briefly the organic compounds present in leachate were extracted by classical solvent extraction technique using 100 mL 1:1 dichloromethane (DCM) and acetone added to 250 mL leachate (Hu et al., 2012). Extraction process was repeated thrice. The extracted organic fraction was filtered through Whatman No. 54 filter paper and then evaporated to dryness at room temperature using a vacuum rotator evaporator. After evaporation, it was dissolved in 2 mL DCM (crude organic extract).

Microorganism and Culture Conditions

An already reported dibenzofuran degrading bacterial strain *Pseudomonas* sp. ISTDF1 (gene bank accession number EU834943) isolated from sludge and sediment sample of Century Pulp and Paper Mill, Lalkuan (79°28′ E longitude and 29°24′ N latitude), Nainital, Uttarakhand, India (Jaiswal et al., 2011). The bacterial strain isolated was maintained in chemostat with minimal salt medium (MSM) containing (gL^{-1}): $Na_2HPO_4.2H_2O$, 7.8; KH_2PO_4, 6.8; $MgSO_4$, 0.2; $NaNO_3$, 0.085; $NH_4(CH_3COO)_3Fe$, 0.01; $Ca(NO_3)_2.4H_2O$, 0.05 with dibenzofuran (1 mM) as carbon source at 30 °C.

Utilization of Recalcitrant Organic Contaminants by Pseudomonas sp. Strain ISTDF1

The bacterial strain was characterized for growth on aromatic pollutants such as Naphthalene, Phenanthrene, Fluorene, Pyrene, Fluoranthene, Pentachlorophenol and Phthalate as a sole carbon source in MSM agar plates with 1 mM concentration at pH 10. This was done to understand the potency of the bacterial strain for use in the treatment of leachate containing many recalcitrant organic compounds.

Treatment of Crude Organic Extract of Leachate by the Bacterial Strain

For this, bacteria was inoculated in MSM containing 2 mL of the crude organic extract as the sole carbon source and incubated at 30 °C in an orbital shaker at 150 rpm. The samples (250 mL) were removed after 0, 24, 48, 120 and 240 hrs and growth of bacterial strain along with degradation of organic compounds (using GC-MS) in leachate was determined.

Sample Extraction and GC-MS Analysis

Bacterial cells were removed by centrifugation at 7000 rpm for 10 min from the collected samples. To the supernatant (250 mL), 100 mL of 1:1 DCM and acetone was added and compounds were extracted by the separatory funnel extraction method, filtered through Whatman No. 54 filter paper and divided into two parts. Both the parts were evaporated to dryness using a vacuum rotator evaporator. One part was reconstituted with 1 mL of DMSO for toxicological analysis while the other was dissolved in 1 mL DCM for GC-MS analysis. The analysis was done using a Shimadzu GC-MS-QP 2010 equipped with a capillary column RTx 5 (dimension 0.25-μm film thickness × 0.25-mm internal diameter × 30 m in length). One microlitre of each extract was analyzed by GC at condition (splitless mode with a split ratio of 10.0; initial temperature 60 °C for 1.0 min; temperature increased from 60 to 320 °C at a rate of 22 °C min^{-1}) (Iida et al., 2002). Data were compared with the inbuilt standard mass spectra library system (NIST-05 and Wiley-8) of GC–MS.

Toxicological Analysis

Cell Culture

The effect of the treatments on the toxicity of the leachate was evaluated by MTT and Comet assay using HuH7 cell line. HuH7 cells were maintained in Dulbecco's Modified Eagle's Medium (DMEM) supplemented with 10 % foetal bovine serum, 1 % antibiotic antimycotic solution in 5 % CO_2 at 37 °C.

Cell Viability

Viability was assessed using MTT assay according to Nwagbara et al. (2007). Cells were seeded in a 96-well surface treated polystyrene tissue culture plates (Corning 3596). After 24 hrs, at 90 % confluency, cells were treated with DMSO (negative control), 1 mM Benzo (α) pyrene (positive control) and with test samples for 24 hrs. All positive control chemicals and test samples were dissolved in DMSO. The final concentration of DMSO in the medium was 0.5 % and all experiments were carried out in three replicates. After 24 hrs of treatment, medium was removed and replaced by fresh medium containing MTT at a final concentration of 0.5 mg/ml and further incubated for 2 hrs. Then, MTT solubilization solution (DMSO) was added into each well and incubated at room temperature for 1 hr for proper solubilization. Absorbance was read at 570 nm and background absorbance at 650 nm was later on subtracted.

Alkaline Single-cell Gel Electrophoresis (Comet Assay)

Comet assay was used to evaluate the genotoxic potential and done according to Singh et al. (1988). Cells were seeded in 12-well plates (Corning 3513). At 90 % confluency, cells were treated with 0.5 % DMSO (negative control), 1 mM Benzo (α) pyrene (positive control) and with test samples for 24 hrs. After treatment, cells were harvested using trypsin followed by mixing with 1 % low melting agarose and embedded on precoated slides. Cells were then denatured with lysis buffer (pH) and incubated overnight at 4 °C. Samples were then electrophoresced at 25 V for 20 min at 4 °C. This was followed by neutralization in 70 % ethanol for 15 min followed by staining with ethidium bromide (2 μg mL^{-1}, 100 μL per slide). The comets were visualized with fluorescent microscope at excitation and emission setting of 518/605 nm. The percentage of DNA in tail, tail moment and olive tail moment (OTM) of 40 randomly selected cells were analyzed from each slide by using CometScore Freeware Software (www.tritekcorp. com).

Results and Discussion

Utilization of Aromatic Contaminants by Pseudomonas *sp. Strain ISTDF1*

Bacterial growth was qualitatively tested on recalcitrant aromatic pollutants such as Naphthalene, Phenanthrene, Fluorene, Pyrene, Fluoranthene, Pentachlorophenol and Phthalate as sole carbon source in MSM agar plates with 1 mM concentration of each at pH 10. The bacteria showed growth over these recalcitrant organic compounds after 48 hrs of incubation as shown in Table 1 even when they were used as sole carbon source showing its potential in utilization and degradation of these compounds.

Growth of Bacterial Strain in MSM Containing Crude Leachate Organic Extract

2 ml of crude extract (containing organic compounds dissolved in DCM) from Okhla landfill leachate was added to the MSM medium inoculated with bacteria

Table 1 Qualitative growth properties of *Pseudomonas* sp. strain ISTDF1 on different carbon sources

Aromatic pollutants	*Qualitative growth*
Naphthalene	+
Phenanthrene	+++
Fluorene	++
Pyrene	++
Fluoranthene	+++
Pentachlorophenol	+++
Phthalate	++

Symbol '+' indicates the number of colonies visibly apparent; '++' indicates number of colonies present on plates were two times more than the +; '+++' indicates number of colonies present on plates were three times more than the +. Qualitative assessment of the ability of strain to grow on MSM agar plate with different aromatic pollutants with 1 mM as the sole carbon source after 48 hrs of incubation at 30 °C. The experiments were repeated three times.

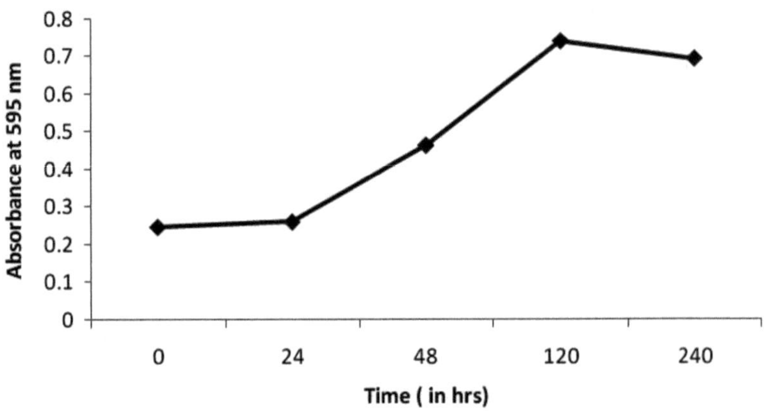

Fig. 1 Growth of *Pseudomonas* sp. ISTDF1 in leachate as sole carbon source in MSM at 30 °C

Pseudomonas sp. ISTDF1. The bacteria was allowed to grow at 30 °C at 150 rpm and growth measurements were taken at different time intervals. There was a steady growth till 120 hrs after which a decline in growth was observed as shown in Fig. 1. A decline in growth may be attributed to the absence of an easy carbon source in the medium required for bacterial growth.

Microbial Degradation of Major Contaminants Present in Landfill Leachate

The inoculated MSM medium containing crude extract was collected at 0 hr (control), 24 hr, 48 hr and 120 hr for extraction of metabolites to be analysed by GC-MS. The organic compounds detected in the 0 hr sample like Nitrobenzene

(RT = 9.89), Brominated dioxin (RT = 10.10), 4,6-Dinitro-1,1,3,3,5-pentamethylindane (RT = 12.21) and Bisphenol A (RT = 13.15) were found to be absent in the extracted metabolite samples of different time intervals. Formation of simpler compounds like 2-Propene (RT = 4.04), 3-Hexen-2-one (RT = 4.37) and Phthalic acid (RT = 10.06) was observed in the 120 hr treated sample. Formation of phthalic acid is a clear cut indication of degradation of organic compounds present in leachate as discussed by Seo et al. (2009). A decrease in the number of peaks with increase in treatment time also indicates gradual mineralization.

Cell Viability

The cell viability derived by means of MTT assay showed that the level of cytotoxicity decreased with increasing duration of bacterial treatment as shown in Fig. 2. The crude extract (0 hr) showed an average absorbance reading higher than the positive control B[a]P indicating that the crude leachate organic extract is less toxic compared to B[a]P. The 24 hr metabolite sample showed an increase in cell viability than the 0 hr untreated sample indicating that the biodegradation of organic compounds in leachate occurred without the formation of toxic intermediates. After 120 hrs of treatment, an appreciable increase in cell viability was observed with the percentage of viable cells reaching upto 81.31 %.

Comet Assay

The results of the single cell gel electrophoresis (Comet assay) with biodegraded (24 hr, 48 hr and 120 hr) and zero hour untreated sample is shown in Figs. 3a and 3b.

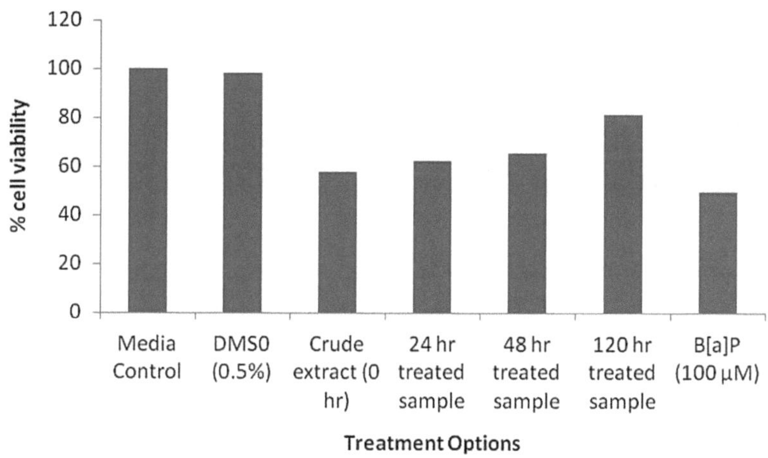

Fig. 2 Evaluation of cytotoxicity of leachate degraded by *Pseudomonas* sp. ISTDF1. **(b)** DMSO (0.5 %) 120 hrs treated 48 hrs treated **(b)** 24 hrs treated Untreated BaP 50 μM **(a)**

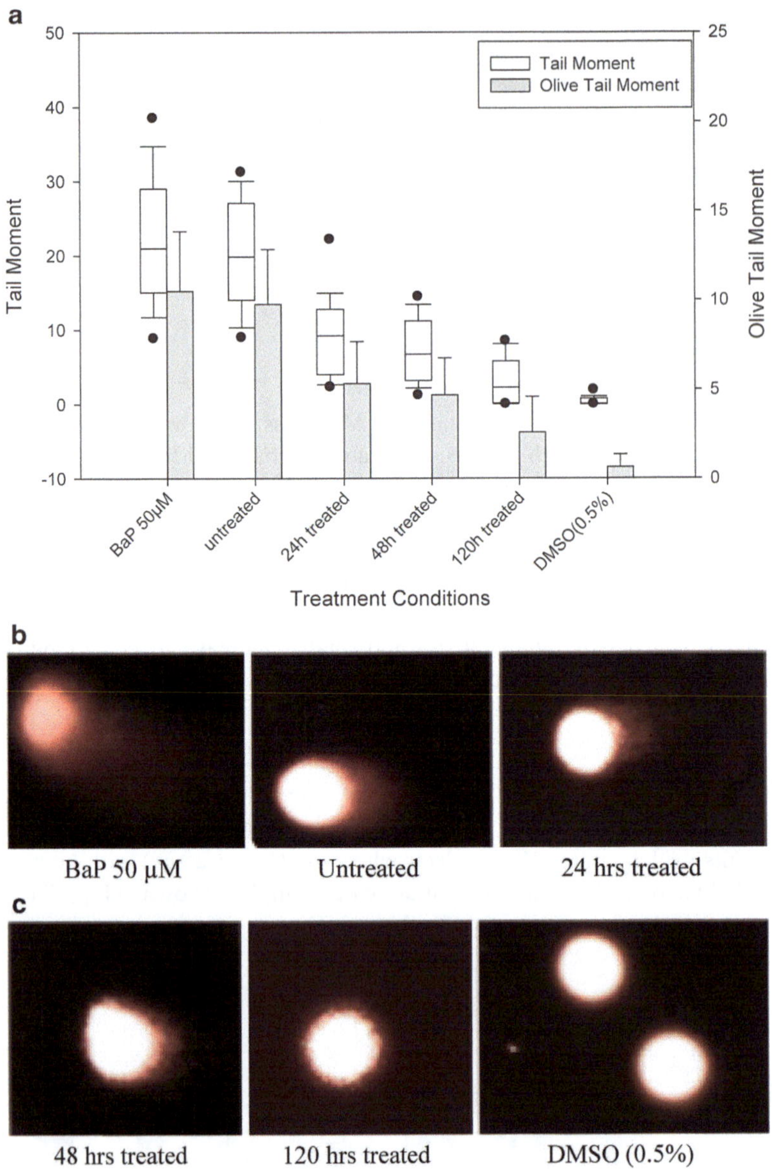

Fig. 3 Genotoxicity of the contaminants before and after treatment with *Pseudomonas* sp. ISTDF1. (**a**) The tail moment and the olive tail moment plotted against different samples. Tail moments of 40 randomly selected comets are presented as quantile box plots. The edges of the box represent the 25th and the 75th percentiles; a solid line in the box presents the median value while dotted line represents mean value. Error bars indicate 90th and 10th percentiles and the black circles indicate outlying points beyond 5th and 95th percentiles. Olive tail moments of same 40 comets are shown in histograms; error bars indicate mean ± standard deviation. (**b**) Images of representative comets from different samples as seen under fluorescent microscope after stained with ethidium bromide

Tail moments of 40 randomly selected comets are presented as quantile box plots. The plot shows that distribution of comets became more homogenous with lower tail moment (3.1706 ± 3.0699) in the 120 h treated sample in comparison to untreated sample (tail moment $= 19.9013 \pm 7.3199$). The olive tail moment data showed a decreasing trend with increasing duration of bacterial treatment. 120 h treated sample resulted in a 5-fold decrease in DNA migration (OTM $= 2.5705 \pm 2.0128$) in comparison to that of the untreated sample (OTM $= 9.7677 \pm 3.0873$).

Conclusion

The present study evaluated the potential of *Pseudomonas* sp. ISTDF1 for bioremediation and detoxification of landfill leachate. The bacterium is capable of degradation of a variety of persistent organic pollutants present in leachate. Significant reduction in both cytotoxicity and genotoxicity in leachate treated with the bacterium for 120 hrs has also been shown through in vitro MTT and Comet bioassays respectively using HuH7 cell line. Biodegradation along with detoxification reveals the potential of the bacterium to be used for leachate treatment.

Acknowledgements This work was supported by research grant from Council of Scientific and Industrial Research (CSIR), New Delhi, India. We thank Mr. Ajai Kumar of Advanced Instrumentation Research Facility (AIRF), JNU, New Delhi for GC-MS analysis and Century Pulp and Paper Mill, Lalkuan, Uttarakhand, for providing effluent and sludge/sediments.

References

Chang, E.E., Miao, Z.F., Lee, W.J., Chao, H.R., Li, L.A., Wang, Y.F., Ko, Y.C., Tsai, F.Y., Yeh, S.C. and Tsou, T.C. (2007). Arecoline inhibits the 2, 3, 7, 8-tetrachlorodibenzo-p-dioxin-induced cytochrome P450 1A1 activation in human hepatoma cells. *Journal of Hazardous Materials*, **146:** 356-361.

Cozzarelli, I.M., Böhlke, J.K., Masoner, J., Breit, G.N., Lorah, M.M., Tuttle, M.L.W. and Jaeschke, J.B. (2011). Biogeochemical evolution of a landfill leachate plume. *Ground Water*, **49:** 663-687.

Gardner, N., Manley, B.J.W. and Pearson, J.M. (1993). Gas emissions from landfills and their contributions to global warming. *Energy and Environmental Management*, **44:** 165-174.

Hu, J., Nakamura, J., Richardson S.D. and Aitken M.D. (2012). Evaluating the effects of bioremediation on genotoxicity of polycyclic aromatic hydrocarbon-contaminated soil using genetically engineered, higher eukaryotic cell lines. *Environmental Science and Technology*, **46:** 4607-4613.

Huset, C.A., Barlaz, M.A., Barofsky, D.F. and Field, J.A. (2011). Quantitative determination of fluorochemicals in municipal landfill leachates. *Chemosphere*, **82:** 1380-1386.

Iida, T., Mukouzaka, Y., Nakamura, K. and Kudo, T. (2002). Plasmid borne genes code for an angular dioxygenase involved in dibenzofuran degradation by *Terrabacter* sp. strain YK3. *Applied Environmental Microbiology*, **68:** 3716-3723.

Jaiswal, P.K., Kohli, S., Gopal, M. and Thakur, I.S. (2011). Isolation and characterization of alkalotolerant *Pseudomonas* sp. strain ISTDF1 for degradation of dibenzofuran. *Journal of Industrial Microbiology and Biotechnology*, **38:** 503-511.

Kapoor, A. and Viraraghavan, T. (1995). Fungal Biosorption – An alternative treatment option for heavy metal bearing wastewaters: A review. *Bioresource Technology*, **53:** 195-206.

Kim, T.J., Lee, E.Y., Kim, Y.J., Cho, K.S. and Ryu, H.W. (2003). Degradation of polycyclic aromatic hydrocarbons by Burkholderis cepacia 2A-12. *World Journal of Microbial Biotechnology*, **19:** 411-417.

Mor, S., Ravindra, K., Dahiya, R.P. and Chandra, A. (2006). Leachate Characterization and Assessment of Groundwater Pollution near Municipal Solid Waste Landfill Site. *Environmental Monitoring and Assessment*, **118:** 435-456.

Mosmann, T. (1983). Rapid colorimetric assay for cellular growth and survival: Application to proliferation and cytotoxicity assays. *Journal of Immunological Methods*, **65:** 55-63.

Nwagbara, O., Darling-Reed, S.F., Tucker, A., Harris, C., Abazinge, M., Thomas, R.D. and Gragg, R.D. (2007). Induction of cell death, DNA strand breaks, and cell cycle arrest in DU145 human prostate carcinoma cell line by benzo[a]pyrene and benzo[a]pyrene-7,8-diol-9,10-epoxide. *International Journal of Environmental Research and Public Health*, **4:** 10-14.

Resnick, S.M. and Gibson, D.T. (1996). Regio- and steriospecific oxidation of fluorene, dibenzofuran and dibenzothiophene by naphthalene dioxygenase from *Pseudomonas* sp. strain NCIB 9816-4. *Applied Environmental Microbiology*, **62:** 4073-4080.

Seo, J.S., Keum, Y.S. and Li, Q.X. (2009). Bacterial degradation of aromatic compounds. *International Journal of Environmental Research and Public Health*, **6:** 278-309.

Singh, N.P., McCoy, M.T., Tice, R.R. and Schneider, E.L. (1988). A simple technique for quantitation of low levels of DNA damage in individual cells. *Experimental Cell Research*, **175:** 184-191.

Slack, R.J. (2005). Household hazardous waste in municipal landfills: Contaminants in leachate. *Science of the Total Environment*, **337:** 119-137.

Tai, H.L., Dehn, P.F. and Olson, J.R. (1994). Use of rat H4IIE and human HepG2 cells as models to investigate the metabolism of 2, 3, 7, 8-tetrachlorodibenzofuran (TCDF). *Toxicologist*, **14:** 271-278.

Talyan, V., Dahiya, R.P. and Sreekrishnan, T.R. (2008). State of municipal solid waste management in Delhi, the capital of India. *Waste Management*, **28:** 1276-1287.

Widziewicz, K., Kalka, J., Skonieczna, M. and Madej, P. (2012). The Comet Assay for the Evaluation of Genotoxic Potential of Landfill Leachate. *The Scientific World Journal*, doi:10.1100/2012/435239.

Identifying Knowledge Gaps in Assessing Health Risks due to Exposures of Nanoparticles from Contaminated Edible Plants

Divya Singh and Arun Kumar

Introduction

Nanoparticles (NPs) have been reportedly found in the environment (Brar et al., 2010; Kumar et al., 2012; Sundaram and Kumar, 2012) and found to pose risks to ecological systems and to the human being (Nowack et al., 2012). During the use of wastewater effluent as potential irrigation water, plants may also receive NPs from water and might accumulate. Further, biosolids (i.e., sludge from wastewater) are also being used as fertilizer and through this route also plant might get exposed to NPs. Considering this pollution route and subsequent potential of exposures to humans, there is a need of testing protocols for assessing risk to human and other communities (Nowack et al., 2012; Kumar, 2012; OECD, 2012).

Understanding effects of NPs in edible crop plants is a matter of importance because NPs are often released into soil environments through wastewater irrigation and biosolids application. The uptake, bioaccumulation, biotransformation, and risks of nanomaterials (NMs) for food crops and human are still not well understood. Very few NMs and plant species have been studied, mainly at the very early growth stages of the plants. Various studies, conducted on NPs and plants, show accumulation of NPs in plants (Rezvani and Zaefarian, 2011). For example, Ag content of shoots exposed to Ag-containing solutions found to be in the range of 0.34 to 2.6 μg/g wet weight (Musante and White, 2010; Lee et al., 2008). Further, NP contents in plant tissues have been observed to increase with increasing NPs concentration in growth media. These findings indicate the possibility of presence of NPs in plant tissues, indicating possible risk of exposure during ingestion of edible portions of plants, irrigated with wastes containing NPs.

D. Singh • A. Kumar (✉)
Department of Civil Engineering, Indian Institute of Technology Delhi, New Delhi, India
e-mail: arunku@civil.iitd.ac.in

N.J. Raju et al. (eds.), *Management of Water, Energy and Bio-resources in the Era of Climate Change: Emerging Issues and Challenges*,
DOI 10.1007/978-3-319-05969-3_19

A review of studies on NPs exposures to plants (Lee et al., 2012; Lin and Xing, 2008; Mazumdar and Ahmed, 2011; Stampoulis et al. 2009) indicates that most of the studies are only on the seedling stage and only few studies have focused on bioaccumulation of NPs in plants. However, none of these studies have conducted human health risk assessment for exposures of NPs from plant, indicating the need for health risk assessment. Towards this, the first question is to know how much information is available and to identify data gaps for focusing research efforts.

The objective of this study was to understand data gaps in quantitative health risk assessment of exposures of NPs from edible plants, which have been irrigated with wastewater (or industrial wastewater) contaminated with NPs. This aspect was illustrated using an example calculation of exposure assessment of silver NPs through consumption of *Sorghum* (*Sorghum bicolor).*

Methodology

A four-step human health risk assessment framework, consisting of hazard identification, exposure assessment, dose–response assessment, and risk characterization steps, was developed in a spreadsheet environment (Gerba, 2000). This risk assessment framework has been previously used for microbial risk assessment (Gerba, 2000; Farber et al., 1996; Bemrah et al., 1998). Parameters required at every step were compiled and literature review was conducted to obtain their values.

Explanation and use of these steps of risk assessment have been described using an example of silver NPs through consumption of *Sorghum* (*Sorghum bicolor*). Figure 1 shows a general schematic of silver NPs in wastewater leading to possible presence in crops which have been irrigated with NP-containing wastewater. Steps include identification of possible sources of NPs in wastewater and their fate in wastewater. NPs in wastewater can finally end up in sludge as well as in effluent water. Wastewater irrigation and sludge application on agricultural land is practiced in many countries, so agricultural land can receive these NPs from either of these two sources. Maximum allowable concentration of NPs in *Sorghum* was estimated using literature data with assumptions and issues during risk estimation were identified. The details of every stage are summarized below.

Hazard Identification

This step identifies hazard posing risk to humans due to NPs. Toxicity of NPs depends on parameters like size, shape, surface area, surface chemistry, porosity, surface charge, reactivity in different matrix, chemical composition, availability, usage, toxicokinetics, toxicodynamics and physicochemical properties (US EPA, 2005).

Exposure Assessment

Scenario Description

The NPs released from different commodities find their way through waste disposal routes into the wastewater effluent and wastewater sludge. Further escape of these NPs into the effluent can contaminate the aquatic and soil environment (Brar et al., 2010). Plants are important components of ecosystem and get in contact with NPs through sludge application and irrigation. So, exposure route in this scenario was considered to be ingestion of food crops irrigated with NPs-containing wastewater (Fig. 1).

Above mentioned NP uptake by plants has basis in reported literature. Study of Lin et al. (2009) tells that transport of C70 occurs simultaneously with the uptake of water and nutrients in the xylem. Another study carried out by Attwood (2012) showed that gold nanoparticles (AuNPs) enters plants through the roots and moved into the vasculature. So this shows that if NPs are present in the soil environment along with other nutrients, there are chances that plant will uptake NPs too. After NP uptake, literature also indicates NP distribution in different parts of plants (Lin and Xing, 2008; Lin et al., 2009; Dimkpa et al., 2012; Lee et al., 2008). Thus, NP taken up by roots has a possibility of getting accumulated in different parts, including edible portions of the plant.

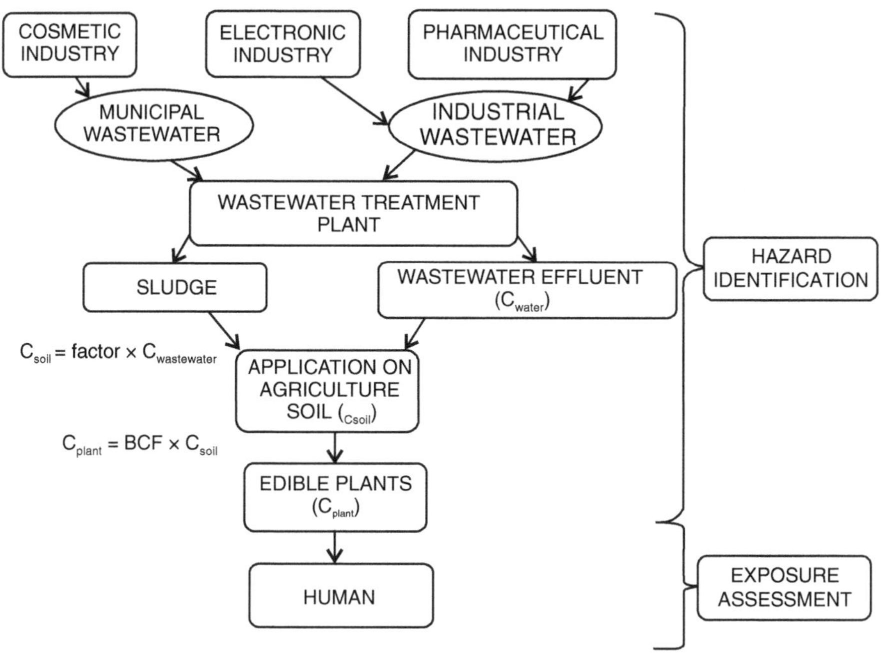

Fig. 1 Schematic showing possible exposure of NPs through consumption of edible plants, irrigated with wastewater effluent, containing NPs (a hypothetical contamination scenario)

Estimation of Average Daily Dose (ADD) of NPs

Average daily dose of NPs is the NPs ingested everyday from oral exposure route. For determining NP uptake during ingestion of edible portion of plant, Eq (1) was used to calculate ADD (mg/kg of body weight/day) as per the USEPA guidelines (USEPA, 1997):

$$ADD = \frac{IR \times ED \times C_{plant}}{BW \times AT} \quad (1)$$

where 'IR' is daily ingestion rate of plants (mg/day), 'ED' is exposure duration for assessing risks (years), 'C_{plant}' is NP concentration in edible portion of the plant (mg/kg dry weight of plant), 'BW' is the body weight (kg) and 'AT' is the averaging time for toxic effects (days). Here C_{plant}, i.e. value of concentration of NP-related metal ion and NPs counts in plant are required. These NP uptake data (i.e., particle number, form, concentration) in edible portion of plant are required for accurate estimate of NP exposed dose. Values of other exposure parameters (i.e., IR, ED, BW, AT) can be obtained from the USEPA (1997) handbook and Gerba (2000) study.

One way of getting C_{plant} data is to monitor NP uptake data in edible portion of plants. As per the literature, not all information is available at one place, if at all available. As per Table 1, most of the studies are focused on qualitative uptake of NPs (Lee et al., 2012; Lin and Xing, 2008; Mazumdar and Ahmed, 2011; Stampoulis et al. 2009; Lee et al., 2008) but all information is missing for a given type of edible portion (Table 2). Although many workers have studied NPs toxicity to other crop plants, such as wheat, spinach, mung bean, tomato (Lee et al., 2011; Khodakovskaya et al., 2009; Tan et al., 2009; Begum et al., 2012; Larue et al. 2012; Lee et al., 2008; Dimkpa et al., 2012). These studies have focused mainly on qualitative information. Quantitative estimates on nanoparticles assimilation are missing. Thus this approach poses problem in estimating ADD value. Another approach for determining C_{plant} is by using mathematical relationship. This approach involves use of NP concentration in water, NP concentration in soil, and NP transfer from soil to edible portion through BCF relationship. For determining NP concentration in soil (C_{soil}) (mg/g), it was assumed to be equal to NP concentration in wastewater ($C_{wastewater}$) (mg/L) (Eq. 2). There exists a gap on knowledge of soil uptake of silver NPs from wastewater during surface irrigation. Here parameter (i.e., "factor") represents transfer of silver NPs from wastewater to soil and indicates adsorption capacity of soil for silver NPs in the wastewater matrix. This parameter was assumed to be 1 (mg/g)/(mg/L). So, here we are assuming that whole of NPs in wastewater is getting transferred in soil without any change. Further, using NP concentration in soil and bioaccumulation (BCF) of silver NPs for *Sorghum* (i.e., mg NP in plant/g dry weight of plant)/(mg NP in soil/g dry weight of soil) (Eq.3), silver NPs concentration in *Sorghum* (C_{plant}) (mg/g) was calculated using Eq. (4).

Table 1 Literature review of studies focusing on nanoparticles-plant interaction for obtaining NP uptake data

Reference/NP/ Focus	*Study focus*	*Availability of uptake information? (quantitative/qualitative)*	*Research gap*
Lee et al. (2012)/Ag NPs	Effect of silver NPs in crop plants *Phaseolus-radiatus* and *Sorghum bicolor*: Media effect on phytotoxicity.	Quantitative bioavailability information through ion released from NPs taken up by plants.	Quantitative available for edible portions of plant. Possible form NP/ion.
Lin and Xing (2008)/ ZnO NPs	Root uptake and phytotoxicity of ZnO NPs.	The root uptake and phytotoxicity were visualized by light, scanning electron, and transmission electron microscopes.	Quantitative data in different plants. Fate in food chain.
Mazumdar and Ahmed (2011)/Ag NPs	Phytotoxicity effect of silver NPs on *Oryza sativa*.	Depositions of silver NPs inside the root cell observed under TEM.	Quantitative uptake mechanisms and data.
Stampouli et al. (2009)/ MWCNTs, Ag, Cu, ZnO, Si)	Assay-dependent phytotoxicity of NPs to plants	The digested plant tissues were analyzed by inductively coupled plasma mass spectroscopy (ICP-MS) for Ag content.	Studies on food chain contamination, potential risk due to NPs uptake.
Lee et al. (2008)/ CuONPs	Toxicity and bioavailability of copper NPs to the terrestrial plants mung bean (*Phaseolusradiatus*) and wheat (*Triticum aestivum*): Plant agar test for water-insoluble NPs.	Uptake of nano-sized copper (nCu) by *Phaseolus radiates* (mung bean) and *Triticum aestivum* (wheat), with reported bioaccumulation factors of 8 and 32 L/kg, respectively.	Toxic effects of NPs.
Dimkpa et al. (2012)/ CuO and ZnO NPs	CuO and ZnO NPs: Phytotoxicity, metal speciation, and induction of oxidative stress in sand-grown wheat.	Cu and Zn, both metals were detected by ICP-MS analysis in digests from plants.	Accumulation in food chain and quantitative data on common edible crops.
Lin et al. (2009)/ CNTs	Uptake, translocation, and transmission of carbon NMs in rice plants	Confirmation of uptake, translocation, and inheritance of C70 by rice using FTIR.	NPs toxicity at genetic level.

Table 2 Toxic effects of Ag NPs to various organs in human

Author	*Title*	*Toxic effect*
Ahmed et al., 2010	Silver nanoparticles applications and human health	These broad applications, however, increase human exposure and thus the potential risk related to their short- and long-term toxicity. A large number of in vitro studies indicate that Ag NPs are toxic to the mammalian cells derived from skin, liver, lung, brain, vascular system and reproductive organs.
Asharani et al., 2009	Cytotoxicity and genotoxicity of silver NPs in human cells	Ag NP-reduced ATP content of the cell which caused damage to mitochondria and increased pro-duction of reactive oxygen species (ROS) in a dose-dependent manner.
Arora et al., 2008	Cellular responses induced by silver NPs: In vitro studies	Oxidative stress. Low doses produced apoptosis and higher dose necrosis.
Panyala et al., 2008	Silver or silver NPs: A hazardous threat to the environment and human health?	Exposure to soluble silver compounds may produce other toxic effects like liver and kidney damage, irritation of the eyes, skin, respiratory and intestinal tract and changes to blood cells.

$$C_{soil} = \text{factor} \times C_{wastewater} \tag{2}$$

$$BCF = C_{plant}/C_{soil} \tag{3}$$

$$C_{plant} = BCF \times C_{soil} \tag{4}$$

Here BCF used in Eq. (3) is based on NP mass in plant obtained from NP mass in soil. Further, BCF information in terms of NP count in plant from NP count in soil is also required to represent BCF. Zhu et al. (2008) reported this aspect by noting NP count value in plant to indicate uptake condition and showed that particles can be translocated throughout the plant and tend to accumulate near root and leaves (Zhu et al., 2008). Thus, BCF in terms of particle count is also required to accurately measure NP uptake in plants and their retention in edible portion of the plant. Thus, overall equation of C_{plant} indicates ion concentration of NP-related metal in edible portion of plant. Further, C_{plant} is also required in terms of NP count in plant. Using these two information (C_{plant} as metal ion concentration and C_{plant} as NP count in plant), risk estimation during NP ingestion can be conducted.

Dose–response Assessment

This step requires information about RfD. Reference dose RfD (i.e., the acceptable dose which does not result in the desired end-point; mg/kg of body weight/day) is based on the critical effect (the effect occurring at the lowest dose); it may not be based on the effect of interest. This information is obtained from toxicity studies. Nowadays, plants are being treated with harmful chemicals like pesticides, insecticides and a significant amount of our food is irrigated with contaminated water. Liver is the organ in human body responsible for the detoxification of harmful chemicals. So, liver is chosen as target organ in this case.

Risk Estimation and Characterization

For non-cancerous contaminant, hazard quotient (HQ) is calculated and for cancerous contaminant, incremental risk of cancer is estimated. HQ is the ratio of the potential exposure to the substance and the level at which no adverse effects are expected. If the HQ is calculated to be equal to or less than 1, then no adverse health effects are expected as a result of exposure. If the HQ is greater than 1, then adverse health effects are possible. The HQ cannot be translated to a probability that adverse health effects will occur and it is unlikely to be proportional to risk. It is especially important to note that an HQ exceeding 1 does not necessarily mean that adverse effects will occur.

$$HQ = \frac{ADD}{RfD} \tag{5}$$

If RfD is not available for a given NP-organ pair, TTD value (i.e., target organ toxic dose) is used. Using TTD value, the overall equation for HQ in terms of all parameters becomes:

$$HQ = \frac{\frac{IR \times ED \times C_{plant}}{BW \times AT}}{TTD} \tag{6}$$

Further, using HQ = 1, maximum allowable value of NP in edible portion of the plant can be estimated (Eq. 7) which can be used to determine maximum allowable contaminant of NPs in plant to develop treatment guidelines.

$$C_{ww} = \frac{HQ \times TTD \times BW \times AT}{IR \times ED \times BCF \times factor} \tag{7}$$

Example Calculation for Exposures of Ag NPs through Ingestion of Sorghum *Seeds*

Sorghum was chosen for an example calculation because uptake and bioaccumulation data are available (Lee et al., 2012). Risk estimation during ingestion of Ag NPs was selected as it has been reported to pose health risks concerns (Table 2). Values of different parameters used for estimating risk value with their justification are presented in Table 3. For determining toxic effects of silver NPs to target organ (liver in this case), data from literature was used. As RfD value of silver NPs for liver (RfD_{ingAg}) was not available (Anandan, 2011), target toxicity dose (TTDs) for liver was used in estimating risks. TTD values are calculated by considering all of the data for a chemical on a given effect and then calculating an 'RfD-like' value for that specific target organ. For the calculation of RfD for liver as a target organ for AgNPs the parameters are summarized in Table 4.

Table 3 Summary of parameters required for estimating exposure risk as per Fig. 1

Parameter	*Description*	*Value/information obtained for the case of Sorghum example calculation*
NP concentration in wastewater ($C_{wastewater}$) (mg/L)	Concentration of NPs frequently used in consumers products and have high chances that sooner they will start coming in wastewater	Literature data (Brar et al., 2010)
Transfer factor from wastewater to soil environment (factor) (mg/g)/(mg/L)	TF = Cp/Cs where Cp (mg/kg-dry) is the elemental concentration in plant and Cs (mg/kg-dry) is its concentration in soil (Uchide et al., 2009)	Transfer of silver NPs from wastewater to soil (i.e., adsorption capacity of soil for silver NP in wastewater matrix) (assumed to be 1, i.e., worst-case scenario)
NP concentration in soil (C_{soil}) (mg/g)	NP concentration available for plant uptake	Calculation using Equation 2
Bioaccumulation factor (BCF)	Ratio of metal concentration in the roots to that in soil (Lorestani et al., 2011)	From literature data BCF as 0.001 kg kg^{-1} of soil (for a concentration range of 0 to 2000 mg/L) (Lee et al., 2012)
NP concentration in edible portion (C_{plant}) (mg/g)	Vegetables and cereals are major components of human diet, being sources of essential nutrients. Concentration of NPs in edible portion is important parameter.	Calculation using Equation 3
Ingestion rate of edible plant	Rate of consumption of vegetable/cereals	Literature data (33.3 mg/day) (Rao et al., 2010)
Exposure duration	Includes acute and chronic type.	One year (USEPA, 1997) (literature data)
Body weight	Weight of person's body.	70 kg (USEPA, 1997) (literature data)
Daily dose of silver NPs ingested through plant (i.e., ADD)	Average daily dose of nanoparticles through ingestion of contaminated plants.	Calculation using Equation 1
Reference dose of NP for a given target organ	Daily oral exposure to the human population that is likely to be without an appreciable risk.	0.1 mg/kg/d for Ag NPs to liver (Anandan, 2011) (literature data)

Table 4 Assumed uncertainty factors and derived RfDs for oral exposure routes for humans for Ag NPs exposures

Parameters	*Definition*	*Value*	*Justification as per the* Anandan (2011) *study*
Data Reference			Kim et al. (2010)
Point-of-departure (POD) (mg/kg/day)	The lower confidence bound on the lowest experimental dose that showed an effect. (USEPA, 1997)	30	NOAEL value available
Uncertainity factors (UFs)	UFs are applied to the POD in determining the RfD. Value range from 1 to 3,000 and depends upon nature of data like whether data is from human or animal studies, duration of exposure, type of POD used etc. depending on the nature of the data used in the assessment (USEPA, 1997).		
UF1 (LOAEL to NOAEL)	A No-Observed-Adverse-Effect Level (NOAEL) is the highest exposure level at which no statistically or biologically significant increases are seen in the frequency or severity of adverse effect between the exposed population and its appropriate control population. (USEPA, 1997).	1	NOAEL value available
UF2 (Exposure duration)	Duration of exposure to the contaminant. This includes chronic and acute type.	10	Chronic toxicity available
UF3 (Interspecies)	Among different species there is difference in their dose, route of metabolism, type of enzymes involved in metabolism (Walton et al., 2001).	10	Assumed animal to human extrapolation
UF4 (Intra individual susceptibility)	This variation among individuals could arise from variation in response over time within subjects, variation of an intrinsic trait among individuals, or a combination of both (Bastain et al., 2003).	1	Only for human adults
UF5 (Data quality)	The assessment results determine the accuracy, completeness, consistency, precision, reliability, temporal reliability, uniqueness and validity of the data.	3	Assumed moderate data quality

The no-observed-adverse-effect-level (NOAEL) for Ag NPs for rats in vivo was obtained as 30 mg/kg bodyweight/day from the study of Kim et al. (2010). Appropriate UFs were applied (Table 4) to obtain an RfD_{ing_Ag} of 0.1 mg/kg/day for Ag NPs. Using TTD value, HQ value and maximum allowable concentration of NPs in wastewater (C_{ww}) were calculated.

Results and Discussion

Identified Gaps at Different Stages of Risk Assessment

Table 5 presents list of identified data gaps in different parameters. As most of the studies of NPs and plants are at their initial stages, only qualitative uptake data using TEM, ICP-MS and EDX analysis are there but quantitative data is missing in almost all the studies.

Table 5 Identified knowledge gaps in parameters

Parameter	*Identified data gaps*
Exposure assessment stage	
NP concentration in wastewater ($C_{wastewater}$) (mg/L)	Some monitored/modelled data available; but not for all NPs
Transfer factor of NPs from wastewater to soil environment (factor) (mg/g)/(mg/L)	Transfer factor values (using adsorption isotherm data); Adsorption capacity determination of different soils to different NPs
NP concentration in soil (C_{soil}) (mg/g)	Monitored data required; if available, approximate calculation not required
Bioaccumulation factor (BCF)	BCF values obtained using metal ions information as well as BCF values obtained using particle count information
Nanoparticles concentration (in terms of metal ions) in plant	Quantitative monitored data required for relevant risk assessment; so that dose can be represented in terms of metal ion content as well as in terms of particle count. Ion differs with NP size in toxicity (Dimkpa et al., 2012; Lee et al., 2008).
Nanoparticles concentration (in terms of particle count) in plant	Monitored data required; especially for the NPs frequently used in consumer products like silver NPs and titanium oxide NPs.
Bioaccumulation factor (BCF) in terms of metal ions	Used during modelling; calculated using NP conc. in plant and in soil
Bioaccumulation factor (BCF) in terms of particle count	Used during modelling; calculated using NP count in plant and in soil
Dose–response assessment	
Point-of-departure (POD) (mg/kg/day)	Not available for all NPs; more toxicity studies on rats are required
Values of Uncertainty factors (UFs) (UF1 to UF5; as per Table 2)	Uncertain due to lack of information and criteria useful in deciding value from the list: UF1 (10 for LOAEL; 1 for NOAEL); UF2 (10 when no chronic data available; 1 when chronic data available); UF3 (10 when no human data is available; 1 for human data); UF4 (1 for human adults); UF5 (1 for high data quality; 3 for moderate data quality; 10 for low data quality) (Schwab et al., 2005)

No study was found so far which can show accumulation data in edible portion of plant. Uptake mechanism is another important issue. Whether NPs are taken up by the plant in the form of ion or particles is still unanswered. For carrying out proper risk assessment what we need is studies on common food crops and high chances of accumulation. Uptake mechanism and toxicity of NPs (Lin and Xing, 2008), ecotoxicological effects of NPs in field and soil condition, risk associated with food chain due to consumption contaminated agricultural species need further studies (Stampoulis et al. 2009). Report highlights the need for more intense evaluation of nanoparticles fate and effects under environmentally relevant conditions (Musante and White, 2010) and to finally determine NP concentration in edible portion of plant.

Example Risk Estimation for Ingestion of Sorghum

Although values of parameters required for estimating hazard quotient value for given scenario was obtained (Table 3), many findings of this study indicated that risk estimation depends mainly on NPs concentration in wastewater. NPs concentration in edible portion of plants, and its uptake rate are either not present or very uncertain. During irrigation, data on fraction of NPs adsorbed on soil and uptake by plants are currently not available. Apart from this, reference dose is another parameter which requires further considerations as it depends on different uncertainty factors as mentioned in Table 4. However, using these literature data and assumptions (Table 4), maximum allowable concentrations of silver NP in *Sorghum* was calculated to be 212 mg silver NPs/kg dry weight of *Sorghum*. It is important to note that above estimate has uncertainties in NPs concentration in plant, reference dose, which need to be characterized properly before making any policy decisions. Out of all uncertain parameters, NPs concentration in edible portion of plant was found to be one of the most important parameters for estimating risk and need to be studied first.

Summary and Conclusions

This study identified knowledge gaps in assessing health risks due to exposures of NPs from contaminated edible plants. Findings of this study are summarized below:

1. Overall review indicated that risk estimation depends mainly on NPs concentration in wastewater, NPs concentration in edible portion of plants, and plant uptake rate. Out of these parameters, NPs concentration in edible portion of plant was found to be one of the most important parameters for estimating risk and need to be studied first.

2. Although qualitative information on NP uptake in edible portion of plant is available, related quantitative data in terms of metal ion concentration as well as in terms of nanoparticles count was found to be lacking. Thus, this aspect needs further detailed study.
3. During example risk estimation, risk estimate (HQ value) was not calculated for exposures of silver NPs from consuming *Sorghum* due to large uncertainty in values of parameters. However, a maximum allowable concentration of silver NPs in *Sorghum* was calculated to be 212 mg silver NPs/kg dry weight of *Sorghum* using toxicity information. This estimate could be used for understanding upper limit of NPs content in edible portion of plant without any risk. However, it is important to note that this estimate has uncertainties which need to be characterized properly before making any policy decisions.

Acknowledgements The authors would like to thank Department of Science and Technology (India) for supporting this study through the grant no: DST/TM/WTI/2 K11/301(G) and Council for Scientific and Industrial Research (CSIR) (India).

References

Anandan, A. (2011). Risk assessment of exposures to mixture of TiO_2 and Ag NPs for humans and Poeciliareticulata (Dissertation). Master of Technology Thesis, Department of Civil Engineering, Indian Institute of Technology Delhi, India.

Arora, S., Jain, J., Rajwade, J.M. and Paknikar, K.M. (2008). Cellular responses induced by silver NPs: In vitro studies. *Toxicology Letters*, **179:** 93-100.

Asharani, V.P., Mun, G.L., Hande, M.P. and Valiyaveettil, S. (2009). Cytotoxicity and Genotoxicity of Silver NPs in Human Cells. *ACS Nano*, **3(2):** 279-290.

Begum, P. et al. (2012). Phytotoxicity of multi-walled carbon nanotubes assessed by selected plant species in the seedling stage. *Applied Surface Science*, **262:** 120-124.

Brar, K., Verma, M., Tyagi, R.D. and Surampalli, R.Y. (2010). Engineered NPs in wastewater and wastewater sludge – Evidence and impacts. *Waste Management*, **30:** 504-520.

Bemrah, N., Sanaa, M., Cassin, M.H., Griffiths, M.W. and Cerf, O. (1998). Quantitative risk assessment of human listeriosis from consumption of soft cheese made from raw milk. *Preventive Veterinary Medicine*, **37:** 121-145.

Birbaum, K., Brogioli, R., Schellenberg, M., Martiona, E., Stark, W.J., Gunther, D. and Limbach, L.K. (2010). No evidence for cerium dioxide nanoparticles translocation in maize plants. *Environ Sci Technol*, **44:** 8718-8723.

Dimkpa, O.C., McLean, E.J., Latta, D.E., Manango, E., Britt, D.W., Johanson, W.P., Boyanov, M.I. and Anderson, A.J. (2012). CuO and ZnO NPs: phytotoxicity, metal speciation, and induction of oxidative stress in sand-grown wheat. *J Nanoparticles Res*, **14:** 1125-1140.

Farber, J.M., Ross, W.H. and Harwig, J. (1996). Health risk assessment of Listeria monocytogenes in Canada. *Intl J Food Microbiol*, **30(1-2):** 145-156.

Gerba, C.P. (2000). Risk assessment. *In:* Environmental Microbiology (Gerba, C.P., Maier, R.M. and Pepper, I.L., Eds.), Academic Press, London.

Khodakovskaya, H. et al. (2009). Carbon Nanotubes are Able to Penetrate Plant Seed Coat and Dramatically Affect Seed Germination and Plant Growth. *American Chemical Society* (ACS Nano), **3(10):** 3221-3227.

Kim, Y.S., Song, M.Y., Park, J.D., Song, K.S., Ryu, H.R., Chung, Y.H. et al. (2010). Subchronic oral toxicity of silver NPs. *Particle and Fibre Toxicol*, **7(20):** 2-11.

Kumar, A. (2012). Making a case for human health risk-based ranking of NPs in water for monitoring purposes. *Environ Sci Technol*, **46:** 5267-5268.

Kumar, P., Kumar, A. and Lead, J.R. (2012). NPs in Indian Environment: Known, unknowns and awareness. *Environ. Sci. Technol.*, **46:** 7071-7072.

Larue, C. et al. (2012). Quantitative evaluation of multi-walled carbon nanotube uptake in wheat and rapeseed. *Journal of Hazardous Materials*, **227-228:** 155-163.

Lorestani, B., Cheraghi, M. and Yousef, N. (2011). Phytoremediation Potential of Native Plants Growing on a Heavy Metals Contaminated Soil of Copper mine in Iran. World Academy of Science, Engineering and Technology, **53:** 377-382.

Lee, W.M., Kwak, J.I. and An, Y.J. (2012). Effect of silver NPs in crop plants Phaseolusradiatus and *Sorghum* bicolor: Media effect on phytotoxicity. *Chemosphere*, **86:** 491-499.

Lee, W.M., An, Y.J. and Kweon, H.S. (2008). Toxicity and bioavailability of copper NPs to the terrestrial plants mung bean (phaseolus radiatus) and wheat (Triticum aestivum): Plant agar test for water-insoluble NPs. *Environ Toxicol Chem*, **27(9):** 1915-1921.

Lin, D. and Xing, B. (2008). Root uptake and phytotoxicity of ZnO NPs. *Environ Sci Technol*, **42:** 5580-5585.

Lin, S., Reppert, J., Hu, Q., Hudson, J., Reid, M.L., Ratnikova, T.A., Rao, A.M., Luo, H. and Ke, P.C. (2009). Uptake, translocation, and transmission of carbon nanomaterial in rice plants. *Small*, **5(10):** 1128-1132.

Mazumdar, H. and Ahmed, G. (2009). Phytotoxicity effect of Silver NPs on Oryza sativa. *Intl J ChemTech Res*, **3(3):** 1494-1500.

Musante, C. and White, J.C. (2012). Toxicity of silver and copper to Cucurbita pepo: Differential effects of nano and bulk-size particles. *EnvironToxicol*, **27(9):** 510-517.

Nowack, B., Ranville, J.F., Diamond, S., Gallego, J.A., Metcalf, C., Rose, J., Horne, N., Koelmans, A.A. and Klaine, S.J. (2012). Potential scenarios for nanomaterial release and subsequent alteration in the environment. *Environ Toxicol Chem*, **31(1):** 50-59.

OECD (2012). Important Issues on Risk Assessment of Manufactured Nanomaterials. *In:* Series on the Safety of Manufactured Nanomaterials No. 33. Organization for Economic Co-operation and Development, Paris, France.

Rezvani, M. and Zaefarian, F. (2011). Bioaccumulation and translocation factors of cadmium and lead in Aeluropuslittoralis. *AJAE*, **2(4):** 114-119.

Schwab, R.W., Hayes, E.P., Fiori, J.M., Mastrocco, F.J., Roden, N.M., Cragin, D., Meyerhoff, R. D., D'Aco, V.J. and Anderson, P.D. (2005). Human pharmaceuticals in US surface waters: A human health risk assessment. *Regul. Toxicol. Pharmacol.*, **42:** 296-312.

Rao, P.P., Basavaraj, G., Ahmad, W. and Bhagavatula (2010). An analysis of availability and utilization of *Sorghum* grain in India. *SAT eJournal*, **8:** 1-8.

Stampoulis, D., Sinha, S.K. and White, J.C. (2009). Assay-Dependent Phytotoxicity of NPs to Plants. *Environ Sci Technol*, **43:** 9473-9479.

Sundaram, B. and Kumar, A. (2012). Engineered nanomaterials-based pollution in India. *Curr Sci*, **110(10)**.

Tan, X.M., Lin, C. and Fugetsu, B. (2009). Studies on toxicity of multi-walled carbon nanotubes on suspension rice cells. *Carbon*, **47:** 3479-3487.

USEPA (United States Environmental Protection Agency) (1997). Exposures Factors Handbook. EPA/600/P 95/002Fa.Cincinnati, Ohio: ORD/NCEA.

USEPA (2005). Nanotechnology White Paper, External Review Draft. United States Environmental Protection Agency, Nanotechnology Workgroup. http://www.epa.gov/osa/pdfs/ EPA_nanotechnology_white_paper_external_review_draft_12- 02-2005.pdf

Vandenhovea et al. (2009). Proposal for new best estimates of the soil-to-plant transfer factor of U, Th, Ra, Pb and Po. *Journal of Environmental Radioactivity*, **100:** 721-732.

Walton, K, Dorne, J.L. and Renwick, A.G. (2001). Uncertainty factors for chemical risk assessment: interspecies difference in the in vivo pharmacokinetics and metabolism of human CYP1A2 substrates. *Food and Chemical Toxicology*, **39:** 667-680.

Conservation Issues and Possible Solutions for Sustainability of Faunal Diversity of Arunachal Pradesh

Anil Kumar

Introduction

Arunachal Pradesh is located in north-east India in the transition zone between the Himalayan and Indo-Burmese region, and has 8.37 million ha geographical area (Sinha, 2008). It is situated between 26° 30′ and 29° 30′ N latitude and 91° 30′ and 97° 30′ E longitude (Fig. 1). The entire territory forms a complex hill system with varying elevations ranging from 50 m in the foot-hills and gradually ascending to about 7000 m. It is mostly hilly and mountainous covered with highly varied and dense vegetation (Kaul and Haridasan, 1987), which support fairly large populations of most taxonomic groups (Zoological Survey of India, 2006b). During last one decade several new species of vertebrates and invertebrates have been discovered from the state (Mishra and Datta, 2007; Sureshan, 2010).

The state of Arunachal Pradesh has a predominantly mountainous landscape with nearly 64% of the total area lying above 1200 m, 13% located between 300–1200 m and 8.87% below 300 m. The topography of Arunachal Pradesh forms a complex hill system of Siwalik and Himalayan origin and is crisscrossed by six major rivers and streams flowing from west to east. These divide the State into five river valleys from the river Kameng (formed by Bichom and Dirang tributaries), the Subansiri river known as Sinyik (Sigen, Sinyum and Kamla tributaries), the Siang river (Nyigong, Siring, Sigong, Sike, Siyom and Yamne tributaries), the Dibang locally known as Sikang or Talon (with Dri, Ahui, Emra and Endra as its tributaries); Lohit river (Noa Dihing, Dichu, Kamlang Tengapani, Melung, Ghalum, Tidding and Dree tributaries) and the Tirap river. These rivers are fed by snow from the Himalayas. Countless deep gorges and fast flowing streams and rivulets form the tributaries of the Siang River

A. Kumar (✉)
High Altitude Regional Centre, Zoological Survey of India, Solan 173 211, Himachal Pradesh, India
e-mail: anilsonta@gmail.com

N.J. Raju et al. (eds.), *Management of Water, Energy and Bio-resources in the Era of Climate Change: Emerging Issues and Challenges*,
DOI 10.1007/978-3-319-05969-3_20

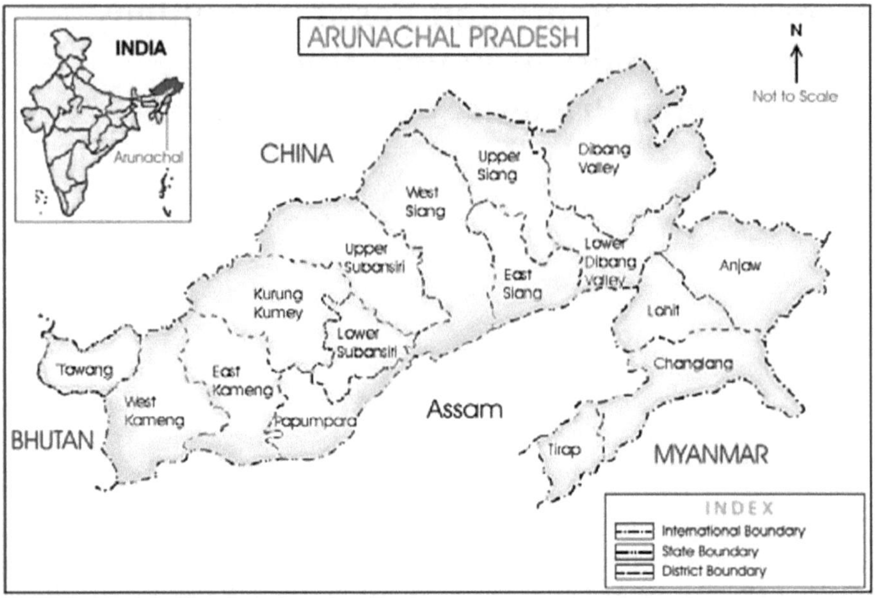

Fig. 1 Map of the Arunachal Pradesh is showing its geographical position and location of districts

(Kalita and Haridasan, 2001; Zoological Survey of India, 2006a). Based on the satellite imagery, a total of twenty one types of vegetation and land cover classes were identified by Roy and Behera (2005). This can be classified into five major categories of vegetation namely tropical forests, sub-tropical forests, temperate forests, sub alpine and alpine forests. Kaul and Haridasan (1987) have classified various forest types with respect to six altitudinal zones, namely: I. Tropical (up to 900 m), II. Sub-tropical (900-1800 m), III. Temperate broad-leaved (1800-2800 m), IV. Temperate coniferous (2800-3500 m), V. Sub-alpine (3500-4000 m) and VI. Alpine (4000–5500 m).

The state has 68,621 km^2 of forest occupying nearly 82% of the total geographic area (83,743 km^2). This includes 54,510 km of closed forest, 14,151 km of open forest, but the true picture of recorded forest cover, appears to be around 51,540 km, comprising 61.5% of the total area of the state (Kaul and Haridasan, 1987; FSI, 2000). Out of this, unclassified forest is 70.25% and the protected forest is merely 0.02% (FSI, 2000). An area of 9527.99 km^2, representing nearly 11.68% of the geographic area of the state, has been brought under Protected Area Network for conservation and development of biodiversity resource, which includes two national parks and eleven wildlife sanctuaries (Sinha, 2008).

Methodology

Present article is based on field surveys conducted by the author (during April 2005 to June 2010; Table 1) especially for the documentation of birds and mammals of Arunachal Pradesh, work undertaken and documented by ZSI scientists/researchers (during last two decades), and published literature by some other workers. Different localities inside the protected areas and outside the area were surveyed extensively throughout the Arunachal Pradesh (except some areas), for the documentation of fauna (Table 1). During the field work, observations on birds were made every day starting from early morning to late evening (6.00 am to 6.00 pm; with few exceptions), with the help of prismatic field binocular (10 × 50) and identification of species was carried out with the help of field guide of Birds of India written by Kazmierczak and Perlo (2000) and a pocket guide to the birds of the Indian Subcontinent by Grimmett et al. (2003). In case of inadequate sightings/doubtful identification, the data was excluded from the present study. Identification of

Table 1 Details of the different areas and localities of Arunachal Pradesh covered in field surveys for the faunal studies

S. No.	*Areas surveyed for faunal studies*	*Dates from- to-*	*Locations covered during the surveys*
1	Pakke Tiger Reserve and adjacent localities	09-11-05 to 21-11-05	Surrounding area of Seijusa and Upper Seijusa, Khari, Bhalukpong and Tipi
2	D'Ering Memorial Wildlife Sanctuary	3-10-2006 to 23-10-2006	Anchalghat, Namsing and Borguli ranges and Mebo area (95°22' E - 95°29' E and 27°51' N - 28°05' N)
3	Sessa Orchid Sanctuary and Eaglenest Wildlife Sanctuary	7-3-2007 to 23-3-2007	Bhalukpong, Tipi, Sessa and Lama Camp, Eaglenest pass, Sundarview and Ramalingam area in Eaglenest Wildlife Sanctuary
4	Tawang Chu Valley and Dirang & surrounding area	23-10-2007 to 02-11-2007	PTSO, Tawang town, Surbhi village, Kitpi village, Jung, and Dirang area.
5	Mouling National Park and adjacent areas	02-06-2008 to 15-06-2008	Ramsing locality (28° 39' 22.8" N and 94° 58' 35.0" E)
6	D'Ering Memorial Wildlife Sanctuary and adjacent areas	11-02-2009 to 18-02-2009	Area between 95°22' E - 95°29' E and 27°51' N - 28°05' N, and area covered namely Anchalghat, Namsing and Borguli ranges.
7	Tawang Chu Valley and adjacent areas	09-05-2009 to 26-05-2009	PTSO, Y-junction lakes, surrounding of Tawang town, Surbhi village, Kitpi village, Jung, Mirba, Mukto, LGG and MJG area
8	Tawang Chu Valley	10-05-2010 to 28-05-2010	-do-
9	Itanagar Wildlife Sanctuary and adjacent areas	During 2006 to 2009 in local surveys	Hati Nullah, Poma village, Basar Nullah, Ganga lake (27°05'39" N, 93°30'15" E)

mammals was carried out with the help of a field guide to Indian mammals (Menon, 2003). Indirect evidences such as pug-marks, dung/pallets, carcass and skins were also considered to infer the occurrence of a species. With the help of local field guide, some villagers were also interviewed. The presence/absence of species were also discussed with forest field staff of the sanctuary and relevant information was incorporated in data. Apart from birds and mammals, specimens of other faunal groups such as fishes, frogs, snakes, reptiles, insects and other invertebrates were collected by supporting staff and deposited to Arunachal Pradesh Regional Centre, ZSI, Itanagar, for further studies.

Results and Discussion

Fauna of Arunachal Pradesh comprised both vertebrate and invertebrate elements. The state is also known for occurrence of endemic and/or rare species of various groups (Figs. 2a and 2b). However, studies are preliminary and limited to a selected number of groups most probably due to remoteness, harsh climate and mostly inaccessible terrain.

Vertebrate Fauna

A comprehensive account of fish fauna of Arunachal Pradesh was produced in 2006. It composed of 143 species, under 61 genera, 21 families and 8 orders

a

Fig. 2a A pair of endangered Hoolock gibbon *Hoolock hoolock*. It is one of the most affected primate species facing conservation threats due to habitat loss, fragmentation and poaching

b

Fig. 2b Two species of gyps vultures (i.e. Himalayan vulture *Gyps himalayensis* and White-rumped vulture *G. bengalensis*) recorded in Arunachal Pradesh. White-rumped vultures are critically endangered and rarely seen in the area

(Zoological Survey of India, 2006a). It includes 50 new records from the state. The family Cyprinidae forms the largest group with 65 species followed by Homolopteridae (17 species), Sissoridae (12 species), Bagridae (seven species), Channidae and Cobitidae (six species each) and the rest with one, two or three species. Recently, an attempt has also been made by ZSI, Itanagar to update the checklist of ichthyofauna of Arunachal Pradesh. Bagra et al. (2009) reported a checklist of 213 species of fishes including 27 species which are yet to be confirmed at the species level. Based on previous studies and present study, 17 species have been added to the checklist, now taking it to 230 species (including few taxa yet to be confirmed at species level) of fishes in total (unpublished project report) (Table 2).

Bordoloi and Borah (2001) reported a checklist of 42 species of amphibians including some unconfirmed species. Recently ZSI, Itanagar encompassed a list of 54 species of amphibians reported from Arunachal Pradesh (unpublished project report). The present documentation of amphibians is based on amphibians collected from different districts of the state and has also taken into account those reported by various workers (Sarkar and Ray 2006; Mathew and Sen, 2010). The comprehensive list contains a checklist of 54 species of amphibian belonging to 22 genera.

Studies conducted by ZSI indicate the distribution of 78 species under 46 genera and 12 families of reptiles; of this two species *viz., Ophites laoensis* and *Xenochrophis punctulata* are the first records to the country. However, Borang and Bhatt (2001) reported a checklist of 113 species belonging to 15 families and 57 genera. Borang et al. (2005) prepared a checklist of the snakes of Arunachal Pradesh. It is mainly based on the zoological expeditions conducted by Dr. Asham Borang and the material examined by Dr. Bharat B. Bhatt, in various museums in

Table 2 Summary of some faunal groups reported so far from Arunachal Pradesh

S. No.	*Taxonomic groups*	*Major faunal elements*	*Number of species*	*References*
Vertebrate Fauna				
1	Pisces	Different species of fishes	230*	Zoological Survey of India (2006a); Bagra et al. (2009)
2	Amphibians	Frogs, toads and Salamander	54	Bordoloi and Borah (2001); Sarkar and Ray (2006); Mathew and Sen (2010)
3	Reptiles	Snakes and lizards	113	Zoological Survey of India (2006a); Borang and Bhatt (2001); Borang et al. (2005)
4	Aves	Different species of birds	>700	Ali and Ripley (1948 and 1987); Singh (1994); Choudhury (2004)
5	Mammals	Different species of mammals	214	Borang (2001); Choudhury (2003)
Invertebrate Fauna				
1	Protozoans	Free-living, parasitic and symbiotic protozoans	72	Das et al. (2006)
2	Crustaceans	Prawns and crabs	12	Kurian and Sebastian (1993)
3	Annelids	Leeches and earthworms	39	Julka (1976 and 1981); Chandra (1991)
4	Trematodes	Endo-parasitic trematodes of amphibians	8	Zoological Survey of India (2006b)
S. No.	*Taxonomic groups*	*Major faunal elements*	*Number of species*	*References*
5	Collembolans	Springtails	27	Zoological Survey of India (2006b)
6	Odonates	Dragonflies and damselflies	92 species/ sub-species	Zoological Survey of India (2006b)
7	Orthopterans	Grasshoppers	56	Singh et al. (2010); Srinivasan and Prabakar (2012).
8	Dictyopterans	Cockroaches	20	Zoological Survey of India (2006b)
9	Thysanopterans	Thrips	15	Singh et al. (2010)
10	Coleopterans	Beetles	106	Singh et al. (2010); Chandra and Gupta (2012)
11	Dipterans	Mosquitoes, black-flies, midges, gnats, horse-flies, robber-flies and hover-flies	271	Zoological Survey of India (2006b)
12	Lepidopterans	Moths and butterflies	397	Chandra and Sambath (2013); Gogoi (2012)

(continued)

Table 2 (continued)

S. No.	*Taxonomic groups*	*Major faunal elements*	*Number of species*	*References*
13	Hymenopterans	Ants, sphecid wasps, potter wasps, yellow wasps and bees	124	Zoological Survey of India (2006b); Singh et al. (2010); Srinivasan and Kumar (2010); Sureshan (2010)
14	Arachnids	Spiders, mites and ticks	114	Zoological Survey of India (2006b)

*Few Taxa yet to be confirmed at species level.

the state. This checklist includes 67 species of snakes belonging to five families and 31 genera. Studies by ZSI reported the restricted distribution of snake (*Boiga gokool*) to the eastern Himalayas and wider distribution of *Trimeresurus monticola* in Southeast Asian countries extending up to Malaysia. The systematic list includes four species of testudines, 19 species of lizards and 55 species of snakes.

The ornithological work done by Ali and Ripley (1948 and 1987) is the key stone in our latest understanding on the diversity and distribution of birds, besides first hand information gathered by some other workers (Godwin-Austen, 1876, 1877; Betts, 1956). Singh (1994) reported a checklist of 519 species of birds of Arunachal Pradesh based on extensive field surveys from 1988 to 1994, carried out by him throughout the state, along with 138 additional species recorded by other workers. Choudhury (2006) listed more than 700 species in his handbook on the birds of Arunachal Pradesh. Many of them are known for their beautiful feathers, songs and flights. As per the records of ZSI, 964 species and subspecies of birds are likely to occur in the state of Arunachal Pradesh (Zoological Survey of India, 2006a), which include 473 species authentically collected and reliably sighted in the field by the scientists of the survey, out of this 416 are resident, 42 are winter visitors, and 11 are summer visitors.

Choudhury (2003) published a book on the mammals of Arunachal Pradesh, which enclosed a checklist of 206 species based upon both primary and secondary sources, along with another 38 species likely to occur or are recorded in adjacent areas. Arunachal macaque *Macaca munzala* was recently discovered species of primate (Mishra et al., 2006). Borang (2001) reported 214 mammalian species from Arunachal Pradesh, belonging to 12 orders, 34 families and 116 genera. However, as per the ZSI records, the mammalian fauna of the state is represented by 105 species and subspecies under 85 genera, 25 families and nine orders.

Invertebrate Fauna

Invertebrate fauna of Arunachal Pradesh mainly comprised different orders of insects, protozoans, crustaceans, ticks, mites and spiders (Table 2). During 1990-92 and in 1999 the scientists of ZSI conducted three faunistic surveys in the districts of Lower Subansiri, West Siang, East Siang, Debang Valley, Lohit and Changlang, and collected diverse groups of free-living and parasitic protozoans. These collections as well as those already known and reported from Arunachal Pradesh, comprised 72 species, out of which 43 species were free-living, 20 species parasitic and nine species symbiotic. Three species of free-living rhizopods namely *Centropyxis sylvatica, Plagiopyxis minuta* and *Nebela tincta* were reported for the first time from India. Sixty two species (40 free-living, 17 parasitic and five symbiotic) constitute first records for Arunachal Pradesh (Das et al., 2006).

Information on crustacean fauna, specially prawns and crabs of Arunachal Pradesh, is still scanty (Kurian and Sebastian 1993). Seven species of Palaemonid prawns have been reported by ZSI. Our knowledge on the taxonomy of annelid fauna of Arunachal Pradesh is confined to the contributions of Moquin-Tandon (1826), Moore (1924), and Chandra (1991) on leeches. The collection comprised seven species of leeches. Total 32 species of earthworms (including 11 new species) have been reported from Arunachal Pradesh (Julka, 1976, 1981). Among endoparasitic organisms, eight species of trematodes of amphibians were identified, belonging to four families, six sub-families, and over six genera. All the species recorded herein constitute the first records of their occurrence in Arunachal Pradesh.

Among insects, the studies of collembolan fauna (springtails) of Arunachal are based on collection made during 1999 from different localities by Zoological Survey of India. So far, 27 species under 15 genera of three families have been reported from the State of Arunachal Pradesh. The order Odonata (dragonflies and damselflies) is an economically important primitive group of insects. Lahiri (1977) first recorded dragonflies from the present Arunachal and it is being considered as the starting point. Later on Lahiri (1985), Lahiri and Sinha (1991), Bhargava (1989), Prasad and Varshney (1995) and Ram and Prasad (1999) reported 92 species and subspecies over 50 genera, 11 families and two suborders, from the state.

Among grasshopers (Order Orthoptera), so far, 56 species of grasshoppers (including previously reported 44 species over two families and 33 genera) have been reported from the state (Singh et al., 2010; Srinivasan and Prabakar 2012). Out of previously reported 44 species, 20 species are distributed over 14 genera and six families. Six species are recorded here for the first time from Arunachal Pradesh. Out of 10 species of family Tettigoniidae, nine species are new records for Arunachal Pradesh (Srinivasan and Prabakar, 2012). So far, 20 species of cockroaches (Order Dictyoptera), distributed over six families and 14 genera, have been reported from the state. Fifteen species of thysanopterans were reported from Kane Wildlife Sanctuary (Singh et al., 2010).

Among beetles, family Scarabaeidae is the largest of the insects order Coleoptera. Biswas and Chatterjee (1985) reported fifty five species of dung beetles from "Namdapha Wildlife Sanctuary" and Biswas (1978) also described four new species from Arunachal Pradesh and Assam. So far 73 species belonging to 14 genera of subfamily Coprinae under the family Scarabaeidae have been identified. Out of 73, twenty two species have been recorded for the first time from Arunachal Pradesh. On the basis of recent studies, the total number of coleopterans has reached up to 106 species (Singh et al., 2010; Chandra and Gupta, 2012). The Diptera are commonly known as (true) flies and include many familiar insects such as mosquitoes, black flies, midges, horse flies, fruit flies, blowflies and houseflies. On the basis of morphological characteristics, order Diptera is divided into three sub-orders namely Nematocera, Brachycera, and Cyclorrhapha. So far, 61 species of Nematocera (mosquitoes, black flies, midges, gnats and related forms), 51 species of Brachycera (horseflies, robberflies and dance flies) and 68 species of Cyclorrhapha (hover flies or flower flies) have been reported by ZSI (Zoological Survey of India, 2006b). Among Order Lepidoptera, 294 species of butterflies belonging to 156 genera have been reported from Dibang Valley of the state (Gogoi, 2012). During September-October 2009, field surveys were conducted by ZSI in Tawang district. Over 250 morpho-species of moths were collected, of which 103 species pertaining to 15 families were identified (Chandra and Sambath, 2013).

Among Hymenoptera, 63 species of ants under 30 genera and seven subfamilies have been reported including the earlier record of 12 species of ants from this state by Tiwari and Maiti (1976) and Tiwari et al. (1977). The members of the family Sphecidae are commonly known as blue mud-dauber, black and yellow mud-dauber, digger or sand or hunting wasps. As per reports, the Sphecidae fauna of the state Arunachal Pradesh composed of 33 species/subspecies belonging to 20 genera under eight subfamilies. Of these, 23 species/subspecies are recorded here for the first time from this state. As per recent reports, the fauna of Vespidae and Apidae of Arunachal Pradesh is composed of 15 species/subspecies belonging to five genera, three subgenera and two subfamilies under the family Vespidae and six species belonging to three genera under the family Apidae. Out of these, three species/subspecies of Vespidae are newly recorded from the state. On the basis of recent publications 13 more species have been added to the hymenoptera fauna of the Arunachal Pradesh (Singh et al., 2010; Srinivasan and Kumar, 2010; Sureshan, 2010).

The studies of the Arachnid fauna of North Eastern parts of India are still fragmentary and no consolidated account has so far been attempted. Bastawade (1985) has reported three species of Scorpions belonging to two families Chaerilidae and Scorpiopsidae from Namdapha Tiger Reserve, Changlang district, Arunachal Pradesh. Bastawade and Pal (1992) have reported the order Schizomida and suborder Cyphophthalmi for the first time from Arunachal Pradesh (Zoological Survey of India, 2006b). Oribatid of crypostigmatid mites form a complex group under the subclass Acarina and are numerically the most abundant soil fauna among the microarthropods inhabiting soil. As per the records of ZSI, a total 35 species belonging to 30 genera and 20 families of oribatid mite have been recorded from different districts of Arunachal Pradesh. Ixodid tick fauna of Arunachal Pradesh was first studied by Nuttall and Warburton (1915). As per recent records, altogether 21 species under eight genera hitherto known from the state have been recorded. Of these, two species viz. *Ixodes kashmiricus* and *I. vespertiliones* have been reported for the first time from Arunachal Pradeh. The spider fauna of Arunachal Pradesh has not yet been studied in a comprehensive manner (Tikader, 1980, 1982). About 350 examples of spider have been studied and identified by ZSI scientists. A total

58 species belonging to 28 genera under 11 families are reported, of those nine families 26 genera 54 species form new records for the state. Out of these, four species four genera under three families are new to science (Zoological Survey of India, 2006b).

Conservation Issues

Arunachal Pradesh is the treasury of mammoth biodiversity and support a sizable number of floral and faunal elements. In spite of such a rich faunal wealth, the state is facing conservation threats as in most other parts of the country. Hunting practices of locals and unplanned developmental activities are the major threats to the key faunal elements. Poaching of large mammals for meat, skin (Fig. 3) and traditional medicines is quite serious conservation issue throughout the state. Most mammal species have been reported to be actively hunted (Fig. 4). Primates, ungulates are mainly killed for meat and carnivores for skin (Selvan et al., 2013a and b). The species such as Black bear *Ursus thibetanus* and Musk deer *Moschus* sp. are hunted for their gall bladders and musk pods respectively (Mishra et al., 2006). Hunting of birds for meat and feathers is also common in most areas of the state. Different communities of local tribes still like it as a daily routine of their life. Snaring of large birds such as galliformes is alarming. However, recently, owing to increasing education level and wildlife awareness, interest in hunting activities is decreasing in some people. Forest degradation (Fig. 5) and developmental activities

Fig. 3 Locals often use skins of large mammals to save themselves from cold. In the picture three local persons are wearing skin of Sambar *Cervus unicolor*

Fig. 4 Birds and mammals hunted for food. It is common practice throughout the state. Present picture was taken in D'Ering Memorial Wildlife Sanctuary. Two species of birds (i.e. Rufous-necked laughing thrush *Dryonastes ruficollis* and one unidentified species; upper right and upper left side individual, respectively) and one mammal species (five individuals) namely Himalayan striped squirrel *Tamiops macclellandi* are seen in the picture

Fig. 5 Cleaning of forest for agriculture and settlements in the adjacent area of D'Ering Memorial Wildlife Sanctuary

have direct and/or indirect impact on biodiversity sustainability. Often, govt. agencies construct the roads for connectivity and development without preventive measures. It is seen that in many cases roads were used for exploitation of natural resources (especially wood) and clearing of land for agriculture and unorganized settlements (Figs. 6a and 6b).

Fig. 6a Landscape view of an area in Tawang district is showing the degradation/loss of vegetation cover due to developmental activities and agriculture expansion

Fig. 6b Extraction of wood is common in many areas of Arunachal Pradesh. Photo was taken near Tom hill area near Dirang, West Kameng district

Shifting cultivation and felling of trees are other major issues. Shifting cultivation is a widely used, tribe-specific agriculture system in Arunachal Pradesh. It is also known as slash and burn agriculture (locally called *jhum*). It is based on the clearing of vegetation, its slashing and burning. During the clearing, usually big logs and wood used for construction work and minor branches and debris are burned to make the land ready for agriculture. After 2–3 years of cropping consequently, land loses its fertility and thus land left for regeneration for about 12–15 years of fallow period (Dash and Pradhan, 1995; Tripathi and Barik, 2003; Murtem et al., 2008). The shifting cultivation is a major occupation for most of the people of Arunachal Pradesh (Murtem et al., 2008). In remote rural villages of the state, it is only means of livelihood, and very closely associated with the tribal culture such as social gathering, traditions and festivals (namely Booriboot, Nyokum, Mopin, Solung etc.). Few decades back, it was believed that it is a time-tested, traditional knowledge based sustainable agriculture system. However, today on the basis of scientific reasons it has been well established that shifting cultivation is environment destructive and faulty land use practice, having a low output as compared to costs. During last few decades, due to increasing population pressure and decreasing *jhum* cycle (period between two successive slashes) from 12–15 years to 3–5 years, it became unsuitable (Tripathi and Barik, 2003). In many cases it has been converted into regular agriculture practice. Wood extraction was very fast in the state till the ban of Supreme Court in 1996. After this large scale commercial exploitation of timber is almost stopped, while illegal and unauthorized felling at minor scale is still seen occasionally.

The topography of Arunachal Pradesh forms a complex hill system and is crisscrossed by six major rivers and streams flowing from west to east. These divide the state into five river valleys from the rivers namely Kameng, Subansiri, Siang, Dibang, Lohit and Tirap. These rivers are fed by snow from the Himalayas (Kalita and Haridasan, 2001). As per estimations, the hydropower potential of the state is about 60,000 MW, and it has been identified as a key area for fast track development by both the government of India and the state government, being regarded as one of the major source of power generation to meet out the country's increasing energy requirements. It is estimated that Arunachal Pradesh has a potential of about 40% of the country's total hydropower generation potential. Recently, the state government has signed Memorandum of Understanding (MoU) with 25 developers for the generation of over 27,000 MW of hydropower potential in the state. At least 88 projects with a capacity of 31,587.5 MW are in the private sector. Currently, however, only one project 405 MW Ranganadi (stage I) is in operation in Arunachal Pradesh. The four big projects currently under construction include 2000 MW Subansiri Lower, 600 MW Kameng, 110 MW Pare and 1750 MW Demwa Lower hydroelectric projects. The economic development in the country in recent years has resulted in widening of the gap between the demand and the supply of power. In order to make power available, the development of hydroelectric power generation is being treated as major solution to bridge the gap between power demand and supply. However, for biodiversity conservation point of view, it is speculated that in future it may create another major threat, owing to possible ignorance of conservation measures. It is speculated that due to construction of major dams, large forest areas will be submerged and infrastructural development will further destroy/ disturb the forest habitats. Engaged manpower (mainly labourers) will directly/ indirectly impose anthropogenic pressure in terms of exploitation of natural resources for their daily need.

Possible Solutions

In most areas of the country, population of most key faunal groups is declining rapidly. Arunachal Pradesh is not untouched from this problem. Therefore, there is a need to conserve and sustain the populations of various faunal groups of this biodiversity hotspot of global importance. As suggested by Choudhury (2010), adequate enforcement of laws, stopping of poaching, review of mega dams, restrictions on arms licenses, building awareness and motivation, and alternatives to shifting cultivation should be implemented. It may possibly be useful to conserve and sustain the faunal diversity of the state.

Government agencies and non-governmental organisations should undertake regular programmes to create awareness for biodiversity conservation. It is now well established that only awareness and education can effectively help to stop the hunting and poaching of animals. With the changing time, the people of Arunachal should adopt the latest agriculture techniques/patterns instead of traditional shifting cultivation. As suggested by Bhuyan et al. (2010) owing to ideal agro climatic conditions of the state, there is a vast scope for cultivation of the non-timber forest products for economic development of the rural people. They can be grown as cash crops in various agro-forestry models such as the agri-silvicultural system, agri-horti-silvicultural system and silvi-pastoral system within the state.

Due to the power requirements of the country, the development of mega hydroelectric power projects can't be avoided now. However, establishment should be commenced with adequate conservation measures such as: 1. The plantation of new trees should be ensured in degraded forest areas to compensate the submerged forests due to dam construction, 2. Priority in benefits should be granted to local displaced people, in terms of employment and sustainable rehabilitation, so that, dependency of residents on forest resources could be minimized, and 3. Developmental activities should be well organised. Construction of houses should be in the form of multistoried flats, so that space for occupation could be reduced.

Acknowledgements The author is grateful to Dr. K. Venkataraman, Director, ZSI for his kind encouragement and support. I would like to thank PCCF & Add PCCF (Wildlife) and DFOs of various protected areas of Arunachal Pradesh, surveyed by ZSI scientists, for official permissions and kind cooperation at various levels. Thanks are due to Officer-in-Charge and staff of Arunachal Pradesh Regional Centre, ZSI, Itanagar for departmental support. Present article is based on mainly faunistic surveys conducted by different ZSI scientists/researchers, some other workers and published literature. Species names, taxonomic status, terminology and nomenclature are taken as reported in original reports/literature.

References

Ali, S. and Ripley, S.D. (1987). Compact handbook of the birds of India and Pakistan. Oxford University Press, New Delhi.

Bagra, K., Kadu, K., Nebeshwar-Sharma, K., Laskar, B.A., Sarkar, U.K. and Das, D.N. (2009). Ichthyological survey and review of the check list of fish fauna of Arunachal Pradesh, India. *Check List*, **5(2):** 330–350.

Bastawade, D.B. (1985). Scorpion fauna of Namdapha Bio-sphere Reserve, Arunachal Pradesh. *Rec. Zool. Surv. India*, **82(1–4):** 259–262.

Bastawade, D.B. and Pal, T.K. (1992). First record of the Arachnid order Schizomida from Arunachal Pradesh. *J. Bombay Nat. Hist. Soc.*, **89(1):** 137.

Betts, F.N. (1956). Notes on birds of the Subansiri area, Assam. *J. Bombay nat. Hist. Soc.*, **53:** 397–414.

Bhargava, R.N. (1989). On a collection of Odonata from West Kameng district, Arunachal Pradesh. *Indian Odonatol.*, **2:** 45–46.

Bhuyan, L.R., Rao, A.N. and Chaudhry, P. (2010). Non-timber forest products of Arunachal Pradesh—Scope, prospects and problems. *Bulletin of Arunachal Forest Research*, **27(1&2):** 47–53.

Biswas, S. (1978). Studies on the Scrab beetles (Coleoptera: Scarabaeidae) of North-East India. Part II: Three new species and two new records from India. *J. Bombay Nat. Hist. Soc.*, **76:** 339–344.

Biswas, S. and Chatterjee, S.K. (1985). Insecta: Coleoptera: Scarabaeidae: Coprinae. *Rec. Zool. Surv. India*, **82(1–4):** 147–177.

Borang, A. (2001). Mammalian fauna of Arunachal Pradesh (checklist & distribution in protected areas). *Arunachal Forest News*, **19(1&2):** 43–82.

Borang, A. and Bhatt, B.B. (2001). Checklist of the reptiles of Arunachal Pradesh (northeast India). *Arunachal Forest News*, **19(1&2):** 132–148.

Borang, A., Bhatt, B.B., Chaudhury, S.B., Borkotoki, A. and Bhutia, P.T. (2005). Checklist of the snakes of Arunachal Pradesh, Northeast India. *J. Bombay Nat. Hist. Soc.*, **102(1):** 19–26.

Bordoloi, S. and Borah, M. (2001). A study on amphibian fauna of Arunachal Pradesh, India. *Arunachal Forest News*, **19(1&2):** 114–116.

Chandra, K. and Sambath, S. (2013). Moth diversity of Tawang District Arunachal Pradesh, India. *Journal of Threatened Taxa*, **5(1):** 3565–3570.

Chandra, K. and Gupta, D. (2012). Pleurostict scarabs (Coleoptera: Scarabaeidae): New distributional records in Arunachal Pradesh, north-east India. *Check List*, **8(5):** 889–893.

Chandra, M. (1991). The Leeches of India – A Hand Book, Zoological Survey of India, Kolkata:

Choudhury, A.U. (2003). The Mammals of Arunachal Pradesh. Regency Publications, New Delhi.

Choudhury, A.U. (2006). A pocket guide to the Birds of Arunachal Pradesh. Gibbon Books and the Rhino Foundation for Nature in NE India, Guwahati.

Choudhury, A.U. (2010). Mammals and birds in Dihang-dibang biosphere reserve, Arunachal Pradesh. *Tigerpaper*, **37(4):** 1–4.

Das, A.K., Nandi, N.C. and Chattopadhyay, P. (2006). Protozoa. *In*: Fauna of Arunachal Pradesh, State Fauna Series 13 (Part-2).

Dash, A.P. and Pradhan, S.K. (1995). Jhumming in Arunachal Pradesh. *In*: Majumdar, D.N. (ed) Further perspective in Shifting Cultivation in North East India, Omson Publication, Guwahati.

Forest Survey of India (2000). The State Forest Report 1999. FSI, Govt of India Press, Dehradun.

Godwin-Austen, H.H. (1876). Fifth list of the birds from the hill ranges of the Northeast Frontier of India. *J. Asiatic Soc. Bengal*, **45(2):** 191–204.

Godwin-Austen, H.H. (1877). Description of three new species of the genera Pellorneum, Actinura and Pomatorhinus lately collected in the neighbourhood of Sadiya, Assam, by Mr MMJ Ogle of the Topographical Survey. *J. Asiatic Soc. Bengal*, **46(2):** 41–45.

Gogoi, M.J. (2012). Butterflies (Lepidoptera) of Dibang Valley, Mishmi Hills, Arunachal Pradesh, India. *Journal of Threatened Taxa*, **4(12):** 3137–3160.

Grimmett, R., Inskipp, C. and Inskipp, T. (2003). Pocket guide to the birds of the Indian subcontinent. Oxford University Press, New Delhi.

Julka, J.M. (1976). Studies on the earthworms collected during the Daphabum expedition in Arunachal Pradesh, India. *Rec. Zool. Surv. India*, **69:** 229–239.

Julka, J.M. (1981). Taxonomic studies on the earthworms collected during the Subansiri expedition in Arunachal Pradesh India. *Records of the Zoological Survey of India*, Occasional Paper No. 26, 1–37.

Kalita, S.N. and Haridasan, K. (2001). Forest and wildlife management in Arunachal Pradesh. *Arunachal Forest News*, **19(1&2):** 26–31.

Kaul, R.N. and Haridasan, K. (1987). Forest types of Arunachal Pradesh—A preliminary study. *J. Econ. Tax. Bot.*, **9(2):** 389–397.

Kazmierczak, K. and Perlo, B. van (2000). A field guide to the birds of India. Om Book Service, Daryaganj, New Delhi.

Kurian, C.V. and Sebastian, V.O. (1993). Prawns and Prawn Fisheries of India. Hindustan Publishing Corporation, Daryaganj, New Delhi.

Lahiri, A.R. (1985). Insecta: Odonata. *In*: Fauna of Namdhapa, Arunachal Pradesh. A proposed Biosphere Reserve. *Rec. zool. Surv. India*, **82(1–4):** 81–97.

Lahiri, A.R. and Sinha, C. (1991). A review of Indian Chlorocyphidae (Insecta: Odonata) with additional notes on taxonomy and distribution for some species and description of a new subspecies. *Rec. Zool. Surv. India*, **89:** 257–268.

Mathew, R. and Sen, N. (2010). Pictorial Guide to the Amphibians of North East India, Zoological Survey of India, Kolkata.

Mishra, C. and Datta, A. (2007). A new bird species from Eastern Himalayan Arunachal Pradesh—India's biological frontier. *Curr. Sci.*, **92(9):** 1205–1206.

Mishra, C., Madhusudan, M.D. and Datta, A. (2006). Mammals of the high altitudes of western Arunachal Pradesh, eastern Himalaya: As assessment of threats and conservation needs. *Oryx*, **40(1):** 1–7.

Menon, V. (2003). A field guide to Indian mammals. Dorling Kindersley (India) Pvt Ltd, Delhi.

Moore, J.P. (1924). Notes on some Asiatic Leeches (Hirudinea), principally from China, Kashmir and British India. *Proc. Acad. Nat. Sci. Phillad.*, **81:** 343–388.

Moquin-Tandon (1826). Monographic de lafamille des Hirudinees, Thesis, Montprllier.

Murtem, G., Sinha, G.N. and Dopum, J. (2008). Jhumia view on shifting cultivation in Arunachal Pradesh. *Bulletin of Arunachal Forest Research*, **24(1&2):** 35–40.

Nuttall, G.H.F. and Warburton, C. (1915). The genus *Haemaphysalis*. *In:* Nuttall et al. (eds) Ticks. A Monograph of the Ixodoidea Part 3.

Prasad, M. and Varshney, R.K. (1995). A checklist of the Odonata of India including data on larval studies. *Oriental Ins.*, **29:** 385–428.

Ram, R. and Prasad, M. (1999). On the collection of Odonata from Arunachal Pradesh, India. *Rec. Zool. Surv. India*, **97:** 113–132.

Roy, P.S. and Behera, M.D. (2005). Assessment of biological richness in different altitudinal zones in the Eastern Himalayas, Arunachal Pradesh, India. *Curr. Sci.*, **88(2):** 321–330.

Sarkar, A.K. and Ray, S. (2006). Amphibia. *In*: Fauna of Arunachal Pradesh. State Fauna Series 13 (Part I).

Selvan, K.M., Gopi, G.V., Habib, B. and Lyngdoh, S. (2013a). Hunting record of endangered Marbled Cat, *Pardofelis marmorata* in the Ziro valley of Lower Subansiri, Arunachal Pradesh, India. *Journal of Threatened Taxa*, **5(1):** 3583–3584.

Selvan, K.M., Gopi, G.V., Habib, B. and Lyngdoh, S. (2013b). Losing threatened and rare wildlife to hunting in Zero valley, Arunachal Pradesh, India. *Current Science*, **104(11):** 1492–1495.

Singh, P. (1994). Recent bird records from Arunachal Pradesh. *Forktail*, **10:** 65–104.

Singh, O.T., Chakravorty. J. and Varatharajan, R. (2010). Entomofauna of Kane Wildlife Sanctuary, Arunachal Pradesh, Northeastern India. *Journal of Threatened Taxa*, **2(13):** 1392–1400.

Sinha, G.N. (2008). Forest and forestry in Arunachal Pradesh. SFRI Information Bulletin No. 27, pp. 1–67.

Srinivasan, G. and Kumar, P.G. (2010). New records of potter wasps (Hymenoptera: Vespidae: Eumeninae) from Arunachal Pradesh, India: Five genera and ten species. *Journal of Threatened Taxa*, **2(12):** 1313–1322.

Srinivasan, G. and Prabakar, D. (2012). Additional records of Tettigoniidae from Arunachal Pradesh, India. *Journal of Threatened Taxa*, **4(14):** 3255–3268.
Sureshan, P.M. (2010). Description of a new species of *Netomocera* Bouček (Hymenoptera: Chalcidoidea: Pteromalidae) from Arunachal Pradesh, India, with a key to world species. *Journal of Threatened Taxa*, **2(12):** 1309–1312.
Tewari, R.N. and Maity, P.K. (1976). Some New Records of Ants from Arunachal Pradesh (Hymenoptera: Formicidae). *Newsl. Zool. Surv. India*, **2(2):** 49–50.
Tiwari, R.N., Guha, D.K. and Maity, P.K. (1977). New Records of Ponerine Ants from Arunachal Pradesh (Hymenoptera: Formicidae). *Newsl. Zool. Surv. India*, **3(3):** 132–133.
Tikader, B.K. (1980). Fauna of India: *Araneae and Spiders*, **1(1):** 1–247.
Tikader, B.K. (1982). Fauna of India: *Araneae and Spiders*, **2(1-2):** 1–536.
Tripathi, R.S. and Barik, S.K. (2003). Shifting cultivation in North-East India. *In*: Bhatt, B.P., Bujarbaruah, K.M., Sharma, Y.P. and Pritam (eds). Proc Approaches for Increasing Agricultural Productivity in Hill and Mountain Ecosystem. ICAR Research Complex for NEH Region, Umiam, Meghalaya.
Zoological Survey of India (2006a). Fauna of Arunachal Pradesh. State Fauna Series 13 (Part-1). Zool Surv. India, Kolkata.
Zoological Survey of India (2006b). Fauna of Arunachal Pradesh. State Fauna Series 13 (Part-2). Zool Surv. India, Kolkata.

Ambient Noise Levels after CNG Implementation in Transport Sector in Delhi

Deepak Singh, Amit Prakash, Amit Kumar, Bhupendra P. Singh, Monika Punia, Sanyogita, Homdutt Sharma, Krishan Kumar, and V.K. Jain

Introduction

Last six decades have seen an increasing awareness in the world towards the surrounding environment. Apart from air, water and soil pollution, noise pollution, which we often ignore, has been perceived to be an important aspect in determining the quality of our environment and life. This environmental menace of noise pollution, mainly associated with urban and industrial areas, has attracted the attention of environmentalists all over the world. Almost all the activities of the modern civilization contribute their shares to the noise environment. A major factor contributing to acoustic environment in cities is the transport noise which originates from road (i.e., vehicular), air and rail traffic. Road traffic noise differs from air, and rail traffic in the way that it is spread throughout the city and continuous in nature. Road traffic noise problems arose in cities due to population increase stemming from accelerated growth, internal migration, and the increasing number of vehicles, which pour into and add to the already overcrowded streets.

Of all the kinds of surface transportation noise, traffic noise is most ubiquitous. Urban road traffic noise is one of the pervasive types of noise pollution and generally considered to be more intrusive than other types of noise such as

D. Singh (✉) • A. Kumar • B.P. Singh • M. Punia • Sanyogita • K. Kumar
School of Environmental Sciences, Jawaharlal Nehru University, New Delhi, India
e-mail: deepaksingh1947@gmail.com

A. Prakash • V.K. Jain
Tezpur University, Tezpur, Assam, India

H. Sharma
Amity University, Noida, Uttar Pradesh, India

N.J. Raju et al. (eds.), *Management of Water, Energy and Bio-resources in the Era of Climate Change: Emerging Issues and Challenges*,
DOI 10.1007/978-3-319-05969-3_21

industrial noise, airport noise and community noise (Unweltbundesant, 2000). It is also intrinsically discontinuous: as a vehicle approaches an observation point, the noise level rises, reaches a maximum level and then decreases as the vehicle moves away. The actual pattern of traffic noise on a main road is complex. Traffic noise is caused by a number of vehicles of different characteristics moving together under variable conditions. The noise level varies enormously according to location, type of location, density of traffic, and time of day.

Noise from motor vehicles depend partly on the vehicles themselves and partly on traffic conditions, the surroundings (vegetation, high rise buildings, etc.) and meteorological parameters (temperature, humidity, wind speed and direction) which are independent of the vehicles. (White and Walker, 1982).

Noise pollution is hazardous to human health in many ways. Noise can lead to physiological, psychological and behavioural changes in human beings. Noise is considered a serious health hazard and inimical to efficiency and performance at place of work. It can also produce many non-auditory effects. It can disturb our work, rest, sleep, communication and also evoke other possibly pathological reactions.

Vehicular traffic is the most dominant source of noise pollution in urban areas. A number of surveys on traffic noise have been conducted in various cities of the world, viz., Tokyo (Yoshida, 1994), Gothenberg (Bjorkman and Rylander, 1997), Beijing (Li et al., 2002) and Messina (Piccolo et al., 2005). In India, surveys on road traffic noise levels have been carried out in Bombay (Dixit et al., 1982), Delhi (Kumar and Jain, 1991; Singh and Jain, 1995; Kumar et al., 1998) and Calcutta (Chakrabarty et al. 1997).

A number of studies have been conducted on the effect, mitigation and perception of traffic noise due to vegetation (e.g., Huddart, 1990; Watts et al., 1999). In India, Tyagi et al. (2006) studied the spectral characteristics of traffic noise reduction by vegetation belts in Delhi and found that attenuation generally increases with frequency.

In the Indian context, empirical models for prediction of road traffic noise have also been developed for few Indian cities e.g. Visakhapatnam (Rao and Rao, 1990) and Delhi (Kumar et al., 1998).

Concerned about the rising pollution levels in Delhi, the Supreme Court on 28 July, 1998, ordered the state government of Delhi to improve the air quality by reducing vehicular pollution in Delhi. The directives of SC to move public transport to CNG has been implemented in Delhi since April 2001. Delhi now boasts of a completely CNG driven bus fleet, three-wheelers (auto-rickshaws) and RTVs. Also significant fraction of taxis on Delhi roads are driven by CNG.

This has prompted us to undertake the present study and developing multiple linear regression models where explanatory variables corresponding to vegetation and traffic density have the real graded values (in percentage) and not 1 or 0.

Methodology and Experimental Details

For the purpose of making noise level measurements, a precision Type 1 integrated sound level meter (RION, NL-10A) was used. The instrument has got the facility of measuring L_{eq} levels for specified intervals of time duration of 10 sec, 1 min, 5 min, 30 min, 1 hr, 8 hrs and 24 hrs. In addition, it also provides manual facility for measuring instantaneous sound pressure levels at slow/fast/impulse modes.

Study Area: Delhi

In the context of metropolitan cities of India, Delhi, of late, has come to be recognised as the fourth most polluted city of the world. The national capital territory of Delhi is an emerging megapolis of India. It covers an area 1483 sq km and had a population 16.8 million based on the 2011 census. Figure 1 gives the vehicular growth in last 15 years in Delhi.

Sampling Sites

In the present study, field experiments were conducted to measure the noise levels at sixty (60) sampling sites spread all over Delhi. The locations of these sites are shown on the map of Delhi. These locations were so chosen as to represent the different zones within an urban area. The sites differ in terms of their land use pattern, vegetation and vehicular traffic density. The 60 sites can be divided into the following five zones viz., (i) Residential (16 sites), (ii) Commercial (16 sites), (iii)

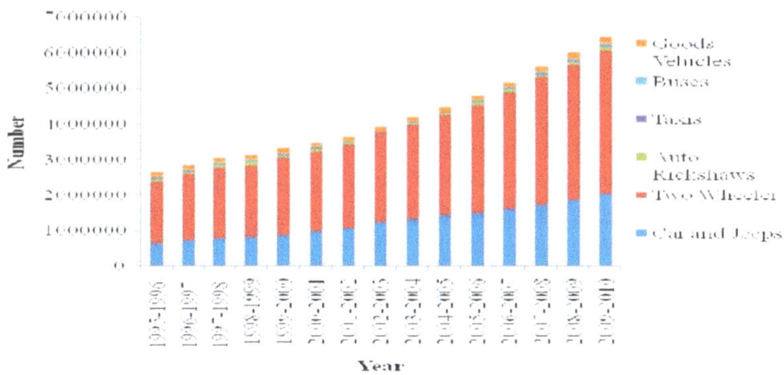

Fig. 1 Vehicular growth in Delhi during 1995–1996 to 2009–2010. (Transport dept - Govt of NCT)

Industrial (5 sites), (iv) Mixed i.e., Residential-cum-Commercial-cum-Industrial (18 sites) and (v) Silence (5 sites). Figure 2 gives the 60 sampling locations in Delhi.

Measurement Details

The measurement of noise levels were made at a distance of five metres from the outer edge of the road. The sound level meter was fixed on a stand about 1.2 m above the ground level and kept perpendicular to the road. At each of the sixty sites, 1 minute L_{eq} values were recorded at an interval of 20 minutes. The total measurement duration was of nine hours (from 9 a.m. to 6 p.m.). Hence a total of 28 readings were taken at each site. The vegetation cover was measured over the circular area of 100 metres radius in terms of percentage cover around each measurement site.

The associated traffic composition was observed by counting the number of vehicles passing the observation point. The vehicles were divided into seven types and were coded for the regression analysis as shown in the bracket after the name of vehicle viz., car/jeep (CJ), two-wheeler (TW), three-wheeler (THW), bus (BUS), taxi (TX), light commercial vehicle (LCV) and rural transport vehicle (RTV). During the model building exercise, these type of vehicles were also categorised into CNG and Non-CNG driven vehicles to evaluate the impact of use of CNG fuel in transport sector on the ambient noise environment of Delhi. Figure 3 shows the CNG driven vehicles viz., RTV (a), Taxi (b), Bus (c) and Three-wheeler (d) in Delhi.

Modelling Procedure

A linear multiple regression model was attempted in the present work. A model equation can be written in the following form:

$$Y = \alpha + \beta_1 X_1 + \beta_2 X_2 + \beta_3 X_3 + \text{--------} + \beta_n X_n + \varepsilon$$

where Y, the dependent variable, represents the noise level, α is a coefficient and $\beta'_i s$ are regression coefficients, X_is is explanatory or independent variables and ε is the stochastic error term. The explanatory variables in the present study are car/jeep (CJ), two-wheeler (TW), three-wheeler (THW), bus (BUS), taxis (TX), light commercial vehicles (LCV), rural transport vehicles (RTV), average temperature (TEMP), average relative humidity (HUM), wind velocity (WV), percentage of vegetation (VEG), existence of flyover (FLY), four dummy variables viz., residential (RES), industrial (IND), mixed (MIX) and silence (SIL) zones. Commercial site has been taken as default variable. The model building exercise was carried out at two levels viz., zone level (at this level five more explanatory variables were added i.e., average temperature, average relative humidity, wind velocity, percentage of

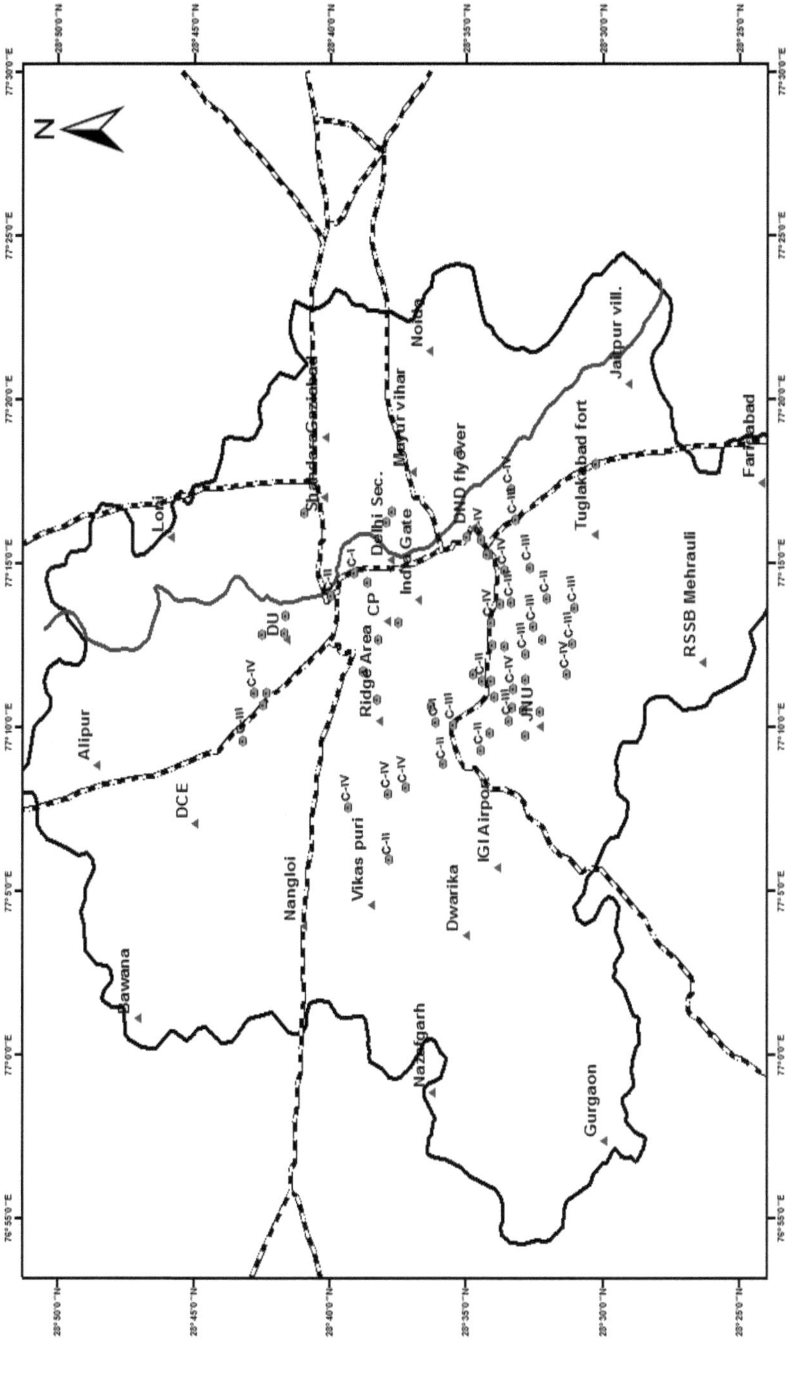

Fig. 2 Sampling sites in Delhi

Fig. 3 CNG driven vehicles in Delhi

vegetation and existence of flyover), and Delhi region level (here four dummy variables viz., residential (RES), industrial (IND), mixed (MIX) and silence (SIL) zones were added. At the zone and Delhi region level, two different types of model were developed, first by taking all the variables into account by clubbing the vehicles into viz., CNG and Non-CNG driven vehicles.

Results and Discussion

From the observations made during various field studies, following models were developed for different zones and entire Delhi. Throughout the data analysis we have taken 10 % level of significance for determination of model coefficients.

Model Development for Residential Zone

An attempt has been made to develop a multivariate regression model for the residential zone as a whole by including vegetation, average temperature, average relative humidity and most dominant wind velocity as independent variables besides the number of different types of vehicles.

The proposed model for residential zone may be represented by the following equation:

$$L_{eq} = 71.294 - 0.040VEG - 0.775WV + 0.164CJ + 0.168TW + 1.737BUS + 2.589RTVs$$

The proposed model for residential zone in terms of CNG and Non-CNG vehicles can thus be represented by the following equation:

$$L_{eq} = 72.195 - 0.059VEG - 0.946WV + 0.074Non - CNG$$

It is interesting to note that CNG driven vehicles cumulatively are not found to contribute significantly to the ambient noise. It is not unexpected since CNG driven vehicles do not ply in significant numbers in residential areas.

Model Development for Commercial Zone

The proposed noise model for commercial zone can thus be represented by the following equation:

$$L_{eq} = 86.091 - 0.064VEG - 0.472TEMP + 0.109HUM - 2.206WV + 0.173CJ + 0.054TW + 0.756\,THW + 1.628BUS + 0.801FLY$$

The model shows flyover has adverse effect on the noise since it may be facilitating large volume of traffic.

The regression model with CNG and Non-CNG vehicles can be given by the equation:

$$L_{eq} = 78.304 - 0.019VEG - 0.025HUM + 0.052Non - CNG$$

Thus basically, the Non-CNG vehicles are the dominant source of ambient noise.

Model Development for Industrial Zone

The proposed model for industrial zone can be represented by the following equation:

$$L_{eq} = 81.248 + 0.435LCV + 0.492CJ$$

The coefficients of different predictors indicate that both LCV and CJ contribute almost equally, towards the ambient noise environment while other variables were found to be insignificant.

The proposed model for industrial zone in terms of CNG and Non-CNG vehicles can be represented by the following equation:

$$L_{eq} = 79.774 + 0.011\text{Non} - \text{CNG}$$

Here it is pertinent to note that in the industrial zone (i) the background level of noise 79.77 is fairly high and (ii) only the Non-CNG vehicles make any contribution towards the ambient noise.

The higher background noise is, in all probability, due to other sources specific to a host of industrial activities viz., machinery noise, industrial workers noise and visitors movements.

Model Development for Mixed Zone

The proposed model for this site may be represented by:

$$L_{eq} = 75.287 - 0.027\text{VEG} + 0.035\text{CJ} + 2.383\text{BUS}$$

The best fit model for this zone, after clubbing vehicles into CNG and Non-CNG driven can be represented by:

$$L_{eq} = 74.886 + 0.029\text{Non} - \text{CNG} + 0.369\text{CNG}$$

It is worth noting that the contribution of CNG driven vehicles in mixed zone is most dominant compared to Non-CNG driven vehicles as evident from their respective regression coefficients 0.369 and 0.029 respectively. This result is in stark contrast to the results obtained for residential and commercial zone where Non-CNG driven vehicles were the dominant sources of noise.

Model Development of Silence Zone

The model proposed for silence zone is:

$$L_{eq} = 65.915 + 0.120\text{TW} + 7.853\text{CJ} + 6.895\text{FLY}$$

The proposed model can be represented by the equation:

$$L_{eq} = 68.571 + 0.077\text{Non} - \text{CNG} + 3.569\text{FLY}$$

In both of the above proposed models, the constant value which represents the background noise are relatively lower as compared to other zones. However, the background noise is still appreciably higher than the standard prescribed (50 dB) for silence zone.

Model Development for Entire Delhi Region

A multivariate model has been developed for Delhi by taking four dummy variables representing residential (RES), industrial (IND), mixed (MIX) and silence (SIL) zones. The commercial (COM) zone has been taken as a default variable. These are in addition to all the other variables taken in case of model development for each zone. Hence a total of 16 variables are considered for developing a model for entire Delhi region.

The proposed model for Delhi is:

$$\begin{aligned} L_{eq} = {} & 72.475 - 1.569RES + 0.954MIX - 0.044VEG - 0.566TEMP \\ & + 0.056CJ + 0.051LCV + 1.552BUS + 1.185THW \end{aligned}$$

The negative coefficient of residential (RES) zone shows that these locations have lower noise level as compared to commercial zone which is taken as default variable. The constant value represents the background noise in Delhi which is 72.48. The positive coefficient for mixed zone (MIX) shows that the noise level here is more than the commercial zone. Vegetation cover (VEG) and temperature (TEMP) show negative coefficients which is expected as they are known to help in attenuating the noise (Tyagi et al., 2006; Kumar et al., 1998; Aylor, 1972a, 1972b). Among the vehicles, buses (BUS) contribute the most followed by three-wheeler (THW), car/jeep (CJ) and LCV. Rest of the variables are not so significant in case of whole of Delhi.

The proposed model can be represented by the equation:

$$\begin{aligned} L_{eq} = {} & 74.966 - 0.057VEG - 1.481RES + 0.954MIX + 0.238CNG \\ & + 0.038Non - CNG \end{aligned}$$

In this case also, RES has negative coefficient and MIX has positive coefficient. The coefficient for CNG and Non-CNG vehicles reveal that both contribute towards the ambient noise of Delhi, though the contribution of CNG vehicles is more than Non-CNG vehicles. Similar inferences were drawn for most of the sites and zones where CNG driven vehicles (buses, three-wheelers etc.) have higher coefficients than the Non-CNG driven vehicles (car/jeep, two-wheelers etc.).

As evident from the proposed model, the value of constant representing average background noise of Delhi is 74.966 dB. This is close to maximum permissible limit for an industrial zone and well above the prescribed limits for residential (55 dB), commercial (65 dB) and silence (50 dB) zones. The traffic density and composition certainly add to this background noise level. Though vegetation is found to play a very positive role in noise attenuation, its low coefficient indicates that thick vegetation cover will be required to attenuate noise appreciably. This result reinforces the earlier inferences about the beneficial role of vegetation in attenuation of noise (Kumar et al., 1998).

Table 1 Comparison of previous study (1999) with current study (2009–10)

		Previous study (1999)			*Current study (2009–10)*			
Type of zone & cocode	*Name of sampling site*	*No. of Veh.*	L_{eq}	L_{eq}/*Veh (a)*	*No of Veh.*	L_{eq}	L_{eq}/*Veh (b)*	*Diff. (b–a)*
Res 7	Delhi Cantt.	42	73.2	1.74	70	76.3	1.09	−0.65
Res 8	Khel Gaon	94	82.22	0.87	61	75.4	1.23	0.36
Res 10	S.N. Market	61	79.35	1.30	59	75.6	1.27	−0.03
Res 11	R.K. Sec-3	55	74.3	1.35	47	73.9	1.59	0.23
Res 12	Raja Garden	58	74.7	1.29	105	79.9	0.76	−0.53
Res 13	R.K. Sec-12	18	77.1	4.28	50	74.7	1.50	−2.78
Com 1	Hauz Khas	65	73.93	1.14	88	78.5	0.90	−0.25
Com 4	Gandhi Nagar	44	73.8	1.68	109	80.9	0.74	−0.93
Com 15	Patel Nagar	80	77.7	0.98	80	79.1	0.99	0.01
Com 16	Dhaula Kuan	64	79.9	1.24	111	80	0.72	−0.52
Ind 3	Azadpur	88	79.25	0.90	162	81.6	0.50	−0.39
Ind 4	Narayana	61	79.45	1.30	139	81.5	0.59	−0.71
Mix 1	Tihar Jail	47	75.1	1.60	71	79.2	1.12	−0.48
Mix 3	Mall Rd	72	79.25	1.10	99	79.7	0.81	−0.30
Mix 4	Olof Palme Mg.	115	79.1	0.69	99	79.5	0.81	0.12
Mix 5	Tara Apart	31	75.6	2.44	89	79.6	0.90	−1.54
Mix 8	N Mandela Rd	25	80.6	3.22	108	78.6	0.73	−2.50
Mix 10	Punjabi Bagh	76	76.85	1.02	124	81.2	0.66	−0.36
Mix 12	Rajendra Nagar	84	82.45	0.98	149	81.8	0.55	−0.43
Mix 13	Nizzamuddin	84	77.97	0.93	143	81.5	0.57	−0.36
Mix 15	Palam Marg	64	76.7	1.20	97	79	0.82	−0.38
Mix 17	Lajpat Nagar	79	79.07	1.00	145	81.6	0.56	−0.44
Sil 1	Rajghat	51	74.9	1.47	140	78.1	0.56	−0.91
Sil 2	AIIMS	88	79.46	0.90	132	81.7	0.62	−0.28

A comparison with the previous study (Srivastava, 1999) has been made to ascertain the impact of CNG introduction in Delhi on the noise environment. The previous study was carried out in the year 1999 (two years before the introduction of CNG in Delhi). Hence a comparison is helpful in understanding the extent of change in the ambient noise that has occurred since then. Out of total 60 sites taken all over Delhi in the present study only 24 sites are common with that of the previous study. Table 1 shows the comparison of previous study (1999) with the current study (2009–10). For comparison purpose, one minute average total vehicle count and the corresponding one minute L_{eq} at these 24 sites are considered. The comparison of L_{eq} values per unit vehicle from the previous study and the present study reveal that the differences at these locations are significant and have lower noise levels in general today in comparison to the past values. With the exception of four places out of 24 places, all other sites have lower L_{eq} per unit number of vehicles. Since 1998, vehicular composition as well as the volume of traffic on road have increased manifold, the L_{eq} values have decreased significantly from the previous values of L_{eq} in Delhi. This may be attributed to the introduction of newer modern technology in the modes of transport as well as because of the introduction of CNG in public transport system. When we look at the model developed for Delhi, it is to be noted that though car/jeep (CJ) and two-wheeler (TW) are two major Non-CNG vehicles contributing to the urban noise, the buses (BUS) and three-wheeler (THW), the two main CNG vehicles have the high regression coefficient values in comparison to car/jeep and two-wheelers. Thus, it can be inferred that buses and three-wheelers enhance the noise level per unit introduction of CNG vehicles as compared to the Non-CNG vehicles. With the usage of advanced and efficient technology in CNG and Non-CNG driven vehicles the noise pollution can further be reduced.

Noise pollution is a significant environmental problem in the urban city centres. Various studies show that the urban road traffic is the root cause of ambient noise. However, many studies such as (Aylor, 1972a, 1972b; Gupta et al., 1986; Tyagi et al., 2006) suggested that other parameters like vegetation cover, road width, presence of flyover, wind velocity etc., also play a significant role in noise pollution. In the present study, an attempt has been made to develop a predictive model for different locations separately as well as to develop a model for five distinct zones namely, residential, commercial, mixed, industrial and silence zones. Further an integrated model for entire Delhi has been developed.

It is evident from the study at individual locations that car/jeep (Non-CNG) plays a significant role at 56 sites out of total 60 observation sites across all the zones. Besides car/jeep (CJ), two-wheeler (TW) is another Non-CNG vehicle, which has significant contribution towards ambient noise level. It contributes significantly at 19 sites out of total 60 sites. Most of these sites, where two-wheeler also plays a significant role, pertain to residential, mixed and commercial zones.

Among CNG driven vehicles, three-wheelers (THW) and buses (BUS) are found to be significant at 23 and 21 sites out of total 60 sites, respectively.

In silence zones, buses (BUS) are not found to play any significant role at any of the five sites, whereas three-wheelers (THW) are found to contribute significantly at one location only.

Similarly, in industrial zones, only at two locations, buses (BUS) contribute to ambient noise whereas, three-wheelers, only at one location. LCVs and RTVs are significant contributor at very few places. RTVs are mostly confined to commercial and mixed zones. They contribute significantly only at one site each in both the zones (commercial and mixed). LCVs show their dominance at all the five industrial sites and only at two sites in mixed zones. The values of constant, which represent the background noise levels, are more than 70 dB at all the locations with the exception of site 1 or Res 1 (residential) and site 60 or Sil 5 (silence) where it is 67.78 and 68.55, respectively. Hence, it can be concluded that at all the sites the background noise exceeds the prescribed permissible limit for various categories of areas.

Close examinations of zone specific models reveal that vegetation has negative coefficient at all the sites. It indicates that at all the sites, vegetation helps in attenuating the ambient noise level. Similar observation has also been reported by Aylor (1972a, 1972b), Kumar et al. (1998) and Tyagi et al. (2006). With the low coefficient value for vegetation, it can be concluded that although green belt around the noise sources reduce the ambient noise but high vegetation density will be required for any substantial noise reduction in Delhi. At residential and commercial zones, wind velocity also plays a significant role in noise reduction as indicated by its negative coefficient at these sites. At sites of other zones, it does not play any significant role. In the last five years, Delhi's road infrastructure has undergone a sea change. A number of flyovers at different traffic intersections have come up. Although these flyovers ease the traffic congestions but increase the number of vehicles at the same stretch of road and hence increase the noise at those locations. It can also be inferred from the regression models of different zones that flyovers contribute significantly, towards ambient noise in silence and commercial zones.

To ascertain the impact of CNG in vehicles in Delhi on its ambient noise environment the vehicle composition was segregated into CNG and Non-CNG vehicles and then regression analysis was carried out. The Non-CNG vehicles are found to dominate in all the five zones while CNG driven vehicles contribute significantly only at mixed zone.

Conclusion

It can be concluded that the conversion of public transport system from petrol/diesel driven to CNG driven modes have a positive effect on noise scenario of Delhi. The models developed for entire Delhi show that the regression coefficient for CNG vehicles is 0.238 in comparison to Non-CNG vehicles, which has only 0.038 as coefficient. This is due to the fact that the number of buses and three-wheelers are noisier than that of the car/jeep which runs on petrol or diesel. But it is to be noted that the CNG buses and three-wheelers are less noisy than that of the diesel or petrol driven buses and three wheelers. This implies that the noise level per unit vehicle decreased significantly, after the induction of CNG vehicles in Delhi. Hence, we can say that CNG vehicles have a positive effect as far as noise pollution scenario in Delhi is concerned. This study has policy implications as far as noise pollution management is concerned. In the present scenario with rapidly increasing population of Delhi, more buses, three-wheelers and other modes of transport will be required in near future. In such a condition, CNG vehicles with rattle-free body and proper sound mufflers can be beneficial for noise pollution control in Delhi. Further, enhancement of vegetation cover in between the roads as well as at the pavements would go a long in protecting the people of Delhi from the adverse impacts of noise.

References

Aylor, D. (1972a). Noise Reduction by Vegetation and Ground. *Journal of the Acoustical Society of America*, **51(1):** 197–205.

Aylor, D. (1972b). Sound Propagation in Relation to Leaf Density, Leaf Width, and Breadth of Canopy. *Journal of the Acoustical Society of America*, **1(1):** 411–414.

Bjorkman, M. and Rylander, R. (1997). Maximum Noise Levels in City Traffic. *Journal of Sound and Vibration*, **205(4):** 513–516.

Chakrabarty, D., Santra, S.C., Mukherjee, A., Roy, B. and Das, P. (1997). Status of road traffic noise in Calcutta metropolis, India. *Journal of Acoustical Society of America*, **101:** 943–949.

Dixit, G.R., Mahadevan, T.N. and Kapoor, R.K. (1982). A Noise Pollution Survey of Bombay. *Scavenger*, **12(2):** 20–25.

Huddart, L. (1990). The use of Vegetation for Traffic Noise Screening. Department of Transport, TRRL Report ISSN 0266–5247 1990. Transport and Road Research Laboratory, Crowthorne.

Kumar, K. and Jain, V.K. (1991). Noise and its Spectral Characteristics at some Traffic Sites in Delhi. *Journal of Acoustical Society of India*, **19(3):** 6–13.

Kumar, K. and Jain, V.K. (1995). A Study of Noise in Various Modes of Transport in Delhi. *Applied Acoustics*, **43:** 57–65.

Kumar, K., Jain, V.K. and Rao, D.N. (1998). A Predictive Model of Noise for Delhi. *Journal of the Acoustical Society of America*, **103(3):** 1677–1679.

Li, B., Tao, S. and Dawson, R.W. (2002). Evaluation and analysis of Traffic Noise from the Main Urban Roads in Beijing. *Applied Acoustics*, **63:** 1137–1142.

Piccolo, A., Plutino, D. and Cannistraro, G. (2004). Evaluation and Analysis of the Environmental Noise of Messina, Italy. *Applied Acoustics*, **66:** 447–465.

Rao, P.R. and Rao, M.G.S. (1990). Traffic Noise Pollution in the City of Visakhapatanam. *Scavenger*, **21(1):** 9–13.

Srivastava, A.K. (1999). M.Phil. Dissertation, School of Environmental Sciences, Jawaharlal Nehru University, New Delhi, India.

Transport Dept Report 2010, Govt of NCT, New Delhi, India.

Tyagi, V., Kumar, K. and Jain, V.K. (2006). A study of the spectral characteristics of traffic noise attenuation by vegetation belts in Delhi. *Applied Acoustics*, **67:** 926–935.

Unweltbundesant (2000). Larmbel, Report of the German Environmental Agency, Berlin.

Watts, G., Chinn, L. and Godfrey, N. (1999). The effects of vegetation on the perception of traffic noise. *Applied Acoustics*, **56:** 39–56.

White, R.G. and Walker, J.G. (1982). Noise and Vibration. John Wiley and Sons.

Yoshida, T. (1994). Estimation of Equivalent Sound Pressure Levels of Community Noise and Road Traffic Noise. *Journal of Acoustical Society of Japan*, **15(1):** 53–57.

Transport of Lindane through Soil Column

M.K. Sharma and C.K. Jain

Introduction

Among a large number of manmade chemicals, organochlorines such as aldrin, dieldrin, DDT and its derivatives (DDD, DDE), HCH (BHC), and PCBs are of great concern due to their highly bioaccumulative nature and toxic biological effects. These chemicals are persistent in nature, biomagnify in the food web and impose various toxic effects in marine organisms (Tanabe et al., 1997). Organochlorine pesticides (OCPs), also existing as typical persistent toxic substances (PTS), have been of increasing concern in the world due to their salient features of persistence, bioaccumulation, and toxicity (Willett et al., 1998; Quan et al., 2003; Wong et al., 2005).

Organic pollutant 'Lindane' is an organochlorinated insecticide and fumigant and highly persistent in groundwater of metropolitan city of Vadodara. It shows a low affinity for soils binding, and may be mobile in soils with especially low organic matter content or subject to high rainfall. It may pose a risk to groundwater contamination. It is very stable in both fresh and salt-water environments and is resistant to photodegradation. The soil of Vadodada city is dull brown in colour. The pattern of pesticides use, structures and their degradation products are also responsible for the transportation (leaching) behaviour of the pesticides. The fine texture of soils in general inhibits pesticides leaching because of either low vertical permeability or high surface area, which enhances the adsorption of pesticides.

Further, column studies of sufficient length and diameter in the laboratory are helpful to study the fate and transport of pollutants under the field conditions. A number of workers have carried out the column experiments to understand the process of transportation of different pesticides through a soil column

M.K. Sharma (✉) • C.K. Jain
National Institute of Hydrology, Roorkee 247 667, Uttarakhand, India
e-mail: mks@nih.ernet.in

N.J. Raju et al. (eds.), *Management of Water, Energy and Bio-resources in the Era of Climate Change: Emerging Issues and Challenges*,
DOI 10.1007/978-3-319-05969-3_22

(Rao and Davidson, 1979; Mouvet et al., 1997; Indelman et al., 1998; Freijer et al., 1998; Lennartz, 1999; Roy et al., 1999; McGechan and Lewis, 2002; Hadiri et al., 2003; Landry et al. 2004; Li et al. 2005; Köhne et al., 2006).

Batch experimental data are often difficult to apply directly to fixed bed column adsorber because isotherms are unable to give accurate data for scale up since a flow column is not at equilibrium. Fixed bed column adsorption experiments are helpful to study the adsorption dynamics. The fixed bed column operation allows more efficient utilization of the adsorptive capacity than the batch process. The adsorbent at the inlet end is contacted continuously by the adsorbate at the initial adsorbate concentration. The most important criterion in the design of fixed bed adsorption systems is the prediction of fixed bed column breakthrough or the shape of the adsorption wave front, which determines the operation life-span of the adsorbent. The design of up flow fixed bed column adsorber includes estimation of shape of breakthrough curves and the appearance of breakpoint (Sarin et al., 2006).

Paknikar et al. (2005) studied the degradation of Lindane from aqueous solution using iron sulphide nanoparticles stabilized by biopolymers and reported that stabilized nanoparticles could degrade Lindane (5 mg/L) with an efficiency of 94 % in 8 h in a subsequent microbiological treatment; residual Lindane, its partially degraded intermediates as well as the stabilizing polymer, could be completely degraded in 1 h.

Kumar and Philip (2006) studied the adsorption and desorption characteristics of Endosulfan on four Indian soils i.e. clayey, red, sandy and composted soil extensively. Adsorption and desorption rates of α- and β-endosufan, calculated from kinetic studies, varied for different type of soils. Maximum adsorption took place in clay soil followed by composted soil and red soil while adsorptions of α- and β-endosulfan were negligible in sand. The binding characteristics of various functional groups were calculated using Scatchard plot. Effect of functional groups was more predominant in clayey soil. Organic matter and pH played a significant role in adsorption and desorption phenomena of endosulfan. Adsorption decreased drastically in clay soil when pH was reduced while desorption was higher at both acidic and alkaline pH ranges compared to neutral pH. Results indicated that α-endosulfan is more mobile compared to β-endosulfan and mobility of endosulfan is the maximum in sandy soil followed by red soil.

In the present investigation, transport behaviour of Lindane in the unsaturated zone was studied by maintaining optimized conditions of the adsorption in the column of length 25 cm and diameter 4″ containing soil of Vadodara city of Gujarat State of India. To achieve the said objectives, batch and column experiments were conducted.

Study Area

The metropolitan city Vadodara is the graceful city of Gujarat State of India. It is bounded by 22°18′ N latitude and 73°16′ E longitude. Vadodara urban agglomeration covers an area of about 140 km^2. Rapid industrial development has taken place in the city. The municipal area is further divided into 28 wards. As per census, 2001, the population of Vadodara metropolitan city is 14,92,398. The disposal site for dumping solid waste is also located at Wadsar. The rivers Jambua, the Surya, the Viswamitri and the Dhadhar, which flow through central part of the district and empty into Gulf of Khambat, are also part of Mahi Basin. Vadodara city is the industrial nucleus of the Gujarat State. The principal industrial areas within Vadodara urban areas are at Makarpura and Nandisari. Many large scale industries are operating continuously. Few industries in the study area are using Lindane as a raw material for producing medicines for vector control diseases. The treated/untreated effluents discharged from some of these industries in the north-eastern part of the study area flow through drains which join the Effluent Channel, which ultimately meets river Mahi while treated/untreated effluents discharged from some of these industries in the southern part of the study area flow through drains which join either river Vishwamitri or river Jambua. Both of these rivers, Vishwamitri and Jambua, meet further downstream in the south end of Vadodara city. The high concentrations of pollutants in these drains/rivers are leaching through soil and ultimately contaminating the groundwater of the city.

The climate of the metropolitan city is moderate tropical type. The temperature of the city varies from 8 °C to 46 °C. The average annual rainfall is recorded as 900 mm. The earliest geological evolution of the basement rocks, exposed in northern and eastern parts, had been controlled by the Precambrian orogenies (Arvalli and Delhi cycles), and the older crystalline rocks ideally show folds, faults and magmatism related to the two orogenies. After Precambrian orogenies, major geological events of Vadodara district were confined to Mesozoic and Cenozoic Eras which can be related with the breaking up of the Gondwanaland and the subsequent northward drift of the Indian sub-continent, involving formation of sediments and Deccan Trap Volcanism with uplifts and subsidence along the two major lineaments—Narmada and Cambay rift system. There is no yield of water upto 50 feet and sandy aquifer was found from 50 to 70 feet.

Material and Methods

Characterization of Soil

The soil of unsaturated zone of the Vadodara city was collected and has been characterized. The physical characteristics of the soil sample of Vadadara city was determined as per the standard methods (USEPA, 1974; Hendershot et al., 1993). The physical parameters determined include pH, density, porosity, content of gravel, sand, clay and silt while chemical parameters determined include calcium, magnesium, nitrate, sulphate, chloride, phosphate, alkalinity and hardness. For measurement of pH, the soil sample (5 g) was shaken with 25 ml distilled water for 60 minutes on mechanical shaker and then sample was filtered by Whatman

42 filter papers. The pH of these filtrate were determined by standard methods. The density and porosity were determined with the help of Multi volume Pycnometer and Geo Pycnometer respectively. The percentage of gravel, sand, clay and silt were determined by sieve analysis and with the help of Master Sizer E System. 50 g sample of the soil was mixed with 250 ml of distilled water and shaken for 60 minutes on mechanical shaker and filtered by Whatman 42 filter paper. The residue of the soil sample was further shaken with 250 ml distilled water separately in the similar way and filtered again. The two filtrates were mixed together to make the total volume of 500 ml of filtrate. The chemical parameters were determined in this filtrate using Standard methods of water analysis (APHA, 1992) for the chemical characteristics of the soil.

Batch Experiment

The batch experiment for adsorption of Lindane on soil of Vadodara for optimizing condition for the column study was carried out for different operating variables viz. pH, initial concentrations, adsorbent dose and adsorbent size. Equilibrium contact time for adsorption of Lindane was found to be 90 minutes; therefore all further adsorption experiments were being carried out for equilibrium time of 90 minutes.

Sorption experiments were conducted in a series of Erlenmeyer flasks of 100 ml capacity covered with a teflon sheet to prevent contamination. 50 ml of Lindane pesticide containing solution (200–2000 μgL^{-1}) was transferred in the flasks together with desired sorbent doses (W_s in gL^{-1}), and placed in a water bath shaker maintained at desired temperature. The pH of the solution was maintained throughout the experiment using dilute HNO_3 and NaOH solutions. The pH was measured before and after the solution had been in contact with the sediment. Aliquots were retrieved periodically and filtered through 0.45 μm membrane filters. The filtrate was analysed for the remaining concentrations of Lindane.

Column Experiment

For transport of pesticides in unsaturated zone of the study area, optimized conditions for adsorption of Lindane on soil were maintained in the column of length 25 cm and diameter 4″ (Fig. 1). For packing the column with the soil, the supporting media i.e. glass wool was packed by hydraulic filling. The slurry of the weighted soil samples was prepared in ultra pure water and it was used to pack the column. The column was kept undisturbed overnight for its full settlement and saturation. The flow of the effluent of the column was controlled by the stopper point at the lower end of the column. The solution of Lindane (10 ml of concentration 10 mg/L) was loaded on to the column and the fractions of 25 ml were collected from outlet of the column. The flow rate of the effluent was maintained constant 0.1 ml/min by

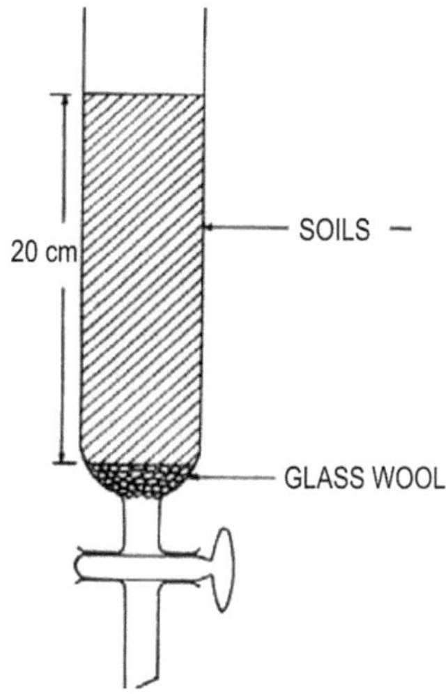

Fig. 1 Soil column

maintaining constant head at upper end of the column by supplying continuous constant supply of distilled water through a bucket having a side tap. The time duration of each 25 ml effluent fraction was also noted.

Analysis of Lindane

The Lindane was analysed in different effluent fractions by gas chromatography. The extraction of pesticides from column effluent fraction was carried out by n-hexane. 25 ml of fraction was shaken with 5 ml of n-hexane in separating funnel for 30 minutes. n-hexane layer was allowed to separate from water for about 10 minutes. The n-hexane layer was separated and the same procedure was repeated three times. Three fractions of extracted n-hexane were mixed together and the total volume of extracted n-hexane was 15 ml. The 15 ml extracted n-hexane was reduced to 1.0 ml by Kuderna-Danish assembly and rotary evaporation apparatus. The concentration was carried out by using water bath at 40 °C temperature. The moisture in concentrated n-hexane (1 ml) was removed by adding anhydrous sodium sulphate (0.02 g) in each extracted n-hexane (1 ml) fractions and keeping them for one hour. All the extracted n-hexane fractions were already clean and clear and therefore no further clean up was carried out.

The qualitative and quantitative analysis of the Lindane in each column fraction was carried out by Aimil Nucon Gas Chromatograph with the ^{63}Ni selective electron capture detector (ECD). This equipment allows the detection of contaminants at trace level concentrations in lower ppb range in the presence of multiple compounds extracted from the matrix to which the detector does not respond. The column used was EQUITY-5, 30 m with internal diameter of 0.25 mm. Nitrogen gas was used as carrier gas at 2.0 ml/min with 28 ml/min as makeup gas. The temperature of the oven was kept at 150 °C with a hold time of 1 minute, then from 150 °C to 200 °C at a rate of 10 °C/minute with a hold time of 1 minute and then from 200 °C to 250 °C at a rate of 1 °C/minute with a hold time of 1 minute and finally to 280 °C at a rate of 10 °C/minute with a hold time of four minutes. The detector was maintained at 285 °C. The standard solution (1 mg/L) of Lindane was prepared in n-hexane. The instrument was calibrated with this standard solution. The chromatogram of the standard Lindane was recorded. The value of the retention time obtained was used for the identification of the Lindane in the extracted column fractions. The peak area obtained by the standard Lindane was used to determine the concentration of pesticide in each fraction. The confirmation of Lindane in the column fraction was achieved by using standard internal addition method. The reproducibility of the results for Lindane was 95 % and above for all column fractions. Further, the mean average reading of an individual sample analyzed in triplicate has been taken for reporting the results. The concentration of Lindane was computed using the formula given by Singh et al. (1987).

Result and Discussion

The leachability of the pesticides in groundwater is controlled by the nature of soil and pesticides itself. The soil nature includes the soil texture, total organic and inorganic matters and biological populations. The physical and chemical properties of the soil of Vadodara city are given in Tables 1 and 2 respectively. The soil of Vadodada city is dull brown in colour. The pattern of pesticides use, structures and their degradation products are also responsible for the transportation (leaching) behaviour of the pesticides. The fine texture of soils in general inhibits pesticides leaching because of either low vertical permeability or high surface area, which enhances adsorption of pesticides (Ali and Jain, 2000). The high organic matter in soils dissolves the pesticides and retards their transportation into the soil. The pH and the temperature of the soils are also important factors for the leachability of pesticides (Hutson and Roberts, 1990; Beltran et al., 1993; Sherma, 1995). However, the mass flow of water through the soil profile is also an important factor of pesticides leaching to groundwater. It has also been reported that some of the bacteria degrade the pesticides into different degradation products (Chen and

Table 1 Physical properties of the soil sample of Vadodara city

Parameters	
pH	7.4
Density (g/cm^3)	2.12
Porosity (%)	12.8
Gravel (%)	16.95
Sand (%)	19.13
Clay (%)	13.43
Silt (%)	50.49

Table 2 Chemical properties of the soil sample of Vadodara city

Parameters	*mg/kg*
Hardness	5000
Alkalinity	1900
Calcium	960
Magnesium	650
Chloride	540
Nitrate	3960
Phosphate	25
Sulphate	250

Table 3 Geochemical characteristics of the soil of Vadodara city

Size fraction (*μm*)	%	*Fe* (*mg/g*)	*Mn* (*mg/g*)	*Organic Matter* (*mg/g*)
0-75	31.8	604.8	4.586	43.1
75-150	19.2	506.6	3.824	42.9
150-210	2.4	478.8	3.268	42.8
210-250	10.6	402.0	2.779	42.8
250-300	22.0	366.2	1.447	42.6
300-425	0.4	334.2	0.820	27.2
425-600	13.6	330.0	0.442	17.1

Mulchandani, 1998). The variety and concentration of the bacteria in soils are also the determining factor for the transportation behaviour of pesticides. The various pesticides have different types of structures with different physico-chemical properties and therefore show different transportation behaviour (Ali and Jain, 2000).

Batch Study

For transport of pesticides in unsaturated zone of the study area, the soil has been characterized for different geochemical phases. The soil has been dried and sieved into different size fraction (Table 3). The organic matter in the different size fractions of the soil was measured in terms of organic carbon. The background Lindane pesticide level in the various fractions of the soil was negligible in the unpolluted zone, compared to the amount of adsorbate added for the adsorption tests. This confirms the absence of any Lindane pesticide particulate attached to the soil particles. The content of iron, manganese and organic matter in various fractions of sediment decreases with increasing particle size indicating the possibility of the two geochemical phases to act as the active support material for the adsorption of Lindane pesticide. Since the soil fraction 250–300 μm has comparatively higher weight percentage, it was considered appropriate to study the comparative adsorption of Lindane pesticide on this fraction and clay and silt fraction

(<75 μm) to demonstrate the importance of the coarser fraction in controlling pesticide pollution. Batch experiment for adsorption of Lindane on soil for optimizing condition for the column study was carried out for different operating variables.

Effect of Operating Variables

Equilibrium Time (t)

The effect of contact time on the adsorption of Lindane on two particle sizes of adsorbents is shown in Figs. 2(a) and 2(b). The asymptotic nature of the plot indicates that there is no appreciable change in the remaining concentration after 90 minutes. This time is presumed to represent the equilibrium time at which an

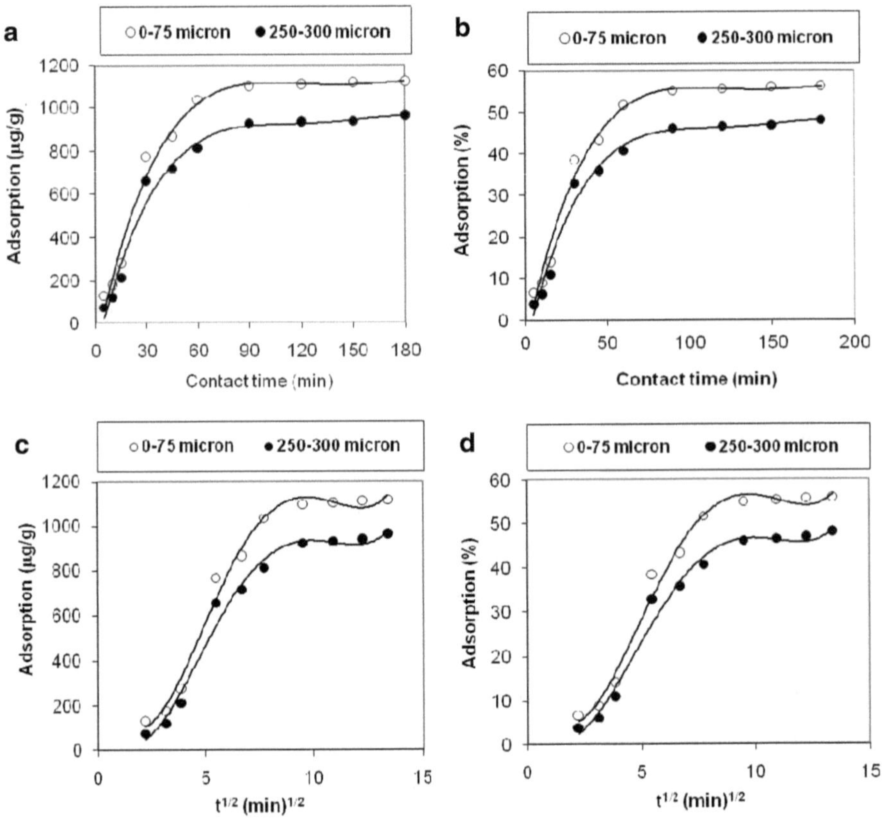

Fig. 2 (**a**) Effect of contact time on adsorption of Lindane. (**b**) Effect of contact time on percent adsorption of Lindane. (**c**) Effect of $t^{1/2}$ on adsorption of Lindane. (**d**) Effect of $t^{1/2}$ on percent adsorption of Lindane

equilibrium concentration is attained. The equilibrium time was found to be independent of initial concentration. All further experiments were, thus, conducted for 90 minutes. Adsorption curves are smooth and continuous leading to saturation, suggesting the possible monolayer coverage of Lindane on the surface of the adsorbent.

According to Weber and Morris (1963), for most adsorption processes, the uptake varies almost proportionately with $t^{1/2}$ rather than with the contact time, *t*. Therefore, plot of Lindane adsorbed, C_t vs $t^{1/2}$, are presented for the two particle sizes of adsorbent in Figs. 2(c) and 2(d). The plots have same general features, initial curved portion followed by linear portion and a plateau. The initial curved portion is attributed to the bulk diffusion, the linear portion to the intraparticle diffusion and the plateau to the equilibrium.

Adsorption Isotherms

The adsorption isotherms of Lindane are shown in Figs. 3(a) and 3(b). It is evident from the isotherm that there is a linear distribution in the range 0 to 1000 μg L^{-1}. Further, for the same equilibration time, the adsorption of Lindane is higher for greater values of initial concentration of Lindane or the percentage adsorption is more for lower concentration of Lindane and decreases with increasing initial concentration. This may be due to the fact that for a fixed adsorbent dose, the total available adsorption sites are limited thereby adsorbing almost the same amount of adsorbate thus resulting in a decrease in percentage uptake of the adsorbate corresponding to an increased initial adsorbate concentration.

A comparison of two plots reveals that the affinity of Lindane is more for <75 μm fraction, i.e., clay and silt as compared to coarser fraction. This may be attributed to the fact that <75 μm fraction contains more iron and manganese

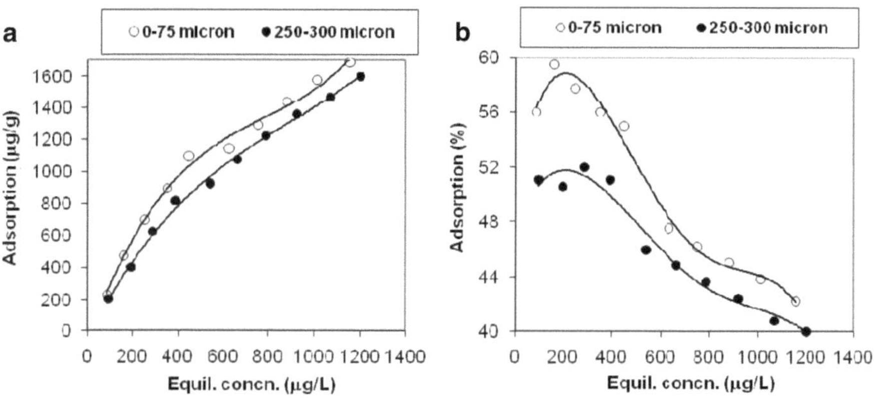

Fig. 3 (**a**) Adsorption of Lindane on the soil of Vadodara city at different concentrations. (**b**) Percent adsorption of Lindane on the soil of Vadodara city at different concentrations

content than that of 250–300 μm fraction indicating the possibility of association of these substrate with clay and silt particles. These findings reveal the possible role of clay and silt components as active support for the adsorption of pesticide.

Effect of pH

The effect of pH on the adsorption of Lindane is shown in Figs. 4(a) and 4(b) for a fixed initial concentration of Lindane (1000 μg/L) and adsorbent dose of 0.5 g/L at a soil particle size of 0–75 and 250–300 μm. Experiments could not be performed at higher pH values due to low solubility of pesticide. A general increase in adsorption with increasing pH of solution has been observed up to the pH value of 6.0 for both the fractions of the soil. From the results it is evident that the pH for maximum uptake of Lindane is 6.0. Further, it is apparent that the adsorption of Lindane rises from 2.0 % at pH 2.0 to 55.0 % at pH 6.0 in the case of clay and silt fraction (0–75 mm) and from 0.4 % at pH 2.0 to 46.0 % at pH 6.0 in the case of coarser soil fraction (250–300 μm). Almost similar behaviour with pH was reported by other authors (Jain and Ram, 1997a, b; Jain and Ali, 2000).

Effect of Adsorbent Dose (W_s)

The effect of adsorbent dose on the adsorption of Lindane is shown in Figs. 5(a) and 5(b) for a fixed initial concentration of 1000 μg/L. It is observed that the adsorption of Lindane per unit weight of adsorbent decreases with increasing adsorbent load. On the other hand percent adsorption increases from 55.0 to 78.6 % for the 0–75 μm fraction with increasing adsorbent load from 0.5 to 2.5 g/L.

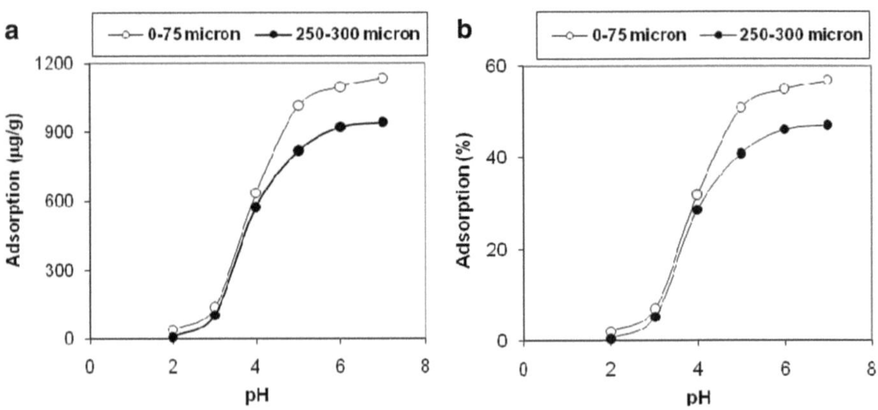

Fig. 4 (**a**) Effect of pH on adsorption of Lindane. (**b**) Effect of pH on percent adsorption of Lindane

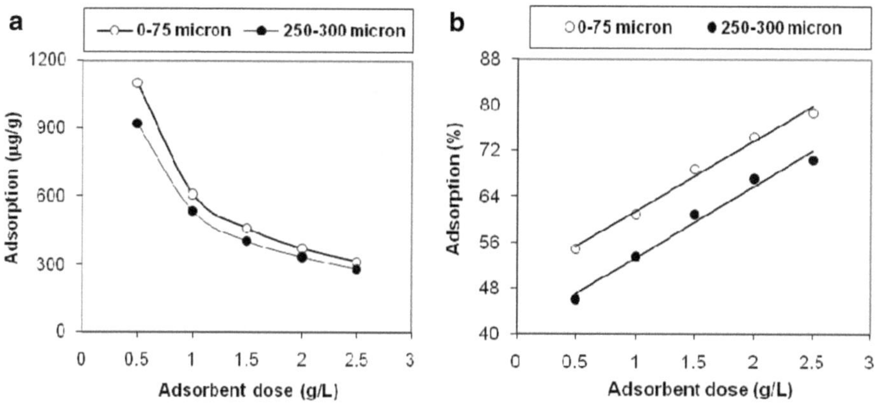

Fig. 5 (**a**) Effect of adsorbent dose on adsorption of Lindane. (**b**) Effect of adsorbent dose on percent adsorption of Lindane

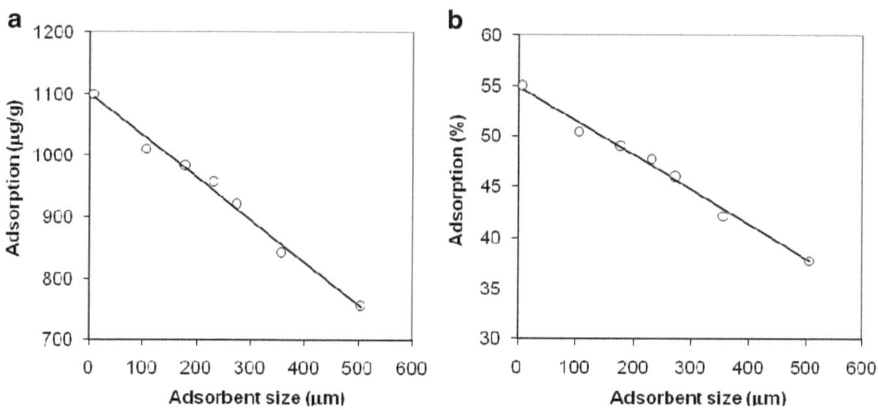

Fig. 6 (**a**) Effect of adsorbent size on adsorption of Lindane. (**b**) Effect of adsorbent size on percent adsorption of Lindane

Effect of Soil Particle Size (d_p)

The effect of particle size of soil adsorbent on the adsorption of Lindane is shown in Figs. 6(a) and 6(b). These plots reveal that for a fixed adsorbent dose, the adsorption of Lindane is higher for smaller adsorbent size. Further, it is observed that the percentage adsorption decreases with increasing geometric mean of adsorbent size. This is because, adsorption being a surface phenomenon, the smaller particle sizes offered comparatively larger surface area and hence higher adsorption occurs at equilibrium. Similar trend with adsorbent dose and size was also reported by other workers (Jain and Ram, 1997a, b; Jain and Ali, 2000). The higher content of iron and manganese in the 0–75 µm soil fraction also accounts for higher adsorption in this fraction, which is the main driving force for the adsorption of Lindane.

Column Study

For transport of Lindane in unsaturated zone of the study area, optimized conditions for adsorption of Lindane on soil as discussed above were maintained in the column of diameter 4″ and of length 25 cm (Fig. 1). 10 ml of 10 mg/L Lindane was loaded in the column and 25 ml effluent fractions were collected maintaining constant effluent rate. The concentration of Lindane was analysed in the collected different effluent column fractions and are given in Table 4.

The diffusion process controls the sorption of Lindane on the soil particles. Initially the uptake of Lindane on soil particles was very high and decreased with increase in flow through the soil bed in the column. The concentration of Lindane in the effluent fraction increased. When the soil bed in the column becomes saturated with Lindane solution, the concentration of the Lindane in the effluent fraction starts decreasing. The results of the experimental study can be summarized as under.

Table 4 Concentration of Lindane in different fractions collected after leaching through soil column

Fraction	*Concentration Lindane (μg/L)*	*Concentration of Lindane (ng)*	*Time taken (minute) for each 25 ml fraction to leach*
1	0.496	12	320
2	1.598	40	340
3	1.618	41	360
4	1.921	48	420
5	2.003	50	435
6	3.177	79	450
7	4.172	104	480
8	5.174	129	480
9	5.335	133	480
10	7.527	188	490
11	10.111	253	540
12	10.301	258	540
13	9.593	240	540
14	8.831	221	540
15	8.559	214	555
16	7.723	193	590
17	7.37	184	605
18	5.722	143	630
19	4.896	122	660
20	2.985	75	660
21	2.435	61	670
22	2.215	55	720
23	1.384	35	780
24	1.244	31	900
25	0.764	19	960

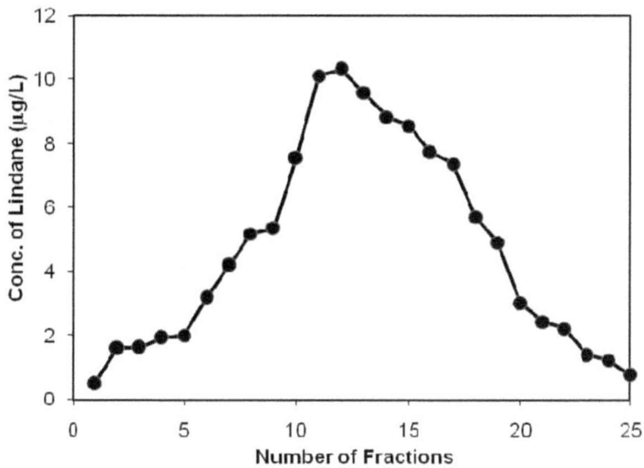

Fig. 7 Transportation of Lindane through soil column

Conditions for the Column Study

- Length of soil column = 25 cm
- pH of mobile phase = 6.0
- Conc. of Lindane loaded = 10 ml of 10 mg/L = 100 µg
- Temperature = 30 °C
- No. of fraction taken = 25
- Flow rate maintained = 0.1 ml/min

Result of Soil Column Experiment

- Maximum concentration of Lindane in effluent column fraction = 10.301 µg/L = 258 ng
- Number of effluent fraction in which maximum conc. of Lindane observed = 12th
- Total weight of the soil taken in the column = 3000 gm/L
- Maximum amount of transportation of Lindane occurred = 3.434 ng/g

It can be inferred from the above results that maximum concentration of Lindane was observed in 12th fraction and maximum amount of transportation of Lindane through the soil of Vadodara city was 3.434 ng/g (Fig. 7). These results can be used to develop a numerical model for simulation of Lindane transport within the vertical soil column.

Conclusion

Lindane is highly persistent in groundwater of metropolitan city of Vadodara, India. It shows a low affinity for soils binding, and may be mobile in soils with especially low organic matter content or subject to high rainfall and may pose a risk to groundwater contamination. The organic carbon content in the soil of Vadodara varies from 1.71 to 4.31 % in different size fractions of Vadodara soil, supporting the ease of mobility of Lindane through the soil (Table 3). Therefore, transport behaviour of Lindane in the unsaturated zone will provide an important input for the study of the migration pattern of Lindane in space and time for future projection. The observed data from the batch and column study can be used to develop a numerical model for simulation of Lindane transport within the vertical soil column.

References

Ali, I. and Jain, C.K. (2000). Transportation behaviour of Lindane in different types of soils. Technical Report TR/BR-7/1999-2000, National Institute of Hydrology, Roorkee, India.

APHA (1992). Standard Methods for the Examination of Water and Waste Waters. American Public Health Association, 18th Edition, Washington, DC.

Beltran, J., Hernandez, F., Lopez, F.J. and Morell, I. (1995). Study of sorption processes of selected pesticides on soils and ceramics, porous cups used for soil solution sampling. *Int. J. Environ. Anal. Chem.*, **58(1&4):** 287–303.

Chen, W. and Mulchandani, A. (1998). The Use of Live Biocatalysts for Pesticide Detoxification. *Trends in Biotech*, **16:** 71–76.

Freijer, J.I., Veling, E.J.M. and Hassanizadeh, S.M. (1998). Analytical solutions of the convection–dispersion equation applied to transport of pesticides in soil columns. *Environ. Modell. Software*, **13(2):** 139–149.

Hadiri, N.E., Ammati, M., Chgoura, M. and Mounir, K. (2003). Behavior of 1, 3-dichloropropene and methyl isothiocyanate in undisturbed soil columns. *Chemosphere,* **52(5):** 893–899.

Hendershot, W.H., Lalande, H., Duquette, M. (1993). Soil Reaction and Exchangeable Acidity. *In*: Carter, M.R. and Gregorich, E.G. (eds). Soil Sampling and Methods of Analysis. Lewis Publishers, London, U.K.

Hutson, D.H. and Roberts, T.R. (1984). Progress in Pesticide Biochemistry and Toxicology. John Wiley and Sons, Chichester U.K.

Indelman, P., Touber-Yasur, I., Yaron, B. and Dagan, G. (1998). Stochastic analysis of water flow and pesticides transport in a field experiment. *J. Contamin. Hydrol.*, **3(1–2):** 77–97.

Jain, C.K. and Ali, I. (2000). Adsorption of cadmium on riverine sediments: Quantitative treatment of the large particles. *Hydrol. Processes*, **14:** 261–270.

Jain, C.K. and Ram, D. (1997a). Adsorption of lead and zinc on bed sediments of the river Kali. *Wat. Res.*, **31(1):** 154–162.

Jain, C.K. and Ram, D. (1997b). Adsorption of metal ions on bed sediments. *Hydrol. Sci. J.*, **42(5):** 713–723.

Köhne, J., Maximilian, Köhne Sigrid and Šimůnek, Jirka (2006). Multi-process herbicide transport in structured soil columns: Experiments and model analysis. *J. Contamin. Hydrol.*, **85(1–2):** 1–32.

Kumar, Madhava and Philip Ligy (2006). Adsorption and desorption characteristics of hydrophobic pesticide endosulfan in four Indian soil. *Chemosphere*, **62(7):** 1064–1077.

Landry David, Dousset Sylvie and Andreux Francis (2004). Laboratory leaching studies of oryzalin and diuron through three undisturbed vineyard soil columns. *Chemosphere*, **54(6):** 735–742.

Lennartz, Bernd (1999). Variation of herbicide transport parameters within a single field and its relation to water flux and soil properties. *Geoderma,* **91(3–4):** 327–345.

Li Kun, Baoshan Xing and William A. Torello (2005). Effect of organic fertilizers derived dissolved organic matter on pesticide sorption and leaching. *Environ Pollut.*, **134(2):** 187–194.

McGechan, M.B. and Lewis, D.R. (2002). SW—Soil and Water Transport of Particulate and Colloid-sorbed Contaminants through Soil, Part 1: General Principles. *Biosystems Engineering*, **83(3):** 255–273.

Mouvet, C., Jeannot, R., Riolland, H. and Maciag, C. (1997). Stability of isoproturon, bentazone, terbuthylazine and alachlor in natural groundwater, surface water and soil water samples stored under laboratory conditions. *Chromosphere*, **35(5):** 1083–1097.

Paknikar, K.M., Nagpal, V., Pethkar, A.V. and Rajwade, J.M. (2005). Degradation of Lindane from aqueous solution using iron sulfide nanoparticles stabilized by biopolymers. *Sci. Technol. Adv. Mater.*, **6(3–4):** 370–374.

Quan, X., Niu, J.F., Chen, S., Chen, J.W., Zhao, Y.Z. and Yang, F.L. (2003). Effect of Fe_2O_3, organic matter and carbonate on photocatalytic degradation of Lindane in the Sediment from Liao River, China. *Chemosphere*, **52:** 1749–1755.

Rao, P.S.C. and Davidson, J.M. (1979). Adsorption and movement of selected pesticides at high concentrations in soils. *Water Res.*, **13(4):** 375–380.

Roy, William R., Krapac, Ivan G. and Chou, Sheng-Fu J. (1999). Chemical Fate and Transport of Atrazine in Soil Gravel Materials at Agrichemical Distribution Facilities. *J. Soil Contamin.*, **8 (3):** 365–387.

Sarin, Vikrant, Singh, T.S. and Pant, K.K. (2006). Thermodynamic and breakthrough column studies for the selective sorption of chromium from industrial effluent on activated eucalyptus bark. *Biores. Techmol.*, **97:** 1986–1993.

Sherma, J. (1995). Pesticides. *Anal. Chem.*, **67:** 1R.

Singh, K.P., Takroo, R. and Ray, P.K. (1987). Analysis of pesticide residues in water. Indian Toxicology Research Centre, Lucknow, India. ITRC Manual.

Tanabe, S., Madhushree, B., Ozturk, A., Tatsukawa, R., Miyazaki, N., Ozdamar, E., Aral, O., Samsun, O. and Ozturk, B. (1997). Persistent organochlorine residues in harbour porpoise (Phocoena phocoena) from the black sea. *Mar. Pollut. Bull.*, **34(5):** 338–347.

US Environmental Protection Agency (1974). Methods for chemical analysis for water and wastes. US EPA Environmental Monitoring and Support Laboratory, Cincinnati.

Willett, K.L., Ulrich, E.M. and Hites, R.A. (1998). Differential toxicity and Environmental fates of hexachlorocyclohexane isomer. *Environ. Sci. Technol.*, **32:** 2197–2207.

Weber, W.J. Jr. and Moris, J.C. (1963). J. Sanit. Eng. Div. ASCE, SA2, 31. *In:* Faust, S.D., Aly, O. M., 1987. Adsorption Processes for Water Treatment. Butterworth, London.

Wong, S., Charles, Scott, Mabury, A. Scott, Whittle, D. Michael, Backus, M. Sean, Teixeira, Camilla, Devault, S. David, Bronte, R. Charles and Derek, C.G. Muir (2004). Organochlorine Compounds in Lake Superior: Chiral Polychlorinated Biphenyls and Biotransformation in the Aquatic Food Web. *Environ. Sci. Technol.*, **38:** 84–92.

Section III
Climate and Natural Resources Management

Effect of Ozone on Biotic Stress Tolerance Potential of Wheat

Usha Mina, Rashmi Aggarwal, Parimal Sinha, Arti Bhatia, and Anshul Fuloria

Introduction

Ozone (O_3) is generated as an air pollutant in the troposphere in a photochemical reaction by the action of sunlight on volatile organic compounds and oxides of nitrogen emitted by vehicles and industry. O_3 concentration in troposphere is rising at an annual rate of 0.5 % (IPCC, 2007) over its background concentration of 10–20 ppb. According to IPCC 4th assessment report (2007), current tropospheric O_3 concentrations over the northern hemisphere in summers are about 30–40 ppb and are expected to rise upto 70 ppb in 2100. Elevated levels of O_3 present in troposphere are phytotoxic and directly affects plants by reacting with apoplastic leaf components and forming reactive oxygen species (ROS) like hydroxyl (OH^-), peroxyl (OH^{2-}) and superoxide (O_2^-) radicals (Fiscus et al., 2005). This oxidative burst causes loss of photosynthetic activity and reduced growth and yield of crops (Fiscus et al., 2005). Economic crop losses due to O_3 were equivalent to $17-$82 million in US, 310 million euros in Netherlands and $ 2 billion in China (Mauzerall and Wang, 2001). In India also O_3 phytotoxic impacts on growth and yield of several crops were reported (Varshney and Rout, 1998, Tiwari et al., 2005; Mina et al., 2010). O_3 also influences plant's susceptibility to biotic stress such as pathogens which causes diseases. Plants have innate mechanisms to protect them from various abiotic and biotic stresses. However the dual stress imposed by O_3 and pathogen affects tolerance of crop and leads to altered host pathogen interaction

U. Mina (✉) • A. Bhatia • A. Fuloria
Centre for Environment Science and Climate Resilient Agriculture, Indian Agricultural Research Institute, New Delhi 110012, India
e-mail: usha_env@iari.res.in

R. Aggarwal • P. Sinha
Division of Plant Pathology, Indian Agricultural Research Institute, New Delhi 110012, India

N.J. Raju et al. (eds.), *Management of Water, Energy and Bio-resources in the Era of Climate Change: Emerging Issues and Challenges*,
DOI 10.1007/978-3-319-05969-3_23

(Fuhrer, 2003). Alteration in pathogenesis potential of pest due to O_3 exposure is of ecological and economical importance.

Enhancement in diseases such as powdery mildew and leaf spot, decrease and no impact on tan spot fungus and other root diseases of wheat crop due to O_3 and pathogen interaction have been reported in literature (Chakraborty et al., 2008). So the disease severity, which is positively related with increase in virulence and aggressiveness of pathogen, is also largely mediated by the host resistance under O_3 exposure. In India reports on impact of O_3 on host-pathogen interactions are very limited.

Wheat (*Triticum aestivum* L.) is a staple food crop of India. Its enhanced productivity post Green Revolution played a key role in ensuring food security (Joshi et al., 2007). Spot Bloch of wheat caused by *Bipolaris sorokiniana* occurs throughout the wheat growing regions and ranks close to leaf rust in destructiveness (Adlakha, 1984). In India it caused as much as 17.5 % yield losses in wheat crop (Saari, 1998). With this background experiments were carried out on wheat crop and its pathogen *B. sorokiniana* with following objectives: (i) To observe the effect of O_3 stress on tolerance potential of wheat; (ii) To assess development of spot blotch disease severity on wheat plants (PBW343) pre-exposed to O_3; and (iii) To observe defense response (Total PR protein and chitinase activity) system of wheat under O_3 and biotic interactions.

Materials and Methods

Plant Material

Untreated seeds of wheat (*Triticum aestivum* L cv. PBW343) were sown in pots (15 cm diameter, containing 1.5 kg sterilized field soil and compost in ratio 3:1 and two seeds per pot), after emergence seedlings were thinned to obtain one plant per pot. All the pots received 300 mg of nitrogen (in the form of urea) and were placed in field under ambient conditions till plants reached tillering growth stage (35-day old). Thereafter the pots were shifted to Open Top Chambers (OTCs), each of 90 × 120 cm (Dia × height) dimension for exposer to different O_3 levels. OTCs were made of transparent polyvinyl sheets with open top.

O_3 Exposure

In OTCs O_3 is supplied by O_3 generator (Standard Appliances, Varanasi, India) by ionizing oxygen (O_2) in the presence of ultra-violet light. The outlet of the O_3 generator was connected to the inlet of the blower of the chamber. The blower assembly mixed the O_3 with ambient air and dispensed the mixture in the connected OTC. O_3 was continuously measured with an automatic UV-absorption analyzer.

Effect of O_3 (abiotic) and O_3 + Pathogen (abiotic + biotic) Stress on Tolerance and Defense Response of Wheat

Wheat plants were exposed to different O_3 levels for 21 days, 8 hours per day. Different O_3 levels selected as treatments were: (i) 'A' Ambient levels with 12–72 ppb O_3 concentration; (ii) 'NF' non-filtered air having O_3 concentration 5-10 % lower than A; (iii) 'EO_3' elevated level having A + 25–30 ppb O_3, and (iv) 'CF' charcoal filtered air with 65-70 % lower than ambient O_3 levels. Average temperature and average relative humidity at the experimental site during the crop growth period were 20 ± 2.5 °C during day, 15 ± 2.0 °C at night and 50 ± 6.0 % respectively. After 21 days exposure one set of exposed plants were maintained under ambient condition to observe their recovery from ozone stress. They were monitored after 14 days of recovery period i.e. on 35th day. Another set was inoculated with the pathogen to observe impact of ozone preexposure on defense response and disease development.

Pathogen Inoculation and Disease Assessment

Wheat plants after 21 days O_3 exposer were prepared for inoculation with isolate BS-75 of *Bipolaris sorokiniana* (Sacc.) Shoem, pathogen of spot blotch disease obtained from the Department of Plant Pathology of the Indian Agricultural Research Institute, New Delhi. Mycelium was grown on potato dextrose agar (PDA, Sigma) for three weeks at 25 °C in the dark. Conidia and hyphae were scraped off from medium with spatula and mixed with sterile distilled water. Conidial suspensions were prepared by filtering the solution through cheesecloth to eliminate mycelial fragments. Inoculum densities were adjusted to $2–2.5 \times 10^3$ conidia per ml and two drops of Tween 20 was added per 100 ml of suspension to improve the contact between inoculum drop and leaf surface. Inoculation was done in evening before sunset. Six tillers of wheat plants in each pot were covered with a transparent polythene bag with a thickness of 0.06 mm and immediately inoculated in the field by spraying 15 ml of the prepared conidial suspension per tiller. Polythene film was removed after 48 hours. The film blocked wind and kept moisture inside while transmitting long wave radiation. Consequently, the wheat surface was kept cooler than the surrounding atmosphere and dew formation on the wheat leaves surface was enhanced. Inoculated plants were monitored for disease development for 14 days. Effect of the abiotic + biotic stress was assessed on inoculated O_3 exposed plants as compared to uninoculated O_3 exposed plants after 14 days of inoculation with respect to following parameters:

O_3 Injury Symptoms

Visual estimates of ozone injury were carried out during and after the exposure. The O_3 injury was recorded for number of leaves exhibiting O_3 induced injury symptoms >5 % of their leaf area.

Air Pollution Tolerance Index (APTI)

The tolerance of wheat plants was determined through APTI which was calculated according to the formula (Escobedo et al., 2008):

$$\text{APTI (mg/g)} = [A(T + P) + R]/10$$

where A represents ascorbic acid; T, total chlorophyll; P, the leaf extract pH; and R, relative leaf water content. These parameters were analyzed as follows.

Chlorophyll Estimation

Total chlorophyll estimation was done according to the method described by Arnon (1949). Three gram of fresh leaf sample from each treatment was blended and then extracted with 10 ml of 80 % acetone and left for 15 minutes for thorough extraction. The liquid portion was decanted into another test-tube and centrifuged at 2500 rpm for three minutes. The supernatant was then collected and the absorbance taken at 645 nm and 663 nm using a spectrophotometer. Calculations were done using the formula:

$$\text{Total Chlorophyll (mg/g fresh wt)} = [20.5 \times D_{645} + 8.02 \times D_{663}] \times V/d \times 1000 \times W$$

where D is absorbance of the extract at 645 and 663 nm; V = total volume of the chlorophyll solution (ml); d = length of light path and W is weight of fresh leave tissue (g).

Leaf Extract pH

Five gram of the fresh leaves was homogenized in 10 ml deionized water. This was then filtered and the pH of leaf extracted determined after calibrating pH meter with buffer solution of pH 4 and pH 9.

Relative Leaf Water Content (RWC)

Relative water content was determined according to the method of Singh (1977) and calculated with the formula:

$$\text{RWC}\ (\%) = [(FW - DW)/(TW - DW)] \times 100$$

where *FW* is fresh weight, *DW* = dry weight, and *TW* = turgid weight.

Fresh weight was obtained by weighing 30 discs of 5 mm diameter cut from the fresh leaves. The discs were then immersed in distilled water overnight till they got saturated, blotted dry and then weighed to get the turgid weight. Next, the leaves discs were oven dried at 70 °C for 4–5 hrs and reweighed to obtain the dry weight.

Ascorbic Acid

Ascorbic acid was estimated by method adapted from Sadasivam and Balasubraminan (1987). Ascorbic acid (AA) was extracted from 5 g leaf tissue in 100 ml (4 %) oxalic acid at 4 °C and centrifuged. Supernatant measuring 5 ml was taken in a 50 ml volumetric flask and 10 ml of oxalic acid was added to it. The flask was swirled and titrated against 2,6-dichlorophenol Indophenol (DCPIP) solution (42 mg sodium bicarbonate + 0.52 mg 2,6-DCPIP made up to 200 ml with distilled water and diluted 1:1 with water) taken in a burette. The end point of titration was the appearance of pink colour. Standard stock solution was prepared by ascorbic acid solution (1 mg/ml) in 4 % oxalic acid solution in a volumetric flask. Stock solution was used to prepare working standards which were titrated against the dye DCPIP. Volume of the dye used was recorded. Amount of ascorbic acid was calculated by formula:

$$\text{Ascorbic acid (mg/g sample)} = (V \times T)/W$$

where *V* is volume of dye in ml used for titration of extract; *T* = AA equivalent per ml of dye; and *W* is weight of leaf material

Disease Development on Wheat Plants Pre-exposed to O_3

Development of *Bipolaris* lesions on plants preexposed to different treatments was monitored during the 14 days post-inoculation period for estimation of disease severity. After 14 days length (l) and width (w) were recorded for each lesion to calculate lesion size. Latent periods were determined by daily microscopic examinations for appearance of conidiophores within lesioned area. The number of days between inoculation and appearance of first mature conidia was recorded for each lesion. Five lesions were cut from each leaf and measured for size. Then placed in

5 ml sterilized distill water and shaken vigorously for one minute. Spores of *B. sorkiniana* in the water were counted with a hemacytometer and the number of spores per 25 mm^2 of spot surface was calculated.

Chitinase Activity

The colorimetric assay of chitinase activity was carried out using a modification (Reissig et al., 1955) of Morgan and Elson's method (Morgan and Elson, 1934). One gram of plant material was homogenized in 3 ml of 0.1 M sodium citrate buffer, pH 5 and centrifuged (10,000 g for 10 min). Supernatant was dialyzed with same buffer and dialyzate used for assay. The assay mixture contained the enzyme extract (1 ml), chitin suspension 4 ml and 3 mg/ml Bovine Serum Albumin (BSA) incubated in water bath at 37 °C for 3 hours. To the reaction mixture 1 ml water was added, boiled and then centrifuged (10,000 g for 10 min). In the supernatant 0.8 M potassium tetraborate was added and boiled for 3 min after which DMAB was added, and the amount of liberated N-acetyl glucosamine (NAG) was determined. The controls included enzyme and substrate blanks as well as internal standards prepared using NAG (0.1-0.4 M range). Absorbance was measured at 585 nm and activity was determined from a calibration curve. One Enzyme Unit (EU) activity is the amount of enzyme which produces 100 μg NAG/ml chitin solution in 1 h at 37 °C at pH 5.2.

PR Protein Estimation

The Total Protein Extraction was first done from wheat leaves using Total Protein Extraction Kit supplied by G-Biosciences. From the extracted total protein, PR proteins were quantitatively estimated by using PR ELISA Kit (catalog No. E0431 pl) supplied by USCN LIFE.

Statistical Analysis

Data obtained from the experiments were pooled and subjected to analysis of variance (ANOVA) and least significant differences were calculated at $p = 0.05$ using MS Excel.

Results

Effect of O_3 Exposure on Wheat

Effect of ozone on wheat were monitored twice, initially after 21 days of exposure and second time after 14 days of recovery period when plants were kept under ambient condition i.e. on 35th day.

O_3 Injury

Ambient O_3 levels at experimental site ranged between 11 and 55 ppb during wheat growing period. Following CF, NF and ambient O_3 levels exposure (21 days), plants did not show significant visible damage on the leaves. However in EO_3 exposed plants old leaves exhibited O_3 injury lesions with 8-10 % of injured area on the leaves. Leaves of six tillers used for inoculation completely lacked any visible damage. On 35th day ozone injury was maximum (23 %) on EO_3 exposed and least on CF exposed plant leaves.

O_3 Exposure and APTI

Exposure to different O_3 levels adversely affected APTI of wheat plants (Fig. 1a). CF exposed plants had highest APTI (10.47) while those exposed to EO_3 had least APTI (10.09) after 21 days of exposure. After 14 days recovery period on 35th day variation in APTI was persisted but it was higher as compared to just after exposure. On 35th day maximum APTI was of CF (10.57) followed by NF, A and EO_3 (8.04) exposed plants respectively.

O_3 and Chitinase Activity

O_3 exposure negatively affected activity of Chitinase enzyme (Fig. 1b). At higher O_3 concentration chitinase activity was lower. Highest chitinase activity was recorded in CF exposure (2.53 EU) followed by NF air, A and least in EO_3 exposure (2.078 EU). On 35th day (after recovery period) variations in chitinase activity under different O_3 levels increased compared to 21st day. EO_3 exposed plants had 27.3 % lower chitinase activity than CF exposed plants.

O_3 and PR Proteins

Exposure of O_3 significantly affected PR protein content in wheat. At higher O_3 concentration lower values of PR protein were recorded. Highest PR protein

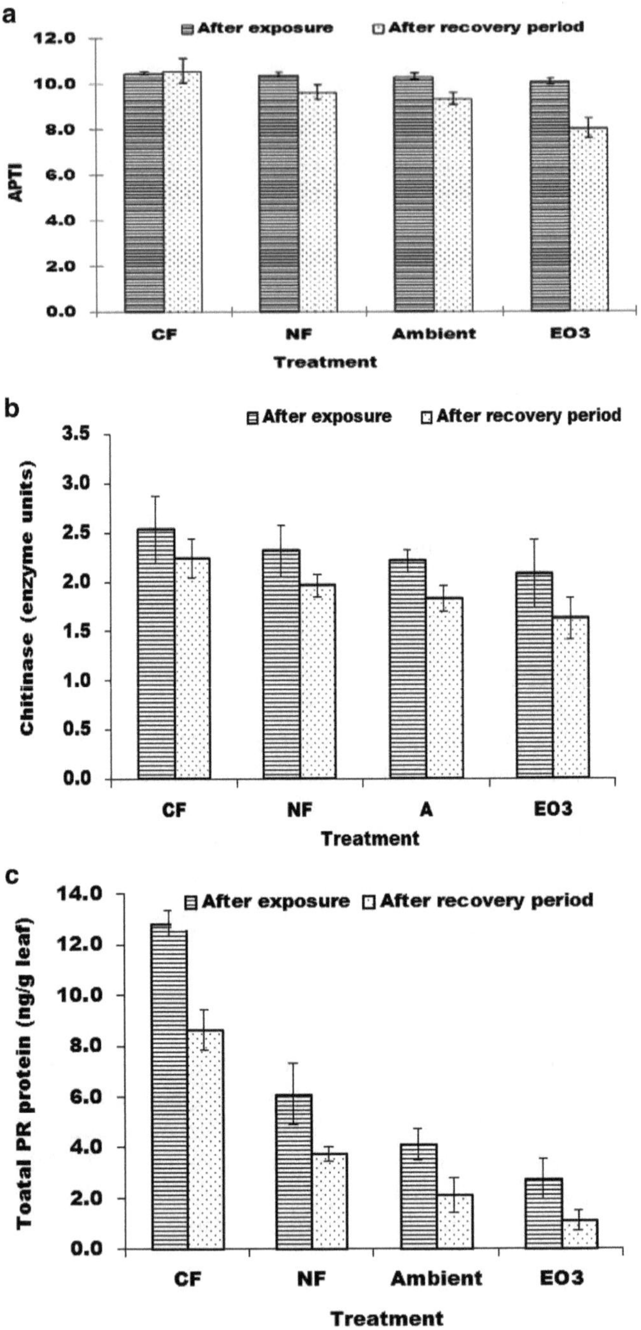

Fig. 1 Effect of ozone exposure on wheat: (**a**) APTI, (**b**) Chitinase activity and (**c**) PR protein content. Vertical bars are equal to SE of the mean (at $p < 0.05$)

content (12.83 ng/g leaf) was in CF exposed plants followed by NF and A exposed plants and lowest (2.74 ng/g leaf) in EO_3 exposed (Fig. 1c). After 14 days recovery period on 35th day PR protein content decreased more in all treatments, however similar trend was observed with highest content in CF (8.65 ng/g leaf) exposed plants and least under EO_3 (1.10 ng/g leaf) exposed plants.

Effect of O_3 Exposure Followed by Pathogen Inoculation Defense Response of Wheat

Disease Severity

Post inoculation the development of spot blotch lesions on wheat plants under different treatment was evaluated as a measure of disease severity (Fig. 2). Different O_3 levels effect on *Bipolaris* growth was variable and highly significant. Number of lesions/leaf was maximum and minimum on plants exposed to EO_3 and CF air respectively. Significant differences in *Bipolaris* lesions size were observed in NF air (27 %), A (50 %) and EO_3 (90 %) exposed plants as compared to CF exposed plants. In the experiment consistent shortening of latent period with increasing O_3 concentration was observed (Fig. 2). Latent period duration was minimum and maximum on plants exposed to EO_3 and CF air respectively. 50 %

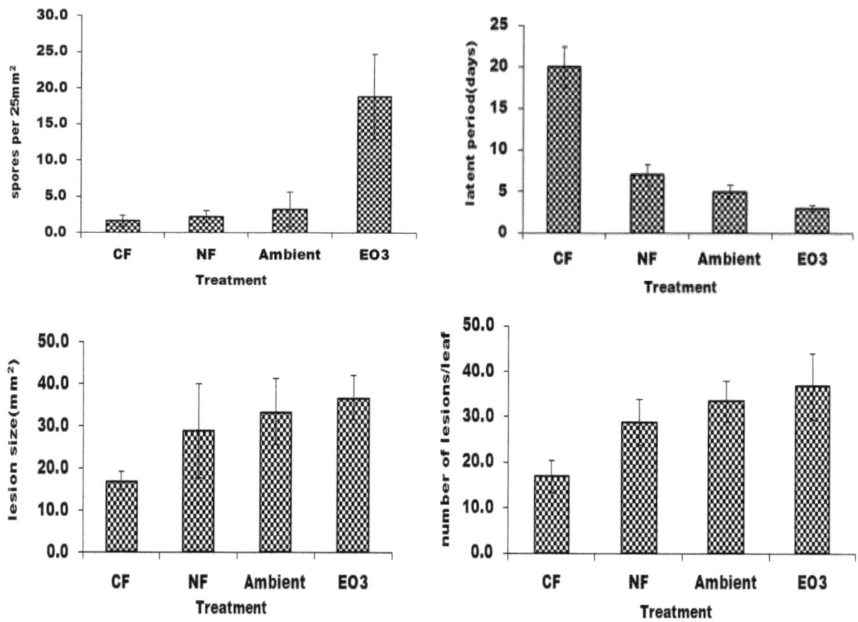

Fig. 2 Effect of ozone exposure on spot blotch disease development. Vertical bars equal to SE of the mean (at $p < 0.05$)

of plants exposed to EO_3 exhibited black spots with conidiophores on leaf surface within seven days of incubation period, whereas only 25 % plants exposed to ambient O_3 levels exhibited black spots with conidiophores even after 10 days of inoculation. In plants exposed to O_3 free air/CF air only on one plant disease symptoms were visible after 20 days of inoculation. Spores/25 mm^2 were highest on the leaves of plants exposed to EO_3 and minimum on plants treated with CF air.

Effect on APTI

The APTI values of O_3 exposed and *B. sorokiniana* inoculated plants were much lower compared to ozone exposed and uninoculated plants. Under CF treatment insignificant change in tolerance occurred leading to maximum latent period and minimum number of lesions per leaf. However the APTI of NF, A and EO_3 exposed and *B. sorokiniana* inoculated plants decreased by 10-11 % and 37 % respectively as compared to exposed but uninoculated ones (Fig. 3a).

Effect on Chitinase Activity

Chitinase activity is reduced more in O_3 exposed and inoculated plants as compared to O_3 exposed and uninoculated plants (Fig. 3b). The reduction in chitinase activity of inoculated plants varied from 6 % in CF exposed and 29 % in EO_3 exposed plants as compared to their respective control plants.

Effect on PR Protein Content

The PR protein content of ozone exposed and *B. sorokiniana* inoculated plants was much lower compared to ozone exposed and uninoculated plants (Fig. 3c). In NF and A pre-exposed diseased plants 30-34 % lesser PR protein was recorded compared to ozone exposed and uninoculated plants. In EO_3 predisposed and diseased plants high reduction of 60.5 % occurred in PR protein content. However under CF pre-exposed and diseased plants PR protein content increased slightly compared to ozone exposed and uninoculated plants.

Discussion

In the present study O_3 exposure decreased chlorophyll, leaf extract pH, ascorbic acid (AA) and water content in wheat plants reflected in reduced APTI. O_3 induced oxidative burst in leaf tissue causes reduced chlorophyll content and negatively affects crop growth and physiology (Fiscus et al., 2005). AA is a strong scavenger of these oxidative free radicals at high pH and acts as a defense compound (Rai

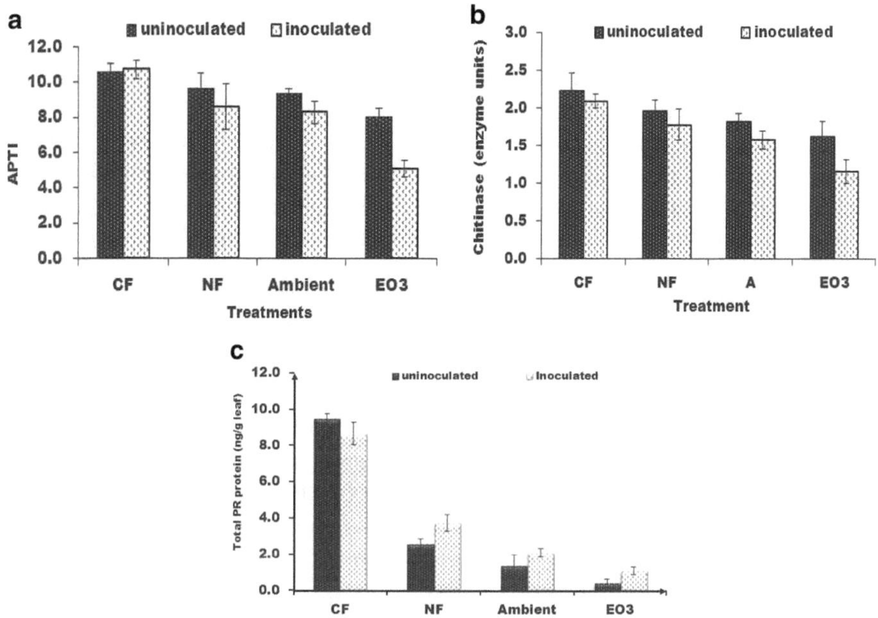

Fig. 3 Defense response of ozone pre-exposed wheat plants to *Bipolaris sorokiniana* inoculation w.r.t ozone preexposed uninoculted plants: (**a**) APTI, (**b**) Chitinase activity and (**c**) Total PR protein content. Vertical bar SE of the mean (at $p < 0.05$)

et al., 2011). Studies reveal that apoplastic AA content increase in O_3 tolerant crops on O_3 exposure (Burkey et al., 2003). Similar to findings of present study, a lower leaf extract pH and low AA content under EO_3 compared to CF was found in wheat by Biswas et al. (2008). Toree (2008) reported variable response of wheat cultivars with respect to AA content and O_3 levels as well as cultivars that exhibited enhancement in AA content were less affected by adverse effects of O_3 compared to sensitive ones. This can be accounted for varied O_3 sensitivity in different wheat cultivars (Barnes et al., 1990; Heagle et al., 2000; Biswas et al., 2008). High water content is also essential for maintaining proper physiological balance and providing resistance to pollutant stress (Liu and Ding, 2008). In tolerant plant species an increase in chlorophyll, relative water content, AA and thus higher APTI occurs in response to pollutant exposure while in sensitive ones like wheat there is reduction in all (Meerabai et al., 2012; Mohammed et al., 2011). In present study low APTI at EO_3 indicated low tolerance potential as well as O_3 sensitivity of PBW343 cultivar of wheat. Disease tolerance and resistance capacity of host plants in presence of O_3 stress can be assessed indirectly through APTI (Escobedo et al., 2008) and induction of important defense compounds such as PR (Pathogenesis Related) proteins which are protective plant proteins specifically induced in pathological or stress related situations (Bowles, 1990).

In present study reduction in chitinase activity and PR protein content in wheat plants under EO_3 as compared to CF was observed. This is contrary to previous studies where increase of PR proteins like chitinase and ß 1–3 glucanase is reported in different crops on O_3 stress exposure (Thalmair et al., 1996). This can be related to varied O_3 sensitivity of different cultivars. For example, near-ambient ozone exposure induced ß-1,3-glucanase activity 22-fold in the O_3 tolerant tobacco line Bel B while in O_3 sensitive line Bel W3 only a fourfold increase in ß-1,3-glucanase activity (Schraudner et al., 1994). Higher induction and stronger chitinase expression against phyto-pathogens in resistant varieties compared to susceptible ones in sugar beet (Nielsen et al., 1993), wheat (Anguelova et al., 2001) and tomato (Lawrence, 2000) was reported. While no difference in the induction timing or amounts of PR-protein in resistant and susceptible cultivars of cotton occurred (Mc Fadden et al., 2001).

After 21 days exposure variations in above parameters under different O_3 treatments was not very significant but on 35th day (after 14-day recovery period) comparatively greater variation and lower values of APTI, chitinase activity and PR protein were recorded in plants exposed to different treatment. O_3 induced negative effect was directly related to its concentration. Lowest APTI, PR protein and chitinase activity were observed in wheat plants exposed to EO_3, followed by A, NF and CF treated plants. Negative effect of O_3 gets more pronounced with increasing age of plant, O_3 concentration and sensitivity of cultivar (Soja et al., 1995). According to Feng et al. (2008) EO_3 (mean concentrations between 31 and 200 ppb) exposure caused significant reduction in wheat growth and yield relative to CF air exposure. The damage is reported to be more in grain filling stage compared to earlier ones (Feng et al., 2008).

Maximum numbers of lesions, lesion size, spores count and short duration latent period of *Bipolaris* were observed on plants exposed to EO_3. However in literature, decreased (Dohmen, 1987), increased (Tiedemann, 1992) and no change (Pfleeger et al., 1999) are reported in number of pathogenic pustules on O_3 exposed plants as compared to control plants. Such variation in response of O_3 predisposed crops to pathogen may be on account of necrotrophic and biotrophic nature of fungal pathogen or O_3 concentration, the stage of plant development (Tiedemann, 1992). *Bipolaris* is necrotrophic pathogen. Dowding (1988) suggested that necrotrophic pathogens (pathogens that use dead tissue as an energy source) would be encouraged by host exposure to O_3 as compared to biotrophic pathogen (pathogen that use live tissue as an energy source) (Pfleeger et al., 1999). In present study necrosis on leaf tips during O_3 exposure has been observed. Availability of necrotic tissue may have promoted growth of *Bipolaris* on O_3 and EO_3 exposed wheat plants as compared to CF exposed ones. Shortening of latent period under EO_3 can apparently be due to stimulatory effect of O_3 on the fungus. Khan and Khan (1999) also reported higher germination of conidia of *Sphaerotheca fuliginea* (powdery mildew fungus) on bottle gourd after O_3 exposure.

In present study O_3 exposer altered wheat response to biotic diseases similar to earlier reports (Manning, 1975; Fuhrer, 2003). Alteration in defense response and disease severity were because of reduced tolerance and enhancement in susceptibility of O_3 exposed wheat to *B. sorokiniana*. Dual stress (O_3 + pathogen) lead to significant reduction in APTI, chitinase activity and PR proteins content as compared to single stress (O_3). Maximum reduction was observed in EO_3 exposed and diseased plants and least in CF exposed disease-free wheat plants. Dual stress of O_3 exposure and *B. sorokiniana* inoculation considerably reduced the tolerance of wheat, hence enhancing its susceptibility to disease. The reduction in APTI was more at higher O_3 concentrations leading to minimum latent period and maximum number of visible lesions on EO_3 exposed plants followed by ambient, NF and CF air exposure. These observations confirm the Tiedemann (1992) report on enhancement in predisposition of wheat to *Septoria nodorum* and *B. sorokiniana* after O_3 exposure. But on the contrary, Pfleeger et al. (1999) reported no enhancement in rust disease severity by exposure of wheat to O_3. Nevertheless O_3 induced reduction in tolerance and defense reaction of wheat as observed in present study substantially supports the increased susceptibility of wheat to *Bipolaris Sorokiniana* at higher O_3 concentrations compared to CF exposure.

Conclusion

O_3 induced negative effects on wheat were directly correlated with its concentration and indirectly with its reduced tolerance and defense response. In present study an enhancement in predisposition of wheat plants to *Bipolaris sorokiniana* was observed. It was supported by observational data on number of lesions, size of lesion, duration of latent period and spore count. Hence owing to continued increase in tropospheric O_3 concentrations the areas of higher ambient concentration of O_3 may experience a higher incidence of spot blotch damage. Wheat being a staple food crop in India, to prevent losses in its productivity due to synergistic effect of O_3 and *Bipolaris* pathogen, proper package of agricultural practices need to be developed and adopted in future.

Acknowledgement Financial support from the Indian Council for Agricultural Research is gratefully acknowledged.

References

Adlakha, K.J., Wilcoxson, R.D. and Raychaudhari, S.P. (1984). Resistance of wheat to leaf spot caused by *Bipolaris sorokiniana*. The American Phytopathological Society. *Plant Disease*, **68:** 320–321.

Anguelova, M.V., Westhuizen, V.D. and Pretorius, Z.A. (2001). Beta-1,3-glucanase and chitinase activities and the resistance response of wheat to leaf rust. *Journal of Phytopathology*, **149:** 381–384.

Arnon, D.I. (1949). Copper enzymes in isolated chloroplasts, polyphenoxidase in beta vulgaris. *Plant Physiology*, **24:** 1–15.

Barnes, J.D., Velissariou, D., Davison, A.W. and Holevas, C.D. (1990). Comparative ozone sensitivity of old and modern Greek cultivars of spring wheat. *New Phytologist*, **116:**707–714.

Biswas, D.K., Xu, H. and Li, Y.G. (2008). Assessing the genetic relatedness of higher ozone sensitivity of modern wheat to its wild and cultivated progenitors/relatives. *Journal of Experimental Botany*, **59:** 951–963.

Bowles, D.J. (1990). Defense-related proteins in higher plants. *Annual Review of Biochemistry*, **59:** 873–907.

Burkey, K.O., Eason, G. and Fiscus, E.L. (2003). Factors that affect leaf extracellular ascorbic acid content and redox status. *Physiologia Plantarum*, **117:** 51–57.

Chakraborty, S., Luck, J., Hollaway, G. et al. (2008). Impacts of global change on diseases of agricultural crops and forest trees. *CAB reviews: Perspective in agriculture, veterinary sciences, nutrition and natural resources*, **3:** 1–15.

De la Torre, D. (2008). Quantification of mesophyll resistance and apoplastic ascorbic acid as an antioxidant for tropospheric ozone in durum wheat (*Triticum durum* Desf. cv. Camacho). *The Scientific World Journal*, **8:** 1197–1209. DOI 10.1100/tsw. 2008.149.

Dohmen, G.P. (1987). Secondary effects of air pollution ozone decreases brown rust disease potential in wheat. *Environmental Pollution*, **43:** 189–194.

Dowding, P. (1988). Air pollution effects on plant pathogens. *In:* Schulte-Hostede, S., Darall, N. M., Blank, L.W. and Wellburn, A.R. (eds). Air Pollution and Plant Metabolism. Elsevier, Amsterdam.

Escobedo, F.J., Wagner, J.E., Nowak, D.J. et al. (2008). Analyzing the cost effectiveness of Santiago, Chile's policy of using urban forests to improve air quality. *Journal of Environment Management*, **86:** 148–157.

Feng, Z.Z., Kobayashi, K. and Ainsworth, E.A. (2008). Impact of elevated ozone concentration on growth, physiology and yield of wheat (*Triticum aestivum* L.): A meta-analysis. *Global Change Biology*, **14:** 2696–2708.

Fiscus, E.L., Booker, F.L. and Burkey, K.O. (2005). Crop loss responses to ozone: Uptake, mode of action, carbon assimilation and partitioning. *Plant, Cell and Environment*, **28:** 997–1011.

Fuhrer, J. (2003). Agroecosystem responses to combinations of elevated CO_2, ozone, and global climate change. *Agriculture, Ecosystems and Environment*, **97:** 1–20.

Heagle, A.S., Miller, J.E. and Pursley, W.A. (2000). Growth and yield response of winter wheat to mixtures of ozone and carbon dioxide. *Crop Science*, **40:** 1656–1664.

Intergovernmental Panel on Climate Change (2007). Summary for Policymakers. *In:* Solomon, S., Qin, D., Manning, M., Chen, Z., Marquis, M., Averyt, K.B. et al. (eds). The Physical Science Basis. Contribution of Working Group I to the Fourth Assessment Report of the Intergovernmental Panel on Climate Change. Cambridge University Press, New York.

Joshi, A.K., Ortiz-Ferrara, G., Crossa, J. et al. (2007). Combining superior agronomic performance and terminal heat tolerance with resistance to spot blotch (Bipolaris sorokiniana) of wheat in the warm humid Gangetic Plains of South Asia. *Field Crop Research*, **103:** 53–61.

Khan, M.R. and Khan, M.W. (1999). Effects of intermittent ozone exposures on powdery mildew of cucumber. *Environmental Experimental Botany*, **42:** 163–171.

Lawrence, C.B., Singh, N.P., Qui, J., Gardner, R.G. and Tuzun, S. (2000). Constitutive hydrolytic enzymes are associated with polygenic resistance of tomato to *Alternaria solani* and may function as an elicitor release mechanism. *Physiological and Molecular Plant Pathology*, **57:** 211–220.

Liu, Y.J. and Ding, H. (2008). Variation in air pollution tolerance index of plants near a steel factory: Implication for landscape-plant species selection for industrial areas. *WSEAS Transactions on Environment and Development*, **4:** 24–32.

Manning, W.J. (1975). Interactions between air pollutants and fungal, bacterial and viral plant pathogens. *Environmental Pollution*, **9:** 87–90.

Mauzerall, D.L. and Wang, X. (2001). Protecting agricultural crops from the effects of tropospheric ozone exposure reconciling science and standard setting. *Annual Review of Energy and the Environment*, **26:** 237–268.

Mc Fadden, H.G., Chapple, R., de Feyter, R. and Dennis, E. (2001). Expression of pathogenesis related genes in cotton stems in response to infection by Verticillium dahliae. *Physiological and Molecular Plant Pathology*, **58:** 119–131.

Meerabai, G., Venkata Ramana, C. and Rasheed, M. (2012). Effect of industrial pollutants on Physiology of Cajanus cajan (L.) – Fabaceae. *International Journal of Environmental Sciences*, **2(4):** 1889–1894.

Mina, U., Kumar, P. and Varshney, C.K. (2010). Response of different growth stages of Potato (*Solanum tuberosum*) to ozone stress. *Phyton*, **49:** 253–266.

Mohammed, K., Kumari, R. and Ramteke, P.W. (2011). Studies on Air Pollution Tolerance of selected plants in Allahabad city, India. *Journal of Environmental Research and Management*, **2:** 042–046.

Morgan, W.T.J. and Elson, L.A. (1934). A colorimetric method for the determination of N-acetylglucosamine and N-acetylchondrosamine. *Biochemical Journal*, **28:** 988–995.

Nielsen, K.K., Mikkelsen, J.D., Dragh, K.M. and Bojsen, K. (1993). An acidic class III chitinase in sugar beet: Induction by Cercospora beticola, characterization, and expression in transgenic tobacco plants. *Molecular Plant-Microbe Interactions*, **6:** 495–506.

Pfleeger, T.G., da Luz, M.A. and Mundt, C.C. (1999). Lack of a synergistic interaction between ozone and wheat leaf rust in wheat swards. *Environmental Experimental Botany*, **41:** 195–207.

Rai, N., Agrawal, R.C. and Khan, A. (2011). Inhibition of DMBA induced mouse skin carcinogenesis by Centella asiatica extract. *Pharmacology* (online), **3:** 536–546.

Reissig, J.L., Strominger, J.L. and Leloir, L.F. (1955). A modified colorimetric method for the estimation of N-acetyl amino sugar. *Journal of Biological Chemistry*, **217:** 959–966.

Saari, E.E. (1998). Leaf blight diseases and associated soil borne fungal pathogens of wheat in south and southeast Asia. *In:* Duveiller, E., Dubin, H.J., Reeves, J. and McNab, A. (eds). Helminthosporium Blights of Wheat: Spot Blotch and Tan Spot. CIMMYT, Mexico.

Sadasivam, S. and Balasubramanian, T. (1987). In: Practical Manual in Biochemistry, Tamil Nadu Agriculture University, Coimbatore, p. 14.

Schraudner, M. et al. (1994). Ambient ozone can induce plant defense reactions in tobacco. Proceedings of Royal Society of Edinburgh Section B, **102:** 55–61.

Singh, A. (1977). Practical Plant Physiology. Kalyani Publishers, New Delhi.

Soja, G. and Soja, A.M. (1995). Ozone effects on dry matter partitioning and chlorophyll fluorescence during plant development of wheat. *Water Air and Soil Pollution*, **85:** 1461–1466.

Thalmair, M., Bauw, G., Thiel, S., Döhring, T., Langebartels, C. and Sandermann, H. Jr. (1996). Ozone and ultraviolet B effects on the defense-related proteins β-1,3-glucanase and chitinase in tobacco. *Journal of Plant Physiology*, **148:** 222–228.

Tiedemann, A.V. (1992). Ozone effects on fungal leaf diseases of wheat in relation to epidemiology. I. Necrotrophic pathogens. *Journal of Phytopathology*, **134:** 177–186.

Tiwari, S., Agrawal, M. and Manning, W.J. (2005). Assessing the impact of ambient ozone on growth and productivity of two cultivars of wheat in India using three rates of application of ethylenediurea (EDU). *Environmental Pollution*, **138:** 153–160.

Varshney, C. and Rout, C. (1998). Ethylene diurea (EDU) Protection against ozone injury in Tomato Plants at Delhi. *Bulletin of Environmental Contamination and Toxicology*, **61:** 188–193.

Isolation and Characterization of Thermo-alkalotolerant *Bacillus* sp. Strain ISTS2 for Carbon Dioxide Sequestration

Smita Sundaram and Indu Shekhar Thakur

Introduction

Global warming marked by spiraling concentration of CO_2 in earth's atmosphere is one of the most serious environmental concerns facing mankind. One of the most effective methods to mitigate the challenge of the rising levels of CO_2 is sequestration of CO_2 by microorganism *inter-amos* others since some microbes are capable of fixing atmospheric carbon dioxide into valuable products.

Carbon capture and storage (CCS) is an approach to mitigate global warming based on capturing carbon dioxide (CO_2) from large point sources. The most important global carbon sink is photosynthetic carbon dioxide fixation by autotrophic biota; however, some bacteria like chemoautotrophs, obligate autotrophs, facultative autotrophs and chemolithoautotrophs can fix atmospheric CO_2 through Rubisco (Kaplan, 1999). Photosynthetic microorganisms and some chemolithotroph can acclimate to a wide range of CO_2 concentration, from as low as 0.001 % to 10 % CO_2 (vol/vol in the air in equilibrium with their environment). Some can even grow in the presence of 40 % CO_2 (Price, 1998). Therefore, the search for bacterial strains for efficient sequestration of carbon dioxide titillates the scientific mind. This research is focused on summarizing how to develop chemolithotrophic bacteria in the chemostat in the presence of different concentration of sodium bicarbonate and gaseous CO_2 as carbon source, activity of carbonic anhydrase (supporting enzyme of Rubisco), development of bacterial CO_2-concentrating mechanisms, production of $CaCO_3$, and correlations with metabolic flexibility in inorganic carbon environments. In this study, a chemolithotrophic bacterium was selected for evaluation of carbon dioxide sequestration. It is more suitable because chemolithotrophic bacteria grow

S. Sundaram • I.S. Thakur (✉)
School of Environmental Sciences, Jawaharlal Nehru University, New Delhi 110067, India
e-mail: isthakur@hotmail.com

N.J. Raju et al. (eds.), *Management of Water, Energy and Bio-resources in the Era of Climate Change: Emerging Issues and Challenges*,
DOI 10.1007/978-3-319-05969-3_24

faster than most algae and cyano bacteria raised for sequestration process. It is non-photosynthetic and can be grown 24 hours a day in rain or sunshine unlike cyano bacteria or algae.

It is well established that a number of algae, cyanobacteria and chemolithotrophic bacteria actively take up and accumulate inorganic carbon (CO_2) intracellularly, which allows cells to photosynthesize efficiently even under atmospheric CO_2. This CO_2 acquisition mechanism is termed as a carbon concentrating mechanism (CCM) and it has been suggested that carbonic anhydrase (CA) plays a key role in the CCM (Kaplan and Reinhold, 1999). CA (EC 4.2.1.1) is a zinc-containing enzyme that catalyzes the reversible dehydration of HCO^-_3 to CO_2. This reaction is known to play important roles in various biological processes such as ion exchange, respiration, pH homeostasis, CO_2 acquisition, and photosynthesis (Badger and Price, 1992). The CA functions to convert an accumulated cytosolic pool of HCO^-_3 into CO_2 within the carboxysome. The generation of CO_2 coupled with a diffusive restriction to the efflux from the carboxysome, possibly imposed by the protein shell, leads to the localized elevation of CO_2 around the active site of Rubisco within the carboxysome. The substrate for carboxysome, HCO^-_3, is accumulated in the cytosol by the operation of a number of active CO_2 and HCO^-_3 transporters. These transporters are located on plasma membrane and exist in both low affinity and high affinity transporter forms (Murray, 2003).

Many of the blue-green algae, and several chemolithotrophic bacteria, contain polyhedral inclusion bodies. In thin section, these inclusions have polygonal profiles (four to six sides); range in size from 50 to 500 nm in diameter (Shively et al., 1973). Basically carboxysome is a polyhedral bacterial micro compartment that enhances carbon fixation by encapsulating the ribulose-1, 5-bisphosphate carboxylase/oxygenase and carbonic anhydrase enzymes. Carboxysomes carry out the final stages of the carbon-concentrating mechanism (CCM), through which autotrophic prokaryotes accumulate inorganic carbon to enhance CO_2 fixation (Todd et al., 2008).

In this report, we present a detailed analysis of the effect of varying levels of dissolved inorganic carbon (DIC) in the growth medium on expression of carboxysomes and over production of carbonic anhydrase which are correlated for sequestration of carbon dioxide in *Bacillus* sp.

Materials and Methods

Sampling Site and Microorganism

The soil sediment was collected from lime containing soil found in JNU New Delhi, India (this soil type had prior history of lime content) and Alwar marble mining site (27.56 N, 76.62 E) for isolation of microbes. The soil collected in clean plastic containers was brought to the laboratory and immediately stored in refrigerator at 4 °C until used for further analysis (Thakur, 1995). 5 g of soil sample was diluted to 25 ml with autoclaved distilled water and centrifuged at 7000 rpm for 10 minutes for isolation of bacteria capable of carbon dioxide sequestration.

Enrichment and Identification of Microorganisms

A chemostat culture was set in a 2-litre glass vessel, effective volume 1 litre, with culture condition as stirring at 150 rpm; temperature at 30 °C; and pH 7.6 in the minimal salt medium (MSM). The composition of MSM (g/l) was: Na_2HPO_4, 7.8; KH_2PO_4, 6.8; $MgSO_4$, 0.2; $NaNO_3$, 0.085; $ZnSO_4.7\ H_2O$, 0.05; $ZnCl_2$, 0.02; Ca $(NO_3)_2.4H_2O$, 0.05 and glucose, 1.00. The 50 ml of supernatant of soil sample containing bacterial cell populations was served as inoculum in the chemostat.

After stabilization of growth pattern, 100 μL of the bacterial community was enriched by MSM with four different concentrations (10 mM, 20 mM, 50 mM, and 100 mM) of sodium bicarbonate and 5 % gaseous CO_2 for evolving CO_2 concentration mechanism. The pH, temperature and rpm were maintained as 7.6, 30 °C and 150 rpm respectively.

The individual colony was picked up, purified and cultured on enzyme induced minimal salt medium having sodium bicarbonate (100 mM). Genomic DNA from the bacterial strain was isolated with the Genome DNA Kit (Qiagen Inc., USA) as described by the manufacturer. The 16S rRNA gene was amplified from genomic DNA by using PCR as discussed by Mishra and Thakur (2010).

Intracellular and Extracellular Enzyme Production

Bacterial strain was grown in MSM as discussed earlier at 37 °C in a rotary shaker for 24–72 h at 200 rpm. To measure the intracellular enzyme activity, cell pellet was washed with distilled water, resuspended in sonication buffer (5 ml) containing 250 mM Tris–HCl buffer and lysozyme 0.005 g and then sonicated for 10 min. After centrifugation at 12,000 rpm, the pellet was discarded; leaving the supernatant to be stored and the protein concentration was determined by Bradford method with bovine serum albumin (Sigma Chemicals) as the standard.

To measure extracellular enzyme activity, bacterial cells grown in MSM were centrifuged (10,000 rpm for 15 min). The supernatant was collected for enzyme assay.

Enzyme Assays

Carbonic anhydrase enzymatic activity was assayed by the Wilbur-Anderson method (Khalifah, 1971). The enzymatic reaction contained 15 mM Tris–HCl buffer (pH 8), 3 mM p-nitrophenyl acetate (p-NPA), 10 to 100 μl enzyme preparations and distilled water, all in a total volume of 3.0 ml. One unit of enzyme activity was expressed as 1 μmol of p-nitrophenyl acetate hydrolysed per minute.

Formation of Carboxysome

Transmission electron microscopy was performed in bacterial cells fixed in glutaraldehyde (1 % solution) and paraformaldehyde (2 %) buffered with sodium phosphate buffer saline (0.1 M, pH 6.8) as discussed by Srivastava and Thakur (2007).

Calcium Carbonate Precipitation

Bacterial product removed from culture media, dried, crushed and scanning electron microscopy was performed as discussed by Mishra and Thakur (2010).

Results and Discussion

Enrichment, Isolation, and Characterization of Bacterium

Bacteria was enriched in MSM at different concentrations of sodium bicarbonate and it was noticed that 15 types of bacterial colonies grew at low concentration of sodium bicarbonate (10 mM) while at higher concentrations such as 20, 50 and 100 mM of sodium bicarbonate the number of colonies reduced to 10, 6 and 2 respectively, reflecting an inversely proportional relationship between higher concentration (stress) of $NaHCO_3$ and growth of bacterial colonies. Bacterial strains surviving at 100 mM concentration on the LB agar plates were isolated on the basis of morphological differentiation of individual colonies and higher carbonic anhydrase activity at different durations and named as S1 and S2. The carbonic anhydrase activity of bacteria grown in 10, 20, 50 mM was not as significant as the bacteria grown on 100 mM concentration of sodium bicarbonate.

In both the strains (S1 and S2) which survived at 100 mM of $NaHCO_3$ only S2 gave more significant carbonic anhydrase activity at different concentration of $NaHCO_3$ (Fig. 1). S2 showed higher activity at 100 mM $NaHCO_3$. Each experiment was performed in three replicates to reduce experimental errors.

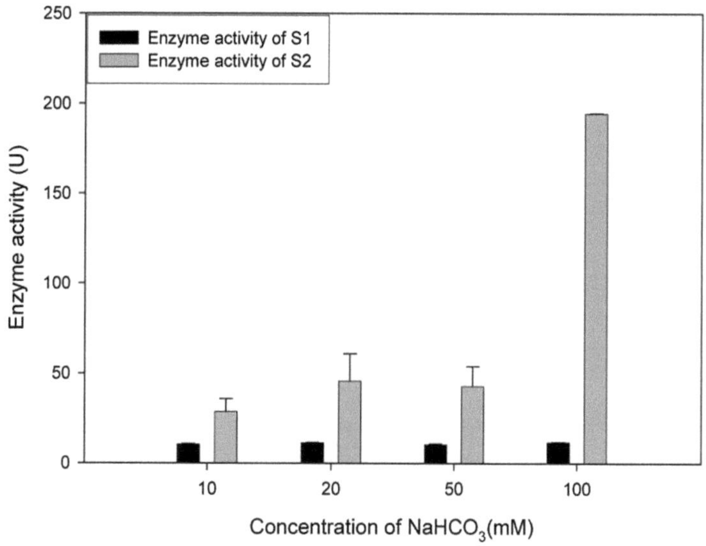

Fig. 1 Production of carbonic anhydrase enzyme activity (U = activity/min) from bacterial isolates S1 and S2 at different duration. Both bacteria grew in MSM containing $NaHCO_3$ (100 mM)

Identification of the Most Efficient CO_2 Sequestering Bacterial Strain by 16S rDNA Sequence Analysis

The bacterial strain (S2) showing higher carbonic anhydrase activity was selected for 16S rDNA sequencing. The resulting sequence was entered into the BLAST nucleotide search programme of the National Center for Biotechnology Information to obtain closely related phylogenetic sequences. Pair-wise alignment giving closest match was chosen and phylogeny tree was drawn using MEGA 3.1 software (Kumar et al., 2004). The sequence after matching was identified as *Bacillus* sp. ISTS2 (accession no. JX473586) having 99 % identity of the 16S rDNA sequence with the corresponding sequences of *Bacillus* sp.

Optimum Growth Condition of Bacterium for CO_2 Sequestration

The growth conditions were optimized with different parameters such as different concentration of sodium bicarbonate, pH, rpm and temperature. Results indicated that the enzymatic activity and optical density of strain S2 increased from 6 to 48 hours and then started to decline (Fig. 2). It was found that S2 was a mesophilic bacterium because it showed significant metabolic activity from temperature range of 25–37 °C.

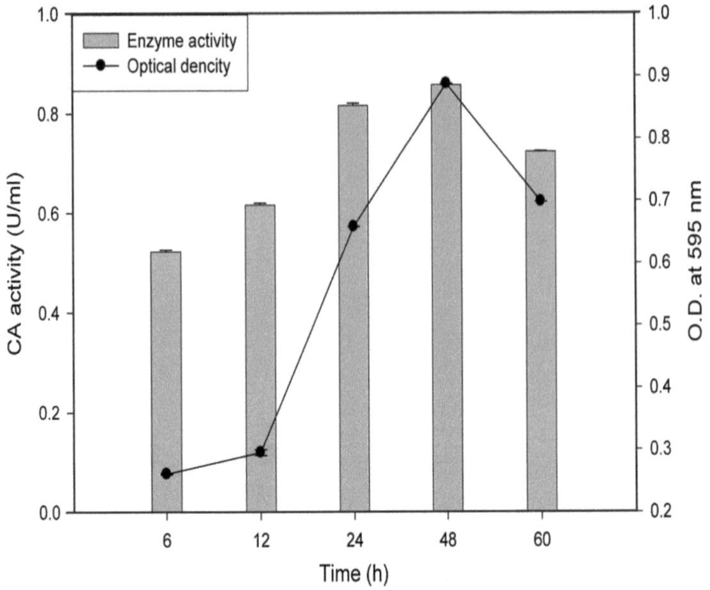

Fig. 2 Time courses for the growth of S2 and activity of CA (U/ml) at MSM containing 100 mM $NaHCO_3$. The values are means of three replicates $\pm$ standard deviations

The effect of pH was also identified as an influencing factor for the sequestration process of carbon dioxide. The initial pH was set from 6 to 11. Strain S2 showed its maximum growth and enzymatic activity at pH 8. Enzymatic activity of carbonic anhydrase was gradually increased from pH 6 to 12. Beyond initial pH 8, strain S2 had shown a sharp decline in growth and hence the reduced enzymatic activity was detected. No bacterial growth and survival was observed at pH lower than 6 and higher than 12.

Initial pH of 7.6 was fixed for the MSM media. Interestingly, a self induced increase in pH was observed at regular intervals of 24 hours. Gradual rise in pH was registered with increasing concentration of $NaHCO_3$ i.e., 10, 20, 50 and maximum pH seen in MSM having 100 mM $NaHCO_3$ (Fig. 3).

Formation of Carboxysomes and $CaCO_3$

Some polyhedral bodies are seen in strain S2 when grown in 5 % gaseous CO_2. This is the first report of presence of carboxysome in a *Bacillus* sp. The number of polyhedral bodies gradually declined with increased carbon concentration. During stress condition along with cell size the number of polyhedral bodies also declined (Fig. 4). Thus it illustrates the innate capacity of the bacteria to evolve into structures which fit well with the demands of the increased presence of CO_2 in the environment facing the bacteria.

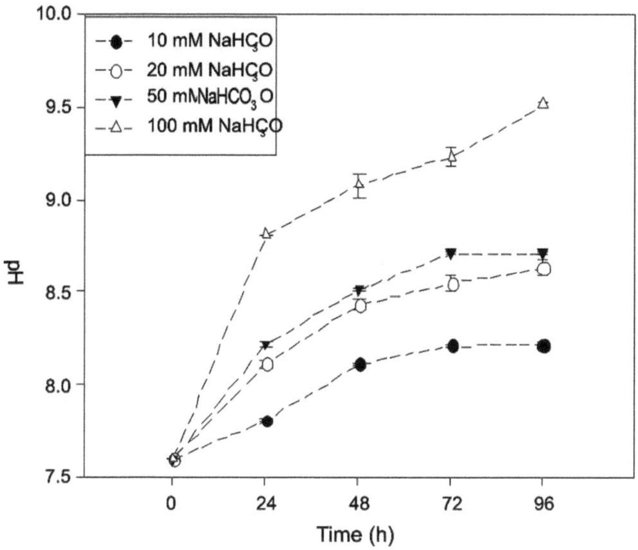

Fig. 3 Changes in the pH at different concentrations of $NaHCO_3$ during bacterial growth at different time intervals

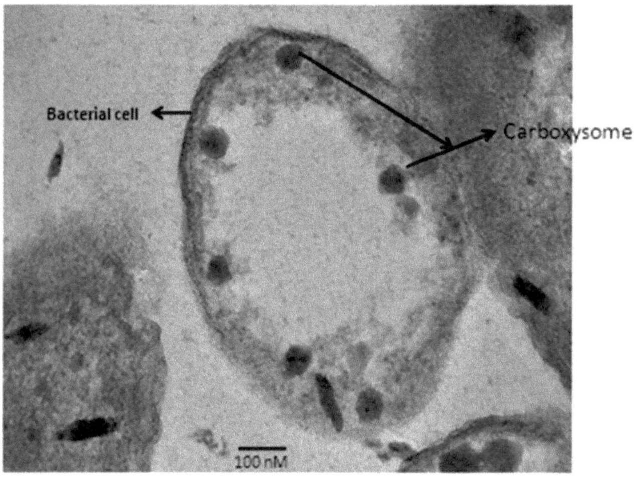

Fig. 4 Transmission electron micrographs of *Bacillus* sp. grown in an MSM containing 5 % of CO_2. Arrowheads indicate carboxysomes. Bar 100 nm

$CaCO_3$ precipitation were observed by S2 strain at different concentrations of $NaHCO_3$ and gaseous CO_2 by scanning electron microscopy (SEM) and confirmed by $CaCO_3$ standard taken from sigma Aldrich (Fig. 5a, b).

Factual outcome of the study pertaining to the presence of carbonic anhydrase and $CaCO_3$ provides a challenging case for utilization of these bacteria in industrial

Fig. 5 Scanning electron microscopy of $CaCO_3$ crystal: (**a**) standard taken from Sigma Aldrich and (**b**) precipitated by *Bacillus* sp. strain ISTS2

settings and meeting the challenges of containing CO_2 and enhancing carbon credits for industries.

Acknowledgements This research work was supported by the research grants of Department of Biotechnology, Government of India, New Delhi, India. The author (SS) thanks Sat Paul Mittal Fellowship organization, Government of India, New Delhi, India, for providing Research Fellowship. We also thank advanced instrumentation facility, Jawaharlal Nehru University, New Delhi, India for providing facility of electron microscopy.

References

Atsumi, S., Higashide, W. and Liao, J.C. (1999). Direct photosynthetic recycling of carbon dioxide to isobutyraldehyde. *Nature Biotechnology*, **27:** 12.

Badger, M.R. and Price, G.D. (1992). The CO_2 concentrating mechanism in cyanobacteria and microalgae. *Physiologia Plantarum*, **84:** 606–615.

Beudeker, R.F., Cannon, G.C., Kuenen, J.G. and Shively, J.M. (1980). Relations between D-ribulose-1,5-bisphosphate carboxylase, carboxysomes, and CO_2 fixing capacity in the obligate chemolithotroph Thiobacillusneapolitanus grown under different limitations in the chemostatArch. *Microbiology*, **124:** 185–189.

Dou, Z., Heinhorst, S., Williams, E.B., Murin, C.D., Shively, J.M. and Cannon, G.C. (2008). CO_2 fixation kinetics of Halothiobacillusneapolitanus mutant carboxysomes lacking carbonic anhydrase suggest the shell acts as a diffusional barrier for CO_2. *The Journal of Biological Chemistry*, **16:** 10377–10384.

Higashide, S.A.W. and Liao, J.C. (2009). Direct photosynthetic recycling of carbon dioxide to isobutyraldehyde. *Nature Biotechnology*, **27:** 12.

Kaplan, A. and Reinhold, L. (1999). CO_2 concentrating mechanisms in photosynthetic microorganisms. *Annual Reviews in Plant Physiology and Plant Molecular Biology*, **50:** 539–559.

Khalifah, R.G. (1971). The carbon dioxide hydration activity of carbonic anhydrase. I. Stop-flow kinetic studies on the native human isoenzymes B and C. *The Journal of Biological Chemistry*, **246:** 2561–2573.

Kumar, S., Tamura, K. and Nei, M. (2004). MEGA3: Integrated software for molecular evolutionary genetics analysis and sequence alignment brief. *Bioinformatics*, **5:** 150–163.

Mishra, M. and Thakur, I.S. (2010). Isolation and characterization of alkalotolerant bacteria and optimization of process parameters for decolorization and detoxification of pulp and paper mill effluent by Taguchi approach. *Biodegradation*, **21:** 967–978.

Murray, R., Badger, M.R. and Price, G.D. (2003). CO_2 concentrating mechanisms in cyanobacteria: Molecular components, their diversity and evolution. *Journal of Experimental Botany*, **54:** 609–622.

Price, G.D., Sueltemeyer, D., Klughammer, B., Ludwig, M. and Badger, M.R. (1998). The functioning of the CO_2 concentrating mechanism in several cyanobacterial strains: A review of general physiological characteristics, genes, proteins and recent advances. *Canadian Journal of Botany*, **76:** 973–1002.

Shively, J.M., Ball, F.L. and Kline, B.W. (1973). Electron Microscopy of the Carboxysomes (Polyhedral Bodies) of Thiobacillusneapolitanus. *Journal of Bacteriology*, 1405–1411.

Spalding, M.H. (2008). Microalgal carbon-dioxide-concentrating mechanisms: Chlamydomonas inorganic carbon transporters. *Journal of Experimental Botany*, **59:** 1463–1473.

Srivastava, S. and Thakur, I.S. (2007). Evaluation of biosorption potency of Acinetobacter sp. for removal of hexavalent chromium from tannery effluent. *Biodegradation*, **18:** 637–646.

Strubel, V., Engesser, K.H., Fischer, P. and Knackmuss, H.J. (1991). 3-(2-Hydroxyphenyl) catechol as substrate for proximal meta ring cleavage in dibenzofuran degradation by Brevibacterium sp strain DPO 1361. *Journal of Bacteriology*, **173:** 1932–1937.

Thakur, I.S. (1995). Structural and functional characterization of a stable, 4-chlorosalicylic acid degrading, bacterial community in a chemostat. *World Journal of Microbial Biotechnology*, **11:** 643–645.

Todd, O.Y., Cheryl, A.K., Sabine, H.G., Cannon, G.C. and Shivel, J.M. (2008). Protein-based organelles in bacteria: Carboxysomes and related Microcompartments. *Nature Reviews Microbiology*, **6:** 681–691.

Watson, G.M. and Tabita, F.R. (1997). Microbial ribulose 1, 5-bisphosphate carboxylase/ oxygenase: A molecule for phylogenetic and enzymological investigation. *FEMS Microbiogy Letter*, **146:** 13–22.

Carbon Footprints of Rice Cultivation under Different Tillage Practices in Rice-wheat System

Divya Pandey, Madhoolika Agrawal, and Jitendra Singh Bohra

Introduction

Continually rising concentration of greenhouse gases (GHGs) in the atmosphere is increasing concerns over how to manage global warming. Quantification of sources and sinks of these gases have been carried out, but there remains incoherence among the estimates due to different nature of sources and processes related therewith. This renders inter-comparison and further utilization of available assessments quite incomparable. Except major point sources like thermal power plants, cement manufacturing, etc. which have been quantified confidently (Matthews et al., 2008), fugitive and sensitive sources/sinks still need proper quantification. Agriculture is one such system which plays dominating role in the global fluxes of CH_4 and N_2O, as their biggest emitter. Substantial inputs of energy, machinery, synthetic fertilizers and pesticides in the modern agriculture bear embodied emissions in addition to much focused direct emissions from soil (Lal, 2004a). At the same time, agricultural soils may reportedly act as considerable carbon sink (Lal, 2004b). Therefore it becomes essential to integrate all inputs-outputs to estimate the actual impact.

For this, a single quantitative indicator which can cover all relevant stages of a cultivation practice at which greenhouse gases are emitted or removed is required. One such indicator is "Carbon footprint" (CF), which refers to the quantity of GHGs expressed in terms of CO_2 equivalents (CO_2-e), emitted into the atmosphere

D. Pandey • M. Agrawal (✉)
Laboratory of Air Pollution and Global Climate Change, Department of Botany, Banaras Hindu University, Varanasi 225001, India
e-mail: madhoo.agrawal@gmail.com

J.S. Bohra
Department of Agronomy, Institute of Agricultural Sciences, Banaras Hindu University, Varanasi 225001, India

N.J. Raju et al. (eds.), *Management of Water, Energy and Bio-resources in the Era of Climate Change: Emerging Issues and Challenges*,
DOI 10.1007/978-3-319-05969-3_25

by an individual, organization, process, product, or event from within a specified boundary (Pandey et al., 2011). Present study is a step in this direction and aims at carbon footprinting of rice cultivation systems under selected tillage managements. The objective was to identify the key points over life cycle of rice cultivation which should be focused for better environmental performance, and to use the calculations for comparing environmental performance of different tillage practices.

Materials and Methods

Tillage Practices under Study

A long-term field experiment on four permutations of tillage and no tillage practices in rice (cv PHB-71)-wheat (cv HD 2824) sequence was conducted during 2003-2009 at the agriculture research farm of the Institute of Agricultural Sciences, Banaras Hindu University, Varanasi. Detailed description of study site, crop raising and maintenance is provided in Pandey et al. (2012). No tillage refers to the practice in which only ripened ears of previous crop are harvested manually which left behind at least 20 cm straw standing on the field. Next crop was sown with the help of no till drill. Minimal soil disturbance took place under this practice. In conventional tillage practice, crops were harvested manually leaving less than 5 cm shoot along with roots in the soil. Soil was tilled up to 30 cm depth during land preparation of the next crop. The four permutations under the study were: (i) conventional tilling and puddling before transplanting of rice and conventional tilling before sowing of wheat (RCT-WCT), (ii) conventional tilling and puddling before rice transplanting and no tilling before sowing of wheat (RCT-WNT), (iii) No tilling before rice sowing and conventional tilling before wheat sowing (RNT-WCT), and (iv) no tilling before sowing of both rice and wheat (RNT-WNT).

Crop Raising

Sowing of rice in RNT-WCT and RNT-WNT was done on 20th June by drilling fine hole of 2.5 cm in the soil and placing seeds into the hole. On the same day, seeds were soaked for the nursery raising of rice to be transplanted in RCT-WNT and RCT-WCT. 21 days old seedlings were transplanted in puddled bed on 16th July. For transplantation, land was prepared by disk ploughing twice and planking through tractor (5036, John Deeri, India with specifications of 35 HP, 2100 rpm, three cylinder direct injection and fuel capacity of 45 litres). Soil was puddled and 10-12 cm water level was maintained for transplanting. Recommended doses of nutrients (150 kg N + 75 kg P_2O_5 + 75 kg K_2O) were applied through 15:15:15 NPK complex fertilizer and urea. Full doses of P and K and 50% of N was applied through 15:15:15 NPK complex at the time of sowing or transplanting. Rest of 75 kg N was applied as urea in two equal splits at critical tillerring and panical initiation stages. Pre-emergence spray of pendimethalin 30 EC at the rate of 1 kg/ha (in RNT-WCT and RNT-WNT) and butachlor 50 EC at the rate of 1.25 kg/ha (in RCT-WCT and RCT-WNT) were done. Mature crops were harvested manually on 30th October.

Post-harvest Operations

The harvest from each experimental plot was bundled separately and dried open in the sunlight for a week. Thereafter threshing was carried out with the help of thresher (Super Shakti 527, S Kumar, India). After threshing, grain weights were recorded and grains from each tillage permutation were processed separately in a grain processing system (GPS) (Agrosaw grain cleaning machine PC-2).

Calculation of Carbon Footprints

All relevant activities from land preparation up to the farm gate (grain processing) were included. For GHG accounting, three tiers were defined on the basis of order of emissions.

Tier1: This tier includes direct GHG emissions from the system. In present study, soil-borne CH_4, N_2O and CO_2, and CO_2 from onsite farm machineries were relevant. Changes in soil C was also a subset of this tier. The detailed methodology of soil-borne GHG emission measurements is described in Pandey et al. (2012). Tier1 farm machineries included tractor and grain processing system (GPS). A record of fuel (diesel) combustion in tractor during each crop was maintained during 2003 to 2008. Average fuel consumption for each hectare of land prepared for rice was calculated. Similarly, fuel consumption in GPS for processing rice grains was recorded annually during 2003-2008 and average fuel consumption per quintal of grain processed by the system was calculated. CO_2 emissions associated with fuel (diesel) consumption in these operations were calculated applying national emission factors reported in INCCA and MoEF (2010). These emission factors assume complete conversion of carbon in the fuel to CO_2 and are based on net calorific values of diesel.

Recalcitrant carbon pool in soil was isolated following Rovira and Vallejo (2002) upto 30 cm soil depth and estimated through elemental analyzer (Flash, 2000; Thermo Fisher, USA). Difference in carbon content between sowing of rice in 2003 and after harvest of rice in October 2011 was attributed equally to number of crops grown during the period. Following the assumption that carbon incorporated in soil ultimately from atmospheric CO_2 and the carbon lost from soil will be converted finally to CO_2, equivalent CO_2 change associated with changes in carbon stock was calculated as:

$$CO_2(\Delta \text{soil}\, C) = (\Delta C \times \text{BD} \times d) \times 10^4 \times 44/12$$

where CO_2 (Δsoil C) is the equivalent CO_2 emitted from or sequestered in soil. The units are (kg ha^{-1}); ΔC is change in soil recalcitrant carbon (kg/kg), BD is bulk density (kg/m^3), and d is depth of sampling (m). When carbon content decreased, CO_2 was emitted and hence was reported as positive value as it contributed positively to carbon footprint and vice versa.

Tier2: This tier covered electricity consumption in thresher and tube well for irrigation. Tube well (33 HP submersible pump, KSP, India) had 55-78 m head giving water at flow rate of 50,000-80,000 litre h^{-1} respectively with an efficiency of 70%. Electricity consumption in the two machines was recorded and embodied CO_2 equivalent emitted was calculated using the CO_2 intensity of electricity generation as reported in INCCA and MoEF (2010).

Tier3: This tier covers other indirect emissions, fertilizers and herbicides in the present study. For fertilizers, emission factors proposed by Tirado et al. (2010) and regarding herbicides, an average value of embodied emissions estimated by Lal (2004a) was utilized.

CO_2 equivalent emissions from all the tiers were added to yield global warming potential (GWP). Yield scaled GWP of the cultivation practices was calculated and reported as CF.

Statistical Analysis

Carbon footprint for the three replicate plots was calculated and their mean and standard errors were calculated. To check differences in mean CF among selected tillage practices, analysis of variance with a post hoc Duncan's test was conducted. Statistical tests were performed with the help of SPSS 10, IBM, USA.

Results and Discussion

CF was highest for RCT-WCT and the lowest for RNT-WNT. Analysis of tiers (Fig. 1) indicated that, Tier1 particularly, emissions of CH_4 was the largest component of CF. Rest of the components (except CO_2) were apparently negligible. It therefore reflected that reduction of CH_4 must be under focus. Rice cultivation being a recognized CH_4 emitter has already undergone rigorous evaluation comparing diverse management options aiming reduction of this gas. Under number of cases, tradeoff of CH_4 with N_2O or CO_2 is observed thereby suggesting that focus should be on net GWP reduction. But in the present study, CH_4 emissions outweighed CO_2 and N_2O under all tillage permutations thus suggesting that the environmental costs of CH_4 reduction practices can be attempted.

Bhatia et al. (2005) calculated net GWP of rice cultivation considering soil-borne CH_4, N_2O and CO_2 to lie in the range 1.186×10^4 to 1.792×10^4 kg CO_2 ha^{-1} with an average of 1.399×10^4 kg CO_2 ha^{-1}. Similarly Pathak et al. (2010) calculated that yield scaled GWP of rice production in India covering only soil-borne GHGs was 1.22 kg CO_2 kg^{-1}. In the present study, GWP of rice cultivation considering only soil-borne emissions ranged from 1.76×10^5 kg CO_2-e ha^{-1} under RCT-WCT to 9.82×10^4 kg CO_2-e ha^{-1} under RNT-WNT. Difference between the GWP estimates of rice cultivation obtained in the present study, and those reported by Bhatia et al. (2005), might be due to different environmental factors, cultivars and agricultural practices. At global scale, CH_4 and N_2O together bear an average GWP of 3.757×10^3 kg CO_2-e ha^{-1}, contribution of CH_4 being 89%, but there is lack of CO_2 measurement reports from rice fields (Linquist et al., 2011). In the present study also, CH_4 appeared as the largest contributor to GWP of rice cultivation (Fig. 1).

Reports on GHG emissions from farm machines and embodied emissions on rice fields is scarce, however few studies have calculated energy intensiveness in some agricultural systems. Lal (2004a), West and Marland (2002) and Kramer et al. (1999) calculated mean annual emissions of 55.73, 98.03 and 120.57 kg CO_2 ha^{-1} respectively from ploughing practices. Corresponding emissions from RCT-WCT in the present work were much higher, while in case of no tillage, negligible emissions. Local farming techniques and conventions, type and efficiency of machines, soil types, as well as the crop also affect the estimates. Land preparation in case of rice cultivation required extra machine work for puddling and working in flooded condition. Unlike tractor, emissions from thresher and GPS were yield dependent.

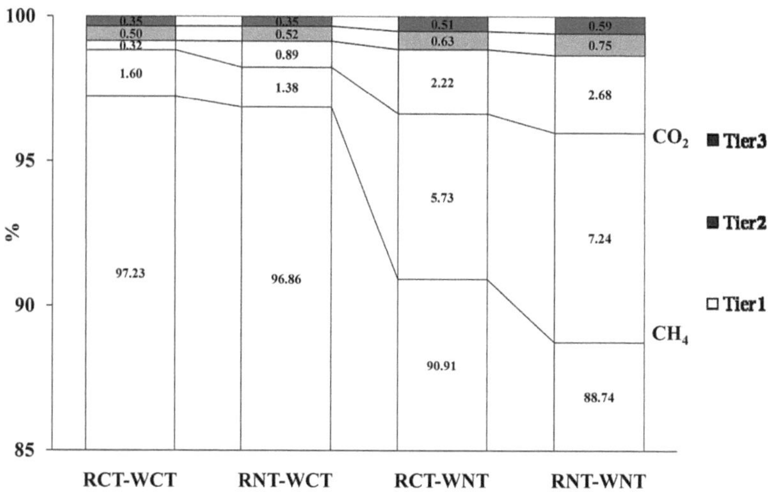

Fig. 1 Percent contribution of different tiers to carbon footprint under different tillage permutations

CO_2 (Δsoil C) was an important component of Tier1 whose direction was sensitive to management options. The reference scenario (RCT-WCT) was a net emitter but reduction in tillage frequency helped carbon accrual in soil, maximum in RNT-WNT followed by RCT-WNT and RNT-WCT. But, carbon sequestration couldn't offset the CF_s to a significant extent but 83 (in RNT-WCT) to 100% (in all except RCT-WNT) of Tier2 and Tier3 emissions taken together.

Among indirect emissions, Tier2 (emissions embodied in electricity) was the largest contributor. Because of high water demand of rice cultivation, irrigation was the major input. Contribution of emissions from onsite farm machineries was negligible yet improvement in efficiency will prove substantial keeping in mind the large area under cultivation. Reduction in tillage frequency suppressed CH_4 emissions with RNT-WNT acting as the lowest emitter. Probable increase in strength and efficiency in methanotrophy at near-surface soil was suggested as a probable cause (Pandey et al., 2012).

Soil carbon status also improved with reduction in tillage frequency. Disruption of aggregates and exposure of organic compounds due to tillage cause its rapid oxidation. Erosion under absence of vegetation cover is a major factor of carbon loss in agricultural soil (Lal, 2004a). Reduction in tillage besides reducing exposed soil surface also reduce soil disturbance and add plant residues. Therefore, soil carbon declined in RCT-WCT while in rest of the practices, soil carbon showed increment with maximum increase in RNT-WNT. West and Marland (2002) estimated carbon sequestration rates in conventional tillage and no tillage to be 256.67 and 733.33 kg CO_2 ha^{-1} yr^{-1} respectively over a 20-year period. In our study, relatively higher sequestration rates of 2739.00 kg CO_2 ha^{-1} yr^{-1} under RNT-WNT, can be due to the reason that the soil was already depleted of the organic carbon than its potential carbon holding capacity. In addition, climate and ecological factors in the region favour a large sink capacity (Lal, 2008). An interesting observation was that, in case of alternately tilled soils, the order of tillage and no tillage had significant influence on carbon sequestration rates. RCT-WNT sequestered approximately twice of the RNT-WCT, the reason being differences in aboveground biomass left after no till practice in the two cases. Shoot weight of rice was greater than wheat. Hence, the tillage practice in which rice residues were left behind added more carbon in soil than wheat residues.

With respect to RCT-WCT, although, RNT-WNT had less emissions (Fig. 2) but environmental conditions could not benefit the RNT-WNT in terms of yield. RNT-WNT produced only 71.67% of the yield obtained under the conventional practice RCT-WCT, while RCT-WNT was the least productive permutation in which yield reduced to 43.33% of RCT-WCT. However, yield improved by the factor of 1.16 under RNT-WCT.

RCT-WCT was found to be the most yielding while RCT-WNT was next to this. RNT-WCT and RNT-WNT were, on the other hand, low yielding practices. This yield factor modulated CF to follow trend different than GWP. While GWP of RCT-WNT was comparable or next to RCT-WCT, relatively lower yield made its CF the highest. The results are partially in consensus with Burney et al. (2010) and Linquist et al. (2011) who concluded that agricultural practices become most carbon efficient near their optimum yields.

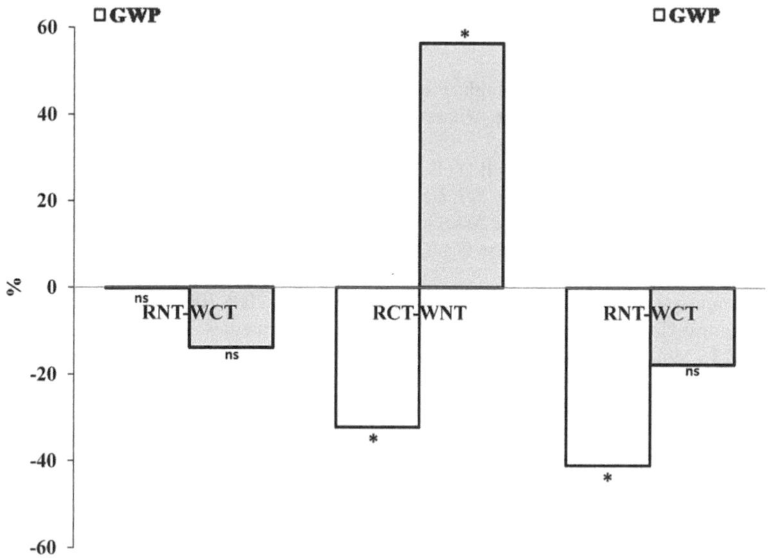

Fig. 2 Cumulative global warming potential and carbon footprint of rice cultivation under different tillage permutations with respect to RCT-WCT. Difference between mean GWP or CF for the practice with those of RCT-WCT are tested by paired sample *t*-test and indicated by levels of significance (*$p < 0.5$, ns – not significant)

Conclusions

No tillage when followed on continuous basis led to significant reductions in CF. RCT-WNT left CF comparable to the reference RCT-WCT, but with reduced economic inputs (tractor for land preparation, labour cost for harvesting and thresher). As already established for rice fields, present study also concluded that strategic reductions in CH_4 emissions is essential, substantial reduction in its emissions will outweigh the environmental costs of the control options. However, CF trends need continuous monitoring for a longer period of time for making any recommendation.

Acknowledgements Divya Pandey is grateful to University Grants Commission, India for research fellowship. Authors thank Head, Department of Botany, Banaras Hindu University for providing necessary facilities. Director, Institute of Agriculture Sciences, Banaras Hindu University is acknowledged for granting permission for carrying out the field experiment at agriculture farm of the Institute. Authors also thank Dr. T. K. Adhya and Dr. Pratap Bhattacharya of Central Rice Research Institute, Cuttack, India for carbon analysis of soil. Financial assistance in the form of research project funded by University Grants Commission, New Delhi is gratefully acknowledged.

References

Bhatia, A., Pathak, H., Jain, N., Singh, P.K. Singh, A.K. (2005). Global warming potential of manure amended soils under rice–wheat system in the Indo-Gangetic plains. *Atmos. Environ.*, **39:** 6976-6984.

Burney, J.A., Davis, S.J. and Lobell, D.B. (2010). Greenhouse gas mitigation by agricultural intensification. *Proc. Natl. Acad. Sci. USA*, **107:** 12052-12057.

Indian Network for Climate Change Assessment (INCCA), Ministry of Environment and Forests (MoEF) (2010). India: Greenhouse Gas Emissions 2007. Ministry of Environment and Forests. Government of India.

Intergovernmental Panel on Climate Change (IPCC) (2007). 2006 IPCC Guidelines for National Greenhouse Gas Inventories, Inst. for Global Environ. Strategies, Hayama, Japan.

Kramer, K.J., Moll, H.C. and Nonhebel, S. (1999). Total greenhouse gas emissions related to the Dutch crop production system. *Agric. Ecosyst. Environ.*, **72:** 9-16.

Lal, R. (2004a). Carbon emission from farm operations. *Environ. Internat.*, **30:** 981-990.

Lal, R. (2004b). Soil Carbon Sequestration Impacts on Global Climate Change and Food Security. *Science*, **304(11):** 1623-1627.

Lal, R. (2008). Carbon sequestration. *Phil. Trans. R. Soc. B.*, **363:** 815-830.

Linquist, B., Van Groenigen, K.J., Adviento-Borbe, M.A., Pittelkow, C. and Van Kessel, C. (2011). An agronomic assessment of greenhouse gas emissions from major cereal crops. *Global Change Biology*, **18(1):** 194-209.

Pandey, D., Agrawal, M. and Bohra, J.S. (2012). Greenhouse gas emissions from rice crop with different tillage permutations in rice–wheat system. *Agric. Ecosyst. Environ.*, **159:** 133-144.

Pandey, D., Agrawal, M. and Pandey, J.S. (2011). Carbon footprints: Current methods of estimation. *Environ. Monit. Assess.*, **178:** 135-160.

Pathak, H., Jain, N., Bhatia, A., Patel, J. and Aggarwal, P.K. (2010). Carbon footprints of Indian food items. *Agric. Ecosyst. Environ.*, **139:** 66-73.

Rovira, V. and Vallejo, V.R. (2002). Labile and recalcitrant pools of carbon and nitrogen in organic matter decomposing at different depths in soil: An acid hydrolysis approach. *Geoderma*, **107:** 109-141.

Tirado, R., Gopikrishna, S.R., Krishnan, R. and Smith, P. (2010). Greenhouse gas emissions and mitigation potential from fertilizer manufacture and application in India. *International J. Agric. Sustain.*, **8(3):** 176-185.

West, T.O. and Marland, G. (2002). A synthesis of carbon sequestration, carbon emissions, and net carbon flux in agriculture: Comparing tillage practices in the United States. *Agric. Ecosyst. Environ.*, **9:** 217-232.

Matthews, S.C., Hendrickson, C.T. and Weber, C.L. (2008). The importance of carbon footprint estimation boundaries. *Environ. Sci. Tech.*, **42:** 5839-5842.

Trend Analysis of Rainfall in Two Contrasting Regional Environments

Tanja Likso and Surender Singh

Introduction

The study of climate change has become the most intriguing aspect of scientific research all over the world. Particular attention has been paid to analyze the regional changes under well debated global warming, i.e. changes in air temperature, with less focus on the impact of climate change on other weather elements including rainfall. While global warming has been attributed to anthropogenic impacts by many scientists, changes in rainfall amounts are still without determining clear causal relationships. Teleconnections viz., El Niño and North Atlantic Oscillation etc. and respective changes in cyclogenesis frequencies are possible causes. The consequences of climate variability and climate change are potentially more significant for the poor in developing countries than for those living in more prosperous nations. Vulnerability to the impacts of climate change is a function of exposure to climate variables, sensitivity to those variables and the adaptive capacity of the affected community. Climate variability can cause abrupt disruptions, such as floods, droughts or tropical storms. These disruptions can take a major toll on a country's economy if a significant part of economic activity is sensitive to the weather and climate.

Under the threat of global warming it is vital to determine the impact that future changes in climate may have on the environment and to what extent any adverse effects can be mitigated. Water resource has become a prime concern for any

T. Likso (✉)
Division for Weather and Climate Obsevation, Meteorological and Hydrological Service,
Grič 3, 10000 Zagreb, Croatia
e-mail: likso@cirus.dhz.hr

S. Singh
Department of AgroMeteorology, Chaudhary Charan Singh Haryana Agricultural University,
Hisar 125004, India

N.J. Raju et al. (eds.), *Management of Water, Energy and Bio-resources in the Era of Climate Change: Emerging Issues and Challenges*,
DOI 10.1007/978-3-319-05969-3_26

development and planning including food production, flood control and effective water resource management. Studies have demonstrated that global surface warming is occurring at a rate of 0.74 ± 0.18 °C over 1906–2005 (IPCC, 2007). Impact of climate change in future is quite severe as given by IPCC reports which signify that there will be reduction in the freshwater availability because of climate change. This has also been revealed that by the middle of 21st century, decrease in annual average runoff and availability of water will project up to 10–30 % (IPCC, 2007). Various researchers have contributed to the study of climate change (Dessens and Bucher, 1995; Serra et al., 2001; Marengo, 2004) with long-term data. Study of different time series data have proved that trend is either decreasing or increasing, both in case of temperature and rainfall. Human interference is also leading to climate change with changing land use from the impact of agricultural and irrigation practices (Kalnay and Cai, 2003). The rainfall received in an area is an important factor in determining the amount of water available to meet various demands, such as agricultural, industrial, domestic water supply and for hydroelectric power generation. Global climate changes may influence long-term rainfall patterns impacting the availability of water, along with the danger of increasing occurrences of droughts and floods.

Rainfall trend analysis, on different spatial and temporal scales, has been of great concern during the past century because of the attention given to global climate change by the scientific community. According to some recent studies, the northwestern Croatia has marked a rainfall increase and on the contrast northwestern India exhibiting decreasing rainfall trend. Impacts of climate change on regional precipitation involve complex processes which are not well understood. Long-term changes in precipitation on global/hemispheric scale are difficult to detect due to large spatio-temporal discontinuities and lack of observations over the oceanic regions. However, recent analyses of rainfall over the Northern Hemispheric land areas since the mid nineteenth century (Houghton et al., 1990) have indicated significant increase of rainfall in mid latitudes and concurrent decrease in low latitudes over the last 30–40 years. It has also been recognized that the global scale climatic changes have widely varying regional manifestations. Regional manifestations of global scale climate change processes are very important because of their close link with the social and economic prosperity of the affected human population.

Predictions of regional scale climate change are still on a loose ground, because of the problems faced by the current climate models in realistically simulating the atmospheric and surface processes with a fine resolution and the air-sea interaction. Even if a reasonable regional scenario can be obtained for a specific part of the world, the complexities in the natural variability themselves make the interpretations highly unstable. Therefore, it is necessary to document and understand the observed climate changes and its variability on a regional scale. Climatic resource endowment plays major role in planning agricultural production in tropical, sub-tropical and even under humid and temperate environment.

Climate interacts intimately with soil quality to determine the yield of a cultivar especially in arid, semi-arid environment, where monsoon rain is the major source of water supply. Climate limits the water supply to the growing crop, especially during pre- and post-monsoon water stresses. Thus, the agroclimatic regional planning applied to agriculture and allied sector is essentially a resource-based planning. The regional approach to climate change/variability is in fact a quest to assess regional vulnerability, based on which appropriate response strategies can be evolved for maintaining ecological balance and sustainable development particularly under monsoon variability in semi-arid tropics and temperate rainy climate. But most of the studies done so far have been based on the anticipated changes in climate related global parameters. Thus, in order to make realistic impact assessment on regional scale, we need transition from global generalities to regional specifics.

With this objective, the present study was aimed to determine the trend of rainfall variations in two contrasting regional environments using Sen´s Estimator of Slope and Mann-Kendall Test. For this purpose, two stations viz., Zagreb-Grič located within northwestern Croatia and Hisar located in northwestern India, have been analyzed for the period 1981–2010 having authentic and realistic data set available. These stations have diverse climates representing semi-arid (Hisar) and temperate rainy climate (Zagreb-Grič). Further, the prime goal of this study is the statistical analyses of annual and monthly rainfall amounts,[1] looking for the presence of linear monotonic trends in the temporal structure of rainfall at two contrasting regional environments. In general, variations in rainfall/precipitation affect water resources system because the water from the atmosphere drives other hydrological processes (e.g. surface flow, groundwater flow). It was also the reason that water from snow is taken into account using the water equivalent. The preliminary results of this study can be used in the long-term planning for water resources management (e.g. irrigation scheduling in crop areas).

Study Domain

The study domain is consisting of two contrasting regional environments viz., Zagreb-Grič[2] (20°37′ N latitude and 86°9′ E longitude and 157 m amsl) in Croatia and Hisar (29°14′ N latitude and 75°68′ E longitude and 214 m amsl) in India (Fig. 1a and b).

Zagreb is the capital and the largest city of the Republic of Croatia in Europe. It is located in the northwest of the country, along the Sava river, at the southern slopes of the Medvednica mountain. Zagreb lies at an elevation of approximately 122 m above mean sea level. According to Köppen climate classification system, most of Croatia (including Zagreb) has a temperate rainy climate. Zagreb has four

[1] For Zagreb-Grič in case of snowfall the water equivalent was used. According to definition (Glickman, 2000) it is the depth of water that would result from the melting of a snow sample.

[2] The main meteorological station Zagreb-Grič is located at Zagreb Upper Town.

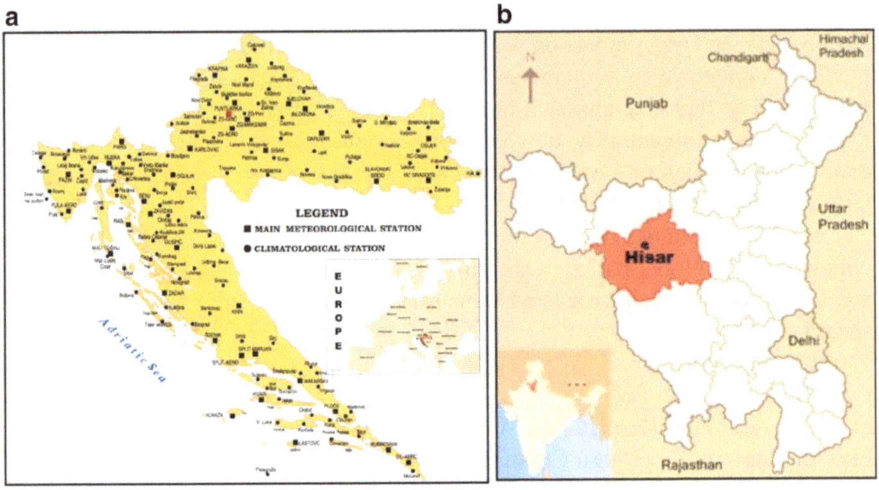

Fig. 1 Study domain (**a**) Zagreb-Grič and (**b**) Hisar

separate seasons: Spring (March to May), Summer (June to August), Fall (September to November) and Winter (December to February). Summers are warm, and winters are cold, without a discernible dry season. The annual variation of precipitation has been classified as the continental type with maximum in warm part of a year (June) and secondary maximum in a late fall. It has been shown that in June, the month with the largest amounts of rainfall in this part of Croatia, cyclones combined with a front give the most abundant rainfall. The average temperature[3] in winter is 1.6 °C and the average temperature in summer is 20.9 °C. At Zagreb-Grič during summer season temperatures rise above 30 °C quite often. The multi-annual average of the number of days with maximum daily temperature $t_{max} \geq 30$ °C is 16.2. This value is obtained from all available data (1861–2011) for Zagreb-Grič and it is referred on the number of such days during the whole year. Snowfall is common in the winter months (December to February) but it can also appear in March and sometimes in April.[4] The highest recorded snow cover ever was 84 cm in February 1895. Fog is more frequent in the cold part of the year, from October to March, than in the warm one. The highest number of foggy days at Zagreb-Grič was recorded in 1920 (99 days in the year). Highest recorded temperature ever was 40.3 °C in July 1950, and lowest was −22.2 °C in January 1942.

The climate of Hisar owes to its continental location on the outer margins of the south-west (SW) monsoon region. It has tropical monsoonal climate and is characterized as arid type of climate. The region has characteristically four seasons during the year viz., Summer (March to May), SW Monsoon (June to September),

[3] The average temperature for Zagreb-Grič here means the multi-annual average (1861–2011) of average monthly temperature for particular season.

[4] The maximum number of days with snow cover ≥ 1 cm for Zagreb-Grič in April is three in 1996.

post-Monsoon (October to November) and Winter (December to February) season. SW monsoon also known as summer monsoon brings rain during last week of June to mid-September. The period from October onward until next June remains almost dry except, few light showers received due to westerly depressions/western disturbances (WDs). The summers are generally quite hot and winters are fairly cool. The main characteristics of climate in the district are its dryness, extremes of temperature and scanty rainfall. Air temperatures during summer are quite high at Hisar, the maximum value of which touched 48.3 °C in May, 1944. However, with the onset of summer monsoon, the temperature ameliorates and falls in the vicinity of 35.0 °C. The lowest temperature recorded at Hisar is −3.9 °C which occurred in January, 1929. The sub-zero temperatures (frost conditions) are not uncommon during winter season. The maximum temperature in summer months between 40 to 44 °C and the minimum temperature between 4 to 6 °C in winter months are normal for the place. The annual average[5] maximum and minimum temperature at Hisar is 31.5 and 16.2 °C, respectively.

Methodology Adopted

Two rainfall data series from two contrasting regional environments (Zagreb-Grič, Croatia and Hisar, India) have been studied. The records were obtained from Meteorological and Hydrological Service of Croatia, Zagreb and India Meteorological Department, New Delhi. The data comprised time series of annual and monthly resolutions for the period 1981–2010. This section is dedicated to giving a brief overview of the methodology used in data analyses. It is well known that homogeneous time series of climatic elements are essential for studies of climatic fluctuations and changes. In order to use time series for climate analysis, it is important to have reliable data without artificial breaks because such inhomogeneities spoil signals of climate trends. Homogeneity testing[6] of the rainfall data was performed by Alexanderson´s Standard Normal Homogeneity Test (Alexandersson, 1986) on monthly and annual basis using AnClim software (v 5.025) developed by Petr Štěpánek (Štěpánek, 2008). In relation to the results of homogeneity testing, both data sets were classified as useful for climate analysis; the Standard Normal Homogeneity Test was not able to locate any significant change-points in mentioned data series. The available metadata do not allow identifying hypothetical changes in instrumentation and location of the measuring stations.

---

[5] The average annual maximum and minimum temperature are calculated on the basis of all months in a year during the period 1970–2000.

[6] The reference series are compared with the test series in order to estimate the relative homogeneity of the test series. The test and reference series are obtained from monthly data on a seasonal and annual basis. The reference stations were selected according to their location, station history and climatic conditions.

Mann-Kendall Test

The Mann-Kendal test is used in cases when data values x_i of time series can be assumed to obey the model (Kendall, 1975).

$$x_{\mathrm{i}} = f(t_{\mathrm{i}}) + \varepsilon_{\mathrm{i}} \tag{1}$$

where *f(t)* is a continuous monotonic increasing or decreasing function of time and the residuals can be assumed to be the same distribution with zero mean. It is assumed that the variance of the distribution is constant in time. Using Mann-Kendall test we want to test the null hypothesis of no trend, H_0, i.e. the observations x_i are randomly ordered in time, against the alternative hypothesis, H_1, where there is an increasing or decreasing monotonic trend. Each data value is compared with all subsequent data values. If a data value from a later time period is higher than a data value from an earlier time period, the statistic S is incremented by 1. On the other hand, if the data value from a later time period is lower than a data value sampled earlier, S is decremented by 1. The net result of all such increments and decrements yields the final value of S (Shahid, 2011). The Mann-Kendall test statistics S is calculated using the formula:

$$S = \sum_{k=1}^{n-1} \sum_{j=k+1}^{n} sgn\ (x_j - x_k) \tag{2}$$

where x_j and x_k are the annual values in years j and k, $j > k$, respectively, and

$$sgn(x_j - x_k) = \begin{cases} 1 & \text{if } x_j - x_k > 0 \\ 0 & \text{if } x_j - x_k = 0 \\ -1 & \text{if } x_j - x_k < 0 \end{cases} \tag{3}$$

For $n \geq 10$, the statistic S is approximately normally distributed with the mean and variance as follows:

$$E(S) = 0 \tag{4}$$

$$VAR(S) = \frac{1}{18}\left[n(n-1)(2n+5) - \sum_{p=1}^{q} t_p(t_p - 1)(2t_p + 5)\right] \tag{5}$$

where q is the number of tied groups and t_p is the number of data in the pth group. The standard test statistic Z is given by

$$Z = \begin{cases} \dfrac{S-1}{\sqrt{VAR(S)}} & \text{if } S > 0 \\ 0 & \text{if } S = 0 \\ \dfrac{S+1}{\sqrt{VAR(S)}} & \text{if } S < 0 \end{cases} \tag{6}$$

The presence of a statistically significant trend is evaluated using the Z value. A positive/negative value of Z indicates an upward/downward trend. To test for either an upward or downward monotone trend (a two-tailed test) at α significance level, H_0 is rejected and a significant trend exists in the observed time series if the $|Z| > Z_{1-\alpha/2}$, where $Z_{1-\alpha/2}$ is obtained from the standard normal cumulative distribution tables.

Sen's Nonparametric Estimator of Slope

To estimate the true slope of an existing linear trend (as change per year) the Sen´s nonparametric method (Sen, 1968) is used. This means that linear model *f(t)* can be described as

$$f(t) = Qt + B \tag{7}$$

where Q is the slope and B is a constant.

In order to obtain the slope estimate Q in equation (7) we should calculate the slopes of all data pairs

$$Q = \frac{x_j - x_k}{j - k} \tag{8}$$
$$i = 1, 2 \ldots, N, \; j > k$$

If there are n values x_j in the time series we obtain as many as $N = n(n-1)/2$ slope estimates Q_i.

The Sen´s estimator of slope is the median of these N values of Q_i. The N values of Q_i are ranked from the smallest to the largest and the Sen´s estimator is

$$Q = \begin{cases} Q_{\frac{n+1}{2}} & \text{if } N \text{ is odd} \\ 0.5\left(Q_{\frac{N}{2}} + Q_{\frac{N+2}{2}}\right) & \text{if } N \text{ is even} \end{cases} \tag{9}$$

A 100(1 – α) % two-sided confidence interval about the slope is obtained by the nonparametric technique based on the normal distribution. The method is valid for n as small as 10 unless there are many ties (Salmi et al., 2002).

At first we compute

$$C_\alpha = Z_{1\frac{\alpha}{2}} \sqrt{VAR(S)} \tag{10}$$

where *VAR(S)* has been defined in equation (5) and $Z_{1-\alpha/2}$ is obtained from the standard normal distribution.

Next $M_1 = (N - C_\alpha)/2$ and $M_2 = (N + C_\alpha)/2$ are computed. The lower and upper limits of the confidence interval, Q_{min} and Q_{max}, are the M_1^{th} largest and the $(M_2 + 1)^{th}$ largest of the N ordered slope estimates O_i. If M_1 and/or M_2 are not a whole numbers, the respective are interpolated. 99 % and 95 % confidence intervals are calculated by a similar procedure.

Results and Discussion

In the non-parametric Mann-Kendall test, trend of rainfall for 30 years from January to December has been calculated (Table 1) for each month individually together with the **Sen's** magnitude of slope (Q) and calculation of percentage (β). In the Mann-Kendall test the *Zc* statistics revealed the trend of the series for 30 years for individual 12 months at Zagreb-Grič from January to December which are 0.560, −0.450, −0.146, 1.035, −1.143, −1.295, 1.669, 1.100, 0.253, 0.138, 1.150 and 1.027, respectively. The trend of above series from January to December at Hisar are −0.100, −0.073, −0.192, −0.086, 0.338, 0.286, −2.785, −3.219, 0.933, −0.053, 0.000 and −0.067, respectively. Trend analysis for both the regions has been done in the present study with 30 years of precipitation data from 1981 to 2010. Sen's slope estimator and Mann-Kendall test has been used for the determination of the trend. At Zagreb-Grič (Fig. 2), *Zc* value in the months of January, April and July to December showed rising trend while it is showing negative trend in February, March, May and June months. Thus *Zc* values for eight months show a positive trend and a negative trend for rest four months. In contrast, at Hisar, only the months of May, June and September exhibited positive trend and rest of the months showed negative trend (Table 2).

In Croatia, Zagreb-Grič region received 870.0 mm annual rainfall exhibiting increasing trend of 4.88 mm per annum as against 410.1 mm annual rainfall with decreasing trend of 4.84 mm per annum at Hisar, India. Annual variability in rainfall was more at Hisar than at Zagreb-Grič owing to comparatively higher values of coefficient of variation (Table 3).

Table 3 is demonstrating the Sen's slope which is also indicating slope magnitude per year for 30 years (Figs. 3 and 4). Highest monthly increasing trend of 1.669 mm during July and of 0.933 mm during September was observed at Zagreb-Grič and Hisar, respectively. On the contrary, maximum decreasing trend of 1.295 mm in June and of 3.219 mm in August was observed at Zagreb-Grič and Hisar, respectively during the whole reference period.

The annual rainfall amount shows large variations between years. In order to obtain smoother curves, 10-year running means were used (Figs. 5 and 6). These

Table 1 Sen's estimator of the monthly rainfall trends (mm/year) at 90 % confidence level

Period (1981–2010)	Stations	
Month(s)	Zagreb-Grič	Hisar
January	0.560	−0.100
February	−0.450	−0.073
March	−0.146	−0.192
April	1.035	−0.086
May	−1.143	0.338
June	−1.295	0.286
July	1.669	−2.785
August	1.100	−3.219
September	0.253	0.933
October	0.138	−0.053
November	1.150	0.000
December	1.027	−0.067

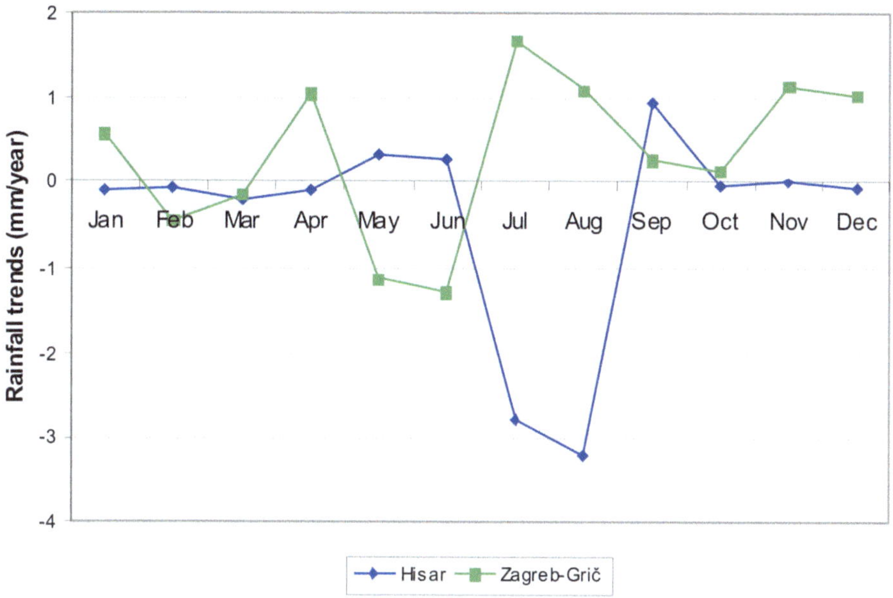

Fig. 2 Monthly rainfall trends at Zagreb-Grič and Hisar

Table 2 Results of the Men-Kendal test for Zagreb-Grič and Hisar in bold values of the standard test statistic Z which exceeds the critical Z value

	Zagreb-Grič			Hisar		
Month(s)	S	VAR(S)	Z	S	VAR(S)	Z
January	55	3141.667	0.963	–51	3137.000	−0.893
February	–44	3140.667	−0.767	–24	3141.667	−0.410
March	–17	3141.667	−0.285	–63	3139.667	−1.106
April	123	3141.667	**2.177**	–36	3133.334	−0.625
May	–76	3141.667	−1.338	33	3140.667	0.571
June	–105	3141.667	**−1.855**	29	3141.667	0.4995
July	190	3141.667	**3.372**	–121	3141.667	**−2.141**
August	43	3141.667	0.749	–151	3141.667	**−2.676**
September	17	3141.667	0.285	81	3141.667	1.427
October	5	3141.667	0.071	–115	3011.000	**−2.078**
November	54	3141.667	0.946	–3	2733.333	−0.038
December	53	3141.667	0.928	–34	3133.000	−0.590
Year	65	3141.667	1.142	–81	3141.667	−1.427

Table 3 Descriptive statistics for the annual time series and linear trends

Measuring station	Period	Mean (mm)	Coeff. of variation	Maximum (mm)	Minimum (mm)	Trends Sen´s slope (mm/year)
Zagreb-Grič	1981-2010	870.0	0.140	1155.1	624.4	4.883
Hisar	1981-2010	410.1	0.260	606.3	213.1	−4.840

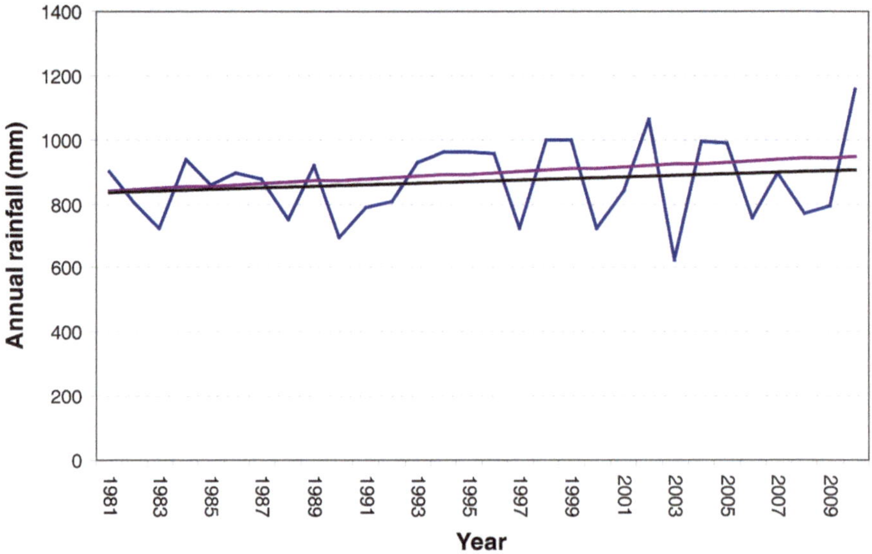

Fig. 3 Annual rainfall amounts at Zagreb-Grič during the period 1981–2010; in black – ordinary regression trend line ($y = 2.4003x + 832.78$), in pink – trend line obtained using the Sen's estimator of Slope ($y = 3.6999x + 841.10$)

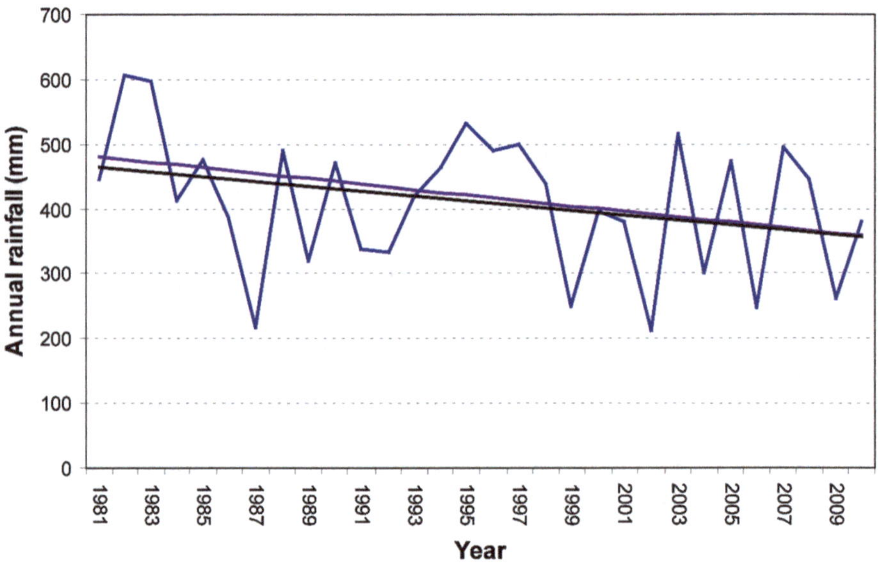

Fig. 4 Annual rainfall amounts at Hisar during the period 1981–2010; in black – ordinary regression trend line ($y = -3.7144x + 467.64$), in pink – trend line obtained using the Sen's estimator of slope ($y = -4.2208x + 480.62$)

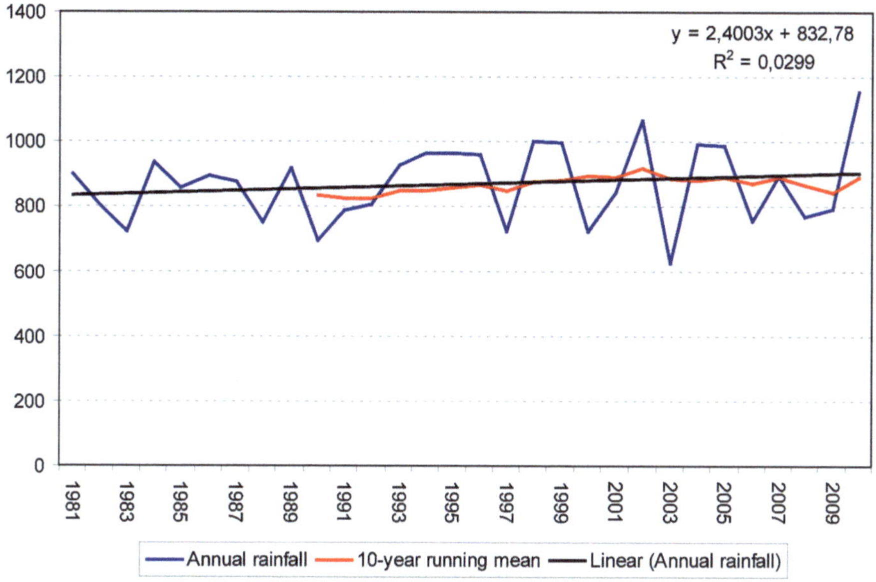

Fig. 5 Annual rainfall amounts, 10-year running mean and linear trend, Zagreb-Grič (1981–2010)

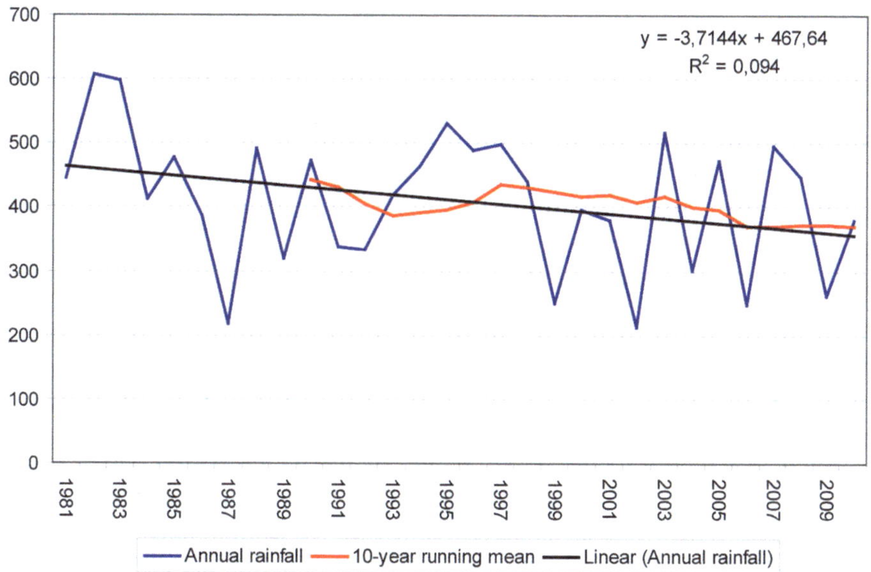

Fig. 6 Annual rainfall amounts, 10-year running mean and linear trend, Hisar (1981–2010)

reveal that at both stations it is possible to distinguish more or less pronounced periods with higher and periods with lower annual rainfall amounts. At Hisar during the period 1991–1995 the drop in rainfall was recorded (see Fig. 6).

From values of the determination coefficients are undeniable that linear trends are not the ideal choice, i.e. they do not provide the best fit to the annual rainfall amounts. According to authors´ opinion a better fit can be obtained using a higher degree polynomial trend line. Despite this fact, the linear trend is one of the basic statistic analyses of time series. It provides reliable information about the overall changes within the time series of some meteorological elements. In this study it reveals the difference in rainfall amounts during the period 1981–2010. The large differences in annual rainfall amounts occurred throughout the study period, especially for Hisar (Fig. 4).

Conclusion

The results of Mann-Kendall test represent both positive and negative trends in both the regions. Some of them are statistically significant at 95 % significance level what is shown in Table 2 (for July, August and October as well as for April, June and July at Hisar and Zagreb Grič, respectively). Rainfall varies in different months for different years. Sen's slope is also indicating increasing and decreasing magnitude of slope in correspondence with the Mann-Kendall test values. There are eight months at Zagreb-Grič and three months at Hisar with increasing trend and *Zc* value along with the increasing slope magnitude. Therefore, it can be concluded that there is evidence of some change in the trend in rainfall of both the regions in these 30 years in different months.

More comprehensive examination should be performed on larger data sets from more weather stations located in the same climatic regions included into this analysis as well as taking into account more weather elements (e.g. temperature, cloudiness, etc.) in order to obtain reliable conclusion about trend of rainfall variations.

The analysis of rainfall for water and agricultural purposes must include information concerning the trends or changes of precipitation, length of the rainy season, the distribution of rainfall amounts through the year, and the risk of dry and wet spells. Further, it can be suggested to look for other weather systems/causes for rainfall variations for a better understanding of climate changes on regional scale which is one of the most serious environmental concerns of our time. Further, study of the regions may reveal other aspects which will be helpful in managing anticipating rainfall and available water resources under the looming threat of climate change and variability.

References

Alexandersson, H. (1986). A homogeneity test applied to precipitation data. *International Journal of Climatology*, **6:** 661–675.

Dessens, J. and Bucher, A. (1995). Changes in minimum and maximum temperatures at the Pic du Midi relation with humidity and cloudiness, 1882–1984. *Atmospheric Research*, **37:** 147–162.

Glickman, T.S. (Managing Editor) (2000). Glossary of meteorology. American Meteorological Society.

Houghton, J.T., Jenkins, G.J. and Ephraum, J.J. (1990). Climate Change, the IPCC Scientific Assessment. Cambridge Univ. Press, New York, Melbourne.

IPCC (2007). Climate Change 2007: Climate change impacts, adaptation and vulnerability. Working Group II contribution to the Intergovernmental Panel on Climate Change FourthAssessment Report. Summary for policymakers, **23.**

Kalnay, E. and Cai, M. (2003). Impact of urbanization and land-use change on climate. *Nature*, **423:** 528–531.

Kendall, M.G. (1975). Rank Correlation Methods, 4th edition.. Charles Griffin, London, U.K.

Marengo, J.A. (2004). Interdecadal variability and trends of rainfall across the Amazon basin. *Theoretical and Applied Climatology*, **78:** 79–96.

Salmi, T., Maata, A., Antilla, P., Ruoho-Airola, T. and Amnell, T. (2002). Detecting trends of annual values of atmospheric pollutants by the Mann-Kendall test and Sen´s slope estimates – the Excel template application Makesens. Finish Meteorological Institute, Helsinki, Finland.

Sen, P.K. (1968). Estimates of the regression coefficient based on Kendal´s tau. *Journal of the American Statistical Association*, **63:** 1379–1389.

Serra, C., Burgueno, A. and Lana, X. (2001). Analysis of maximum and minimum daily temperatures recorded at Fabra observatory (Barcelona, NE Spain) in the period 1917–1998. *International Journal of Climatology*, **21:** 617–636.

Shahid, S. (2011). Trends in extreme rainfall events of Bangladesh. *Theoretical and Applied Climatology*, **104:** 489–499.

Sirois, A. (1998). A brief and biased overview of time series analysis or how to find that evasive trend. *In:* WMO report No. 133 WMO/EMEP workshop on advanced statistical methods and their application to air quality data sets. Helsinki, September 1998, 14–18.

Štěpánek, P. (2008). AnClim – Software for time series analysis. Dept. of Geography, Faculty of Natural Sciences, MU, Brno. http://www.climahom.eu/AnClim.html

Regional Climate Modelling over the Himalayas

A.P. Dimri and P. Maharana

Introduction

The western Himalayas and adjoining north Indian region receives almost 1/3rd of its annual precipitation during winter season (December, January and February – DJF) due to eastward moving low-pressure synoptic weather systems, called Western Disturbances (WDs) in Indian meteorological parlance (Pisharoty and Desai, 1956; Rao and Srinivasan, 1969; Singh, 1979; Kalsi, 1980; Kalsi and Haldar, 1992; Dimri and Mohanty, 2009). Mohanty et al. (1998) carried out a brief survey of the frequency and track of the WDs and their impact on the quantum of precipitation and other climatological parameters at some sites in the north India. Further the Himalayan mountain ranges, having different altitudes and orientations, cause the prevailing weather conditions to be complex. Surface weather elements like precipitation and temperature are highly dependent upon local topography and local atmospheric circulations (Dimri, 2004).

It is very important to mention here that these regions of the Himalayas are very poorly monitored and having very sparse observations. Prediction of these important parameters is thus a crucial problem for local man and machine and thus could be handled with dynamical and statistical downscaling up to certain extent (Dimri et al., 2002; Mohanty and Dimri, 2004; Dimri and Mohanty, 2007). Topographical heterogeneity and land-use variability of the Himalayas generate numerous meso/micro scale circulations in the cascading narrow valleys and ridges, which dominate in determining the precipitation and temperature patterns over the region. Topography steers atmospheric circulations by strong thermodynamical and dynamical forcings and land surface exchanges momentum, energy, water and chemical constituents with the atmosphere (Dimri and Ganju, 2004; Dimri, 2009). As a

A.P. Dimri (✉) • P. Maharana
School of Environmental Sciences, Jawaharlal Nehru University, New Delhi 110067, India
e-mail: apdimri@hotmail.com

N.J. Raju et al. (eds.), *Management of Water, Energy and Bio-resources in the Era of Climate Change: Emerging Issues and Challenges*,
DOI 10.1007/978-3-319-05969-3_27

result, the land atmosphere interactions have a substantial impact on weather and climatic patterns and on the water and the energy cycles of the climate systems (Dickinson, 1995; Pielke and Avissar, 1990). These interactions occur through an array of complete dynamical, physical, biological and hydrological processes, many of which are highly non-linear in nature (Atkinson, 1981; Avissar and Verstraete, 1990). In many cases, land surfaces are characterized by pronounced spatial heterogeneity illustrated by complex patterns of topography and soil water content (Quinn et al., 1991; Rodriguez Iturbe et al., 1995), satellite imageries (Loveland et al., 1991) or vegetation microscale characteristics (Avissar, 1993).

Numerous simulation studies over the United States (Giorgi et al., 1993a), eastern Asia and Japan (Hirakuchi and Giorgi, 1995), and Europe (Marinucci and Giorgi, 1992; Jones et al., 1995) show that regional climate models (RCMs) reproduce well the observed present day mean climate when the model is forced by boundary conditions that accurately represent the observed synoptic scale atmospheric conditions. However, winter season integrations over the complex Himalayan region are not studied extensively. In addition, a model parameterization-based skillful reproduction of precipitation and temperature over a region does not verify that a model can successfully simulate precipitation and temperature over the regions having most complex and variable topography as of western Himalayas (WH). In addition to this, most of the RCM studies focus on mean climate conditions and less is known about interannual and interseasonal variability. Therefore, objective is to examine winter-time interannual variability of precipitation and temperature over the WH. In addition, comparison of modelled and observed interannual variability will help in assessing a range of meso/micro scale atmospheric conditions. For doing so long simulations are needed to yield statistically meaningful results and then compare the simulated and observed precipitation and temperature anomalous for specific months and seasons (Luthi et al., 1996; Jenkins and Barron, 1997; Walsh and McGregor, 1997). By comparing particularly observed anomalies, one can directly test if the RCM (RegCM3; Giorgi et al., 1993) has the correct sensitivity to the specific forcings.

Therefore, winter-time (DJF) interannual variability is examined over the western Himalayas by analyzing 22 years (1980–2002) simulation with the Abdus Salam International Center for Theoretical Physics (ICTP) - RegCM3 driven by the National Centre for Environmental Prediction (NCEP), US, boundary conditions. Model and experimental design and observations used for validation is presented in the next section. Comparison of RegCM3 simulated and observed precipitation and temperature to assess interannual variability is discussed in section 'Results and Discussion'. Broad findings and salient features are presented in the last section as conclusions.

Model and Experimental Design

The regional cliamte model (RCM) used in the present work is the version of RegCM3 developed by Giorgi et al. (1993 a, b) with some of the augmentations discussed by Giorgi and Shields (1999). The dynamical core of the RegCM3 is equivalent to the hydrostatic version of the fifth-generation Pennsylvania State University-National Center for Atmospheric Research (NCAR), US, Mesoscale Model (MM5). For the present simulation the standard model configuration is used with 23 σ levels, with medium resolution planetary boundary layer (PBL) scheme with five lowest 1.5 km of the atmosphere, at approximately 40, 110, 310, 730 and 1400 m above surface (Giorgi and Bates, 1989) and the model top at 50 hPa. The physical parameterization employed in the simulations include the comprehensive radiative transfer package of the NCAR Community Climate Model version 3 (CCM3) (Kiel et al., 1996), the non-local boundary scheme by Holtslag et al. (1999) and mass flux cumulus cloud scheme of Grell (1993). Land surface processes are described via Biosphere-Atmosphere Transfer Scheme or BATS (Dickinson et al., 1993). A simplified explicit cloud water scheme is used which prognostically calculates precipitation and cloud water for radiation calculations (Giorgi and Shields, 1998).

Model domain of grid size 60 km × 60 km is considered in the present study as shown in Fig. 1(a). The computational domain is considered from Mediterranean Sea to India using a Lambert conformal projection, centered over central Asia. This domain size is chosen so that flow pattern of the winter-time synoptic weather system WDs, which is the primary source of winter-time precipitation, could also be understood. However, discussion is focused mainly over the WH and two geographical regions. The topography for control grids is obtained from a 30" (about 1 km) resolution global datasets produced by the U.S. Geological Survey (USGS) (Fig. 1(b)). The landuse distribution for the control experiments is also obtained from 30" landuse dataset produced by USGS (Loveland et al., 1991). A version of this dataset is already available in the form of BATS surface types. From the 30" dataset, fractional cover of different surface types is calculated for each cell of the different model grids, and the grid cell is then assigned the surface type with the largest fractional cover and henceforth considering landuse and soil type. Further, lateral meteorological boundary conditions for the simulations are obtained from analyses of observations by the NCEP (Kalnay et al., 1996) and therefore the model results can be directly compared with the observations for the simulated period. Soil temperatures are initialized with the temperature of the bottom model level and soil water content is initialized as a function of vegetation type (Giorgi and Bates, 1989).

Six months simulation, starting from 01 October to 31 March, is made for each year (1980–2002) which encompasses full winter season. This particular period is chosen for the study, as enormous amount of precipitation in the form of snow is received over the WH. Hence, most of the discussion is pertaining towards surface parameter—precipitation and temperature—variabilities during this period. Observed seasonal (DJF) precipitation and surface air temperature needed for the evaluation of the simulations are obtained from 0.5° resolution global land datasets developed by the Climate Research Unit (CRU) of the University of East Anglia (New et al., 2000; Willmott and Matuura, 2001) and station data from the Snow and Avalanche Study Establishment (SASE), Chandigarh, India.

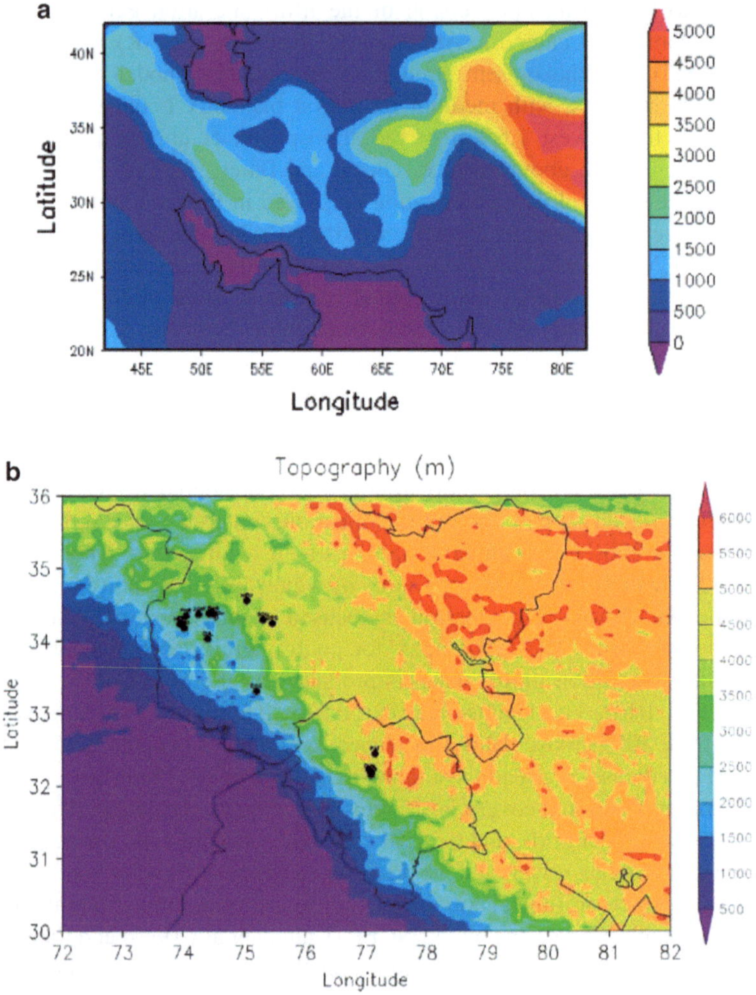

Fig. 1 (**a**) Model domain used in the present study; (**b**) Detailed topographic (m) map of the Himalayan region

Results and Discussion

Observed and model simulation based winter-time interannual variability of precipitation and temperature over the WH is illustrated. In addition, a detailed analysis and intercomparison of the model simulation with the corresponding observed reanalysis is presented and discussed. Further, comparison is drawn at two stations, viz., Bahang (latitude 32°16′33″, longitude 77°09′03″, altitude 2192 m) and Gulmarg (latitude 34°30′00″, longitude 74°29′00″, altitude 2800 m), situated in WH region. These stations are chosen in such a way that they represent different climatic and geographic conditions of the region and have recorded data. Statistics and associated error analysis, in terms of root mean square error (RMSE) and correlation coefficient (CC), is carried out to assess model's skill.

Precipitation

Figure 2 shows bias in RegCM3 simulated seasonal averaged (DJF) precipitation with CRU observed data over the Himalayan region for winter (DJF) during the period 1991–92 to 1999–2000 only. Model simulated regional spatial distribution of the precipitation over the Himalayan region as affected by the mountain systems of the region is presented. Figure 2 shows that model was able to simulate well the variability associated in the nature of precipitation up to certain extent and hence seasonal averaged precipitation is simulated with overestimation in the model physics. This drift in the model simulation may be because of the fact that though the topographically induced cold season precipitation maxima are reproduced, but the corresponding peak precipitation values are somewhat overestimated. It may be attributed to the fact that in CRU observations representation of density of high elevation and low elevation stations may not be that homogeneous so that fine resolution resolvable scale circulations are not reproduced. This is evidently a problem related to relatively coarse resolution of observed data set over the complex topographical Himalayan region. Apart from that model was able to simulate well the anomalous high (DJF, 1990) and low (DJF, 1996) seasonal precipitation.

Further, model statistics are presented in Fig. 3 which shows that model has strength to simulate precipitation. It illustrates area averaged RMSE and CC between seasonal averaged observed (CRU) and model simulated precipitation. It can be seen that model was not only able to capture seasonal variability in the precipitation with least RMSE but also showed strong correlation in between simulated and observed precipitation. In addtion to that, Fig. 4 illustrates anomaly of seasonal averaged observed (CRU, WM and SASE) and model simulated precipitation field at Bahang and Gulmarg. It shows that model simulations-based winter-time interannual variability is close to the observed seasonal precipitation variability. Apart from simulating well the precipitation distribution, model in general could correctly bring out the precipitation variability except during anomalous high and low seasonal precipitation.

Overall the regional topographical forcing is captured by the model, as evident by the fact that most of the topographically induced precipitation is reproduced well by the model simulation. The reason of model overestimate is that the dominant precipitation process is mostly of resolvable scale nature and is induced by topographic uplift within the westward moving cyclonic systems (WDs). As a result, precipitation is mostly forced by the topographical gradients and the underpresentation of these gradients by the relatively smooth topography leads to overestimation of the precipitation maxima. This estimation can be corrected by

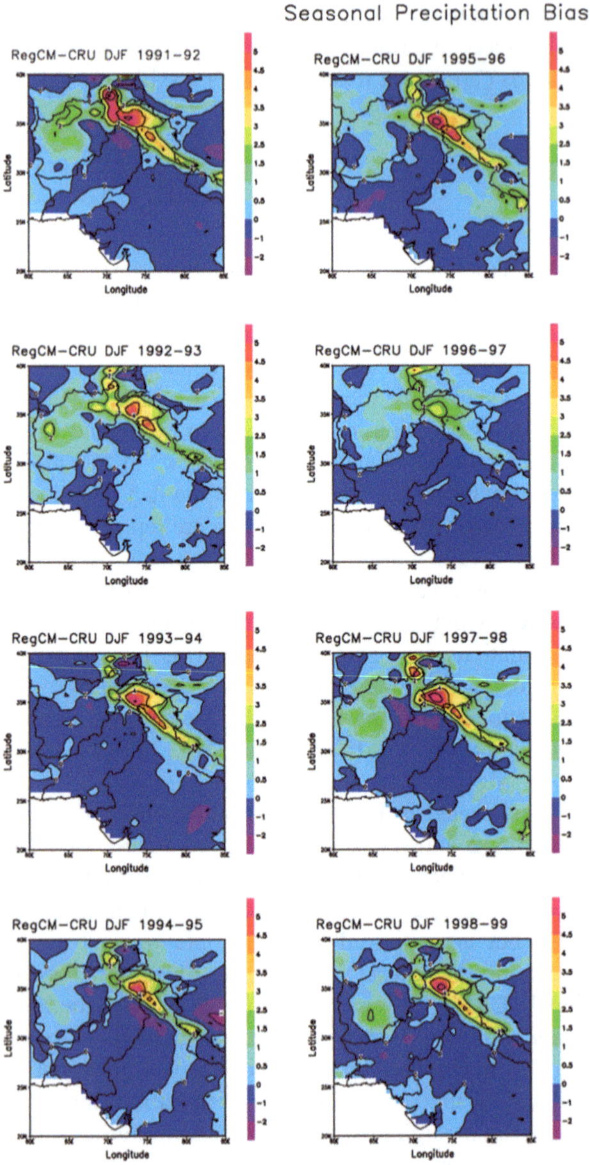

Fig. 2 Bias in winter-time (DJF) seasonal averaged model simulated and observed precipitation (mm/day)

subgrid topographic disaggregation of precipitation and since the subgrid scale variability does affect precipitation formation, a relatively low sensitivity of precipitation for the different experiments can be expected (Dimri, 2009). This is especially the case for winter season, when the precipitation generation mechanism are mainly dynamical in nature. These results are indeed evident from the comparison of the winter precipitation fields in Figs. 2 and 4.

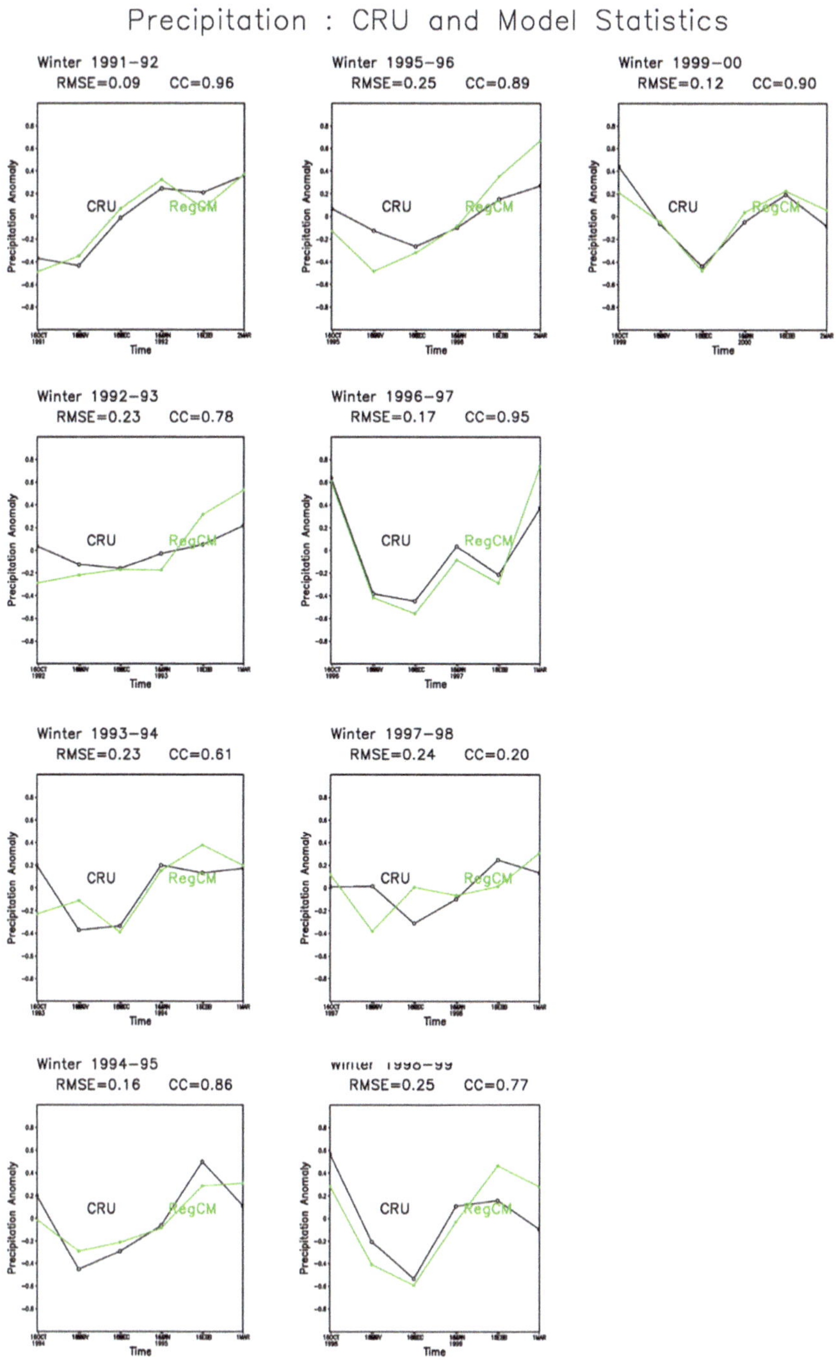

Fig. 3 Area averaged model statistics in precipitation simulation in terms of root mean square error (RMSE) and correlation coefficient (CC)

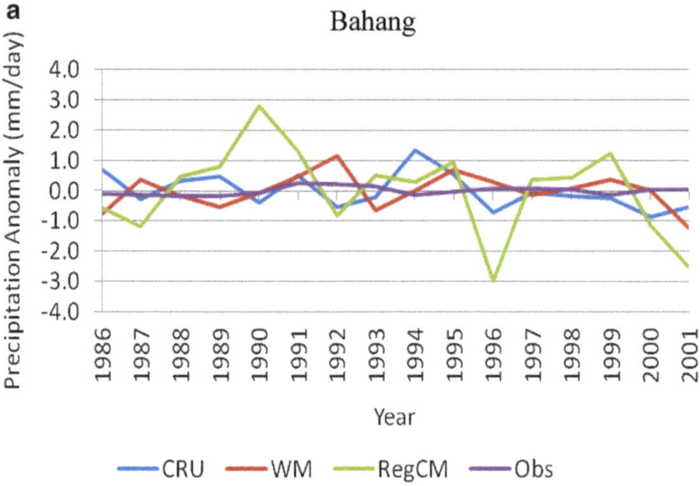

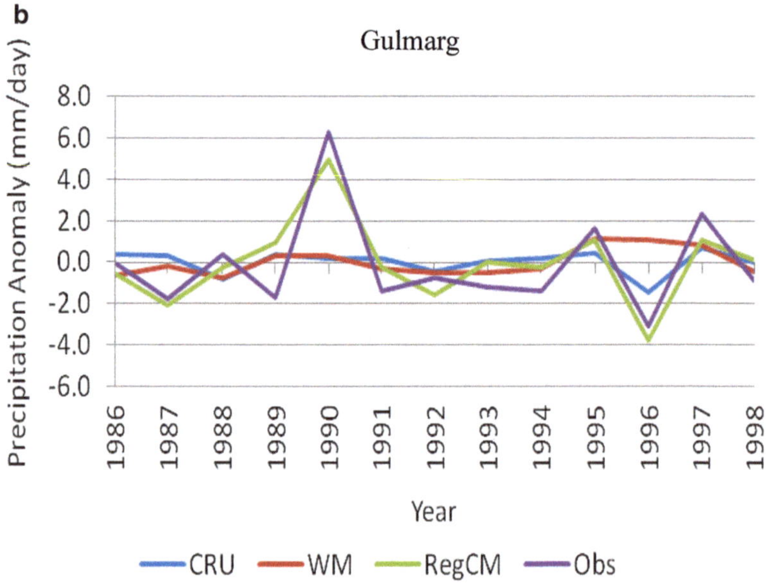

Fig. 4 Comparison of seasonal averaged model simulated and observed precipitation anomaly (mm/day) at (**a**) Bahang and (**b**) Gulmarg

Small differences across the simulation are essentially due to the internal model variability (Giorgi and Bi, 2000). Simulation experiments show high intensity of precipitation maxima, lying along the orientation of the Himalayas, which are not shown in observed (CRU) datasets. This difference may lie in the fact that topography and vegetation cover is very fast changing within a km over the Himalayan region whereas observed dataset of CRU is presented at 0.5° resolution. Here one point is to be noted that due to smoothening of model topography, meso/micro scale circulations, which are dominant in complex topography of the Himalayas, may be overlooked during simulation procedures. Due to topographic smoothening low density of high elevation points is considered in the model. Apart from this, smoothening of topography and landuse in simulation experiments reduce the impact of orographic lifting due to steep gradient in the Himalayan region. Most of the time orographic lifting is the predominant mechanism to modulate the precipitation amount in complex topographical region of the Himalayas. Also, smoothening of landuse type, which is rapidly changing within a km in the Himalayan region, may lead to over/under estimate the precipitation amount. Overall the simulations could capture seasonal precipitation pattern and their variability.

A very important point in variability of the precipitation amount is due to the substantiable sensitive snow forming processes in the Himalayan region. As already told, during winter most of the precipitation is in the form of snow over the Himalayan region. As a result, the response of the snow can be highly nonlinear, since snow formation and melting are regulated by the processes that are essentially step function of temperature thresholds. In the comparison of simulated and observed precipitation some important points need to be considered. First, while the model calculates the snow depths in terms of the liquid water equivalent, the original observed data (SASE) are recorded in centimetres of snow. To obtain equivalent liquid water depth we scaled the snow depth by a factor of 1/3, which is roughly a characteristic of the density of the aging snow (Dickinson et al., 1993). Admittedly a large uncertainty is implicit in this assumption. Second, the station density is irregular in space and it includes a relatively small number of high elevation stations. Third, observed snow depth at a station is strongly affected by processes such as snowdrift and snow sheltering by upwind obstacles, which are not included in the model. For these reasons the comparison with observations is necessarily limited in scope and mostly aimed at providing qualitative indications of the model behaviour.

It is evident from the figures that the spatial variability of precipitation increases substantially with the resolution of the land surface. This is because the temperature produced is such that precipitation can be in the form of snowfall over the higher subgrid peaks and rainfall over the subgrid valleys even though the coarse grid precipitation is only in the form of rainfall or snowfall. As a result, snow tends to accumulate over the high resolution peaks and melt more effectively over the correspoding valleys. This can be expected from the complex Himalayan topography as illustrated in Fig. 1. Indeed even with the limitations discussed, the observations clearly show that snow depth is characterised by a pronounced finescale variability.

Surface Air Temperature

Figure 5 shows bias in simulated seasonal averaged (DJF) temperature with CRU data over the Himalayan region of complex topography. In winter, Fig. 5, seasonal temperature patterns over the Himalayan region are reproduced well in model

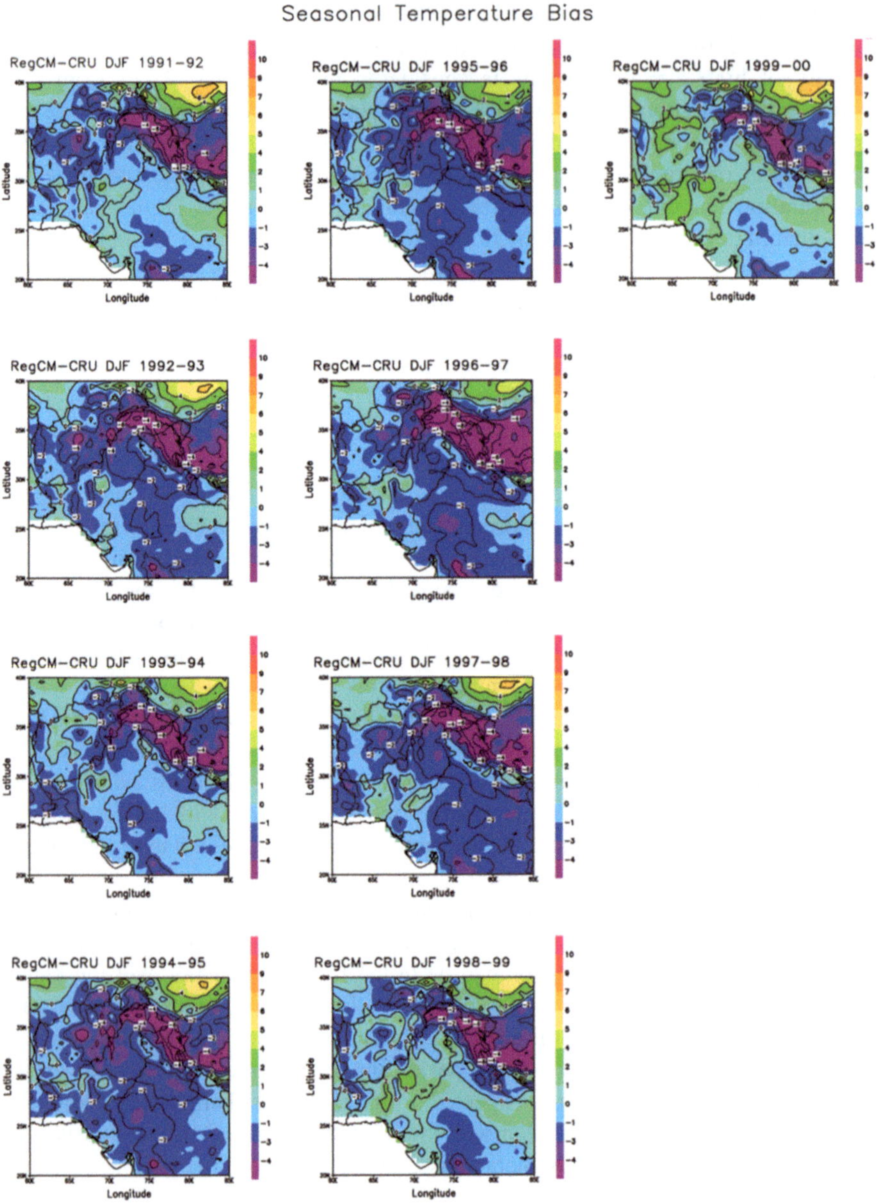

Fig. 5 Same as Fig. 2, but for temperature (°C)

simulation. Over the complex mountaineous region model shows cold bias. It is likely that the model bias is artificially enhanced by a temperature underestimation in the model simulation induced by the relatively low density of high elevation stations. Also note that the model simulations tend to show finer scale structure than the reanalysis observations, in particular over the main mountaineous ranges. The reanalysis observations do not show finescale structure because they are provided with 0.5° resolution. Further Fig. 6 shows area averaged RMSE and CC between simulated results and observed value of tempearture. Some of the differences between model results and observations, Fig. 6, appear mostly attributable to the finer structure of the simulated fields, particularly over the Himalayan region.

It can be seen that model could simulate well the temperature values over the Himlayan region. Effect of topographic heterogeneity and landuse variability induces decrease in temperature spatial variability. Further comparison with anomaly of realtime seasonal averaged observed (CRU, WM and SASE) and simulated surface air temperature at Bahang and Gulmarg is presented in Fig. 7. Seasonal temperature variability is reproduced with reasonable satisfaction over the Himalayan region. Also, simulation results do not reproduce the similar temperature variability as observed at all the places across the Himalayas. Some of this may be attributed to the fact that observational data is an integration of multiple types/ sources of data and control simulation results are affected by smoothening of model topography. Whereas, effect of subgrid scale heterogeneity shows pronounced effect on temperature variability due to surface subgrid parameterization over the region (Dimri, 2009). The temperature anomaly interannual variability does not show similar structure which can be improved upon by using subgrid parameterization scheme by enhancing the initial landuse information in the model simulation. It shows that though model underestimates temperature but temperature anomaly interannual variability of warm and cold region needs to be studied further over the heterogeneous Himalayan region. Zone of negative tempratures are in corroboration with the fact that finer resolvable scale circulations are well depicted in the model simulations. Further, comparison of seasonal averaged observed (CRU, WM and SASE) and model simulated surface air temperature shows that as extreme cold temperatures approach model was able to capture well the temporal variability of temperature over the region. Not only this, spatial variability of cold temperature region is also well depicted in the simulation.

At few of the seasons model simulation could not reproduce the finescale temperature distribution as observed. This is due to the fact that heterogeneity in topography and variability in landuse is complex to be represented in the model with finer scale. Also, due to smoothening of topography in simulation, fine resolution resolvable scale circulations are not reproduced. Further, comparison shows that model has tendency to underestimate the cold region temperatures and overestimate the warm region temperatures by few degrees. Again the model biases can be due to the relatively low density of high elevation stations in the observed data sets. Coming down to the resolvable scale features simulated by the model it is seen that finer scale distribution of surface air temperatures are brought out. Simulations are able to represent the temporal and spatial variability with finer details. It can be noticed that topographical distribution is well resolved to bring out the finer scale surface circulation over the region which is comprised of valleys and hill ranges.

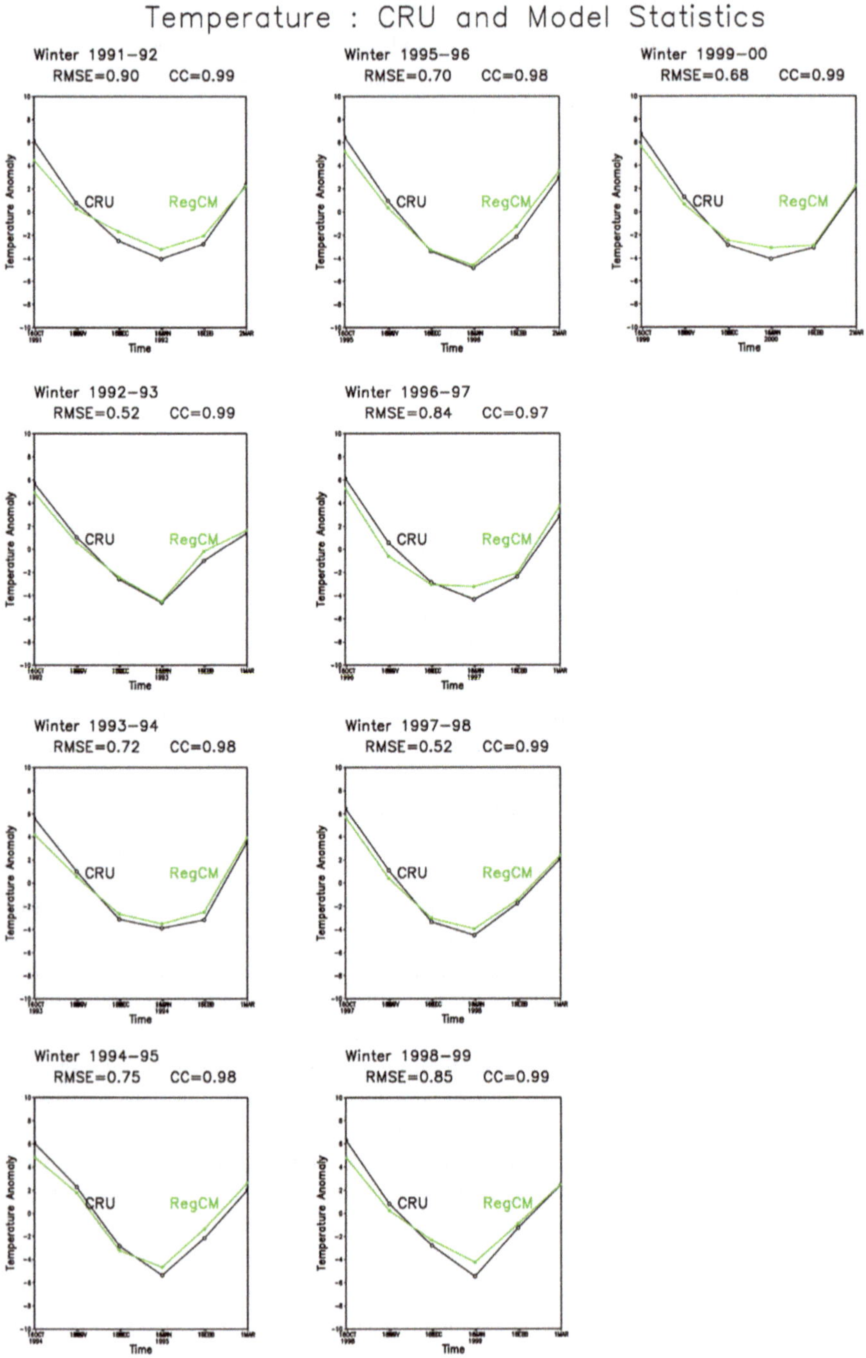

Fig. 6 Same as Fig. 3, but for temperature (°C)

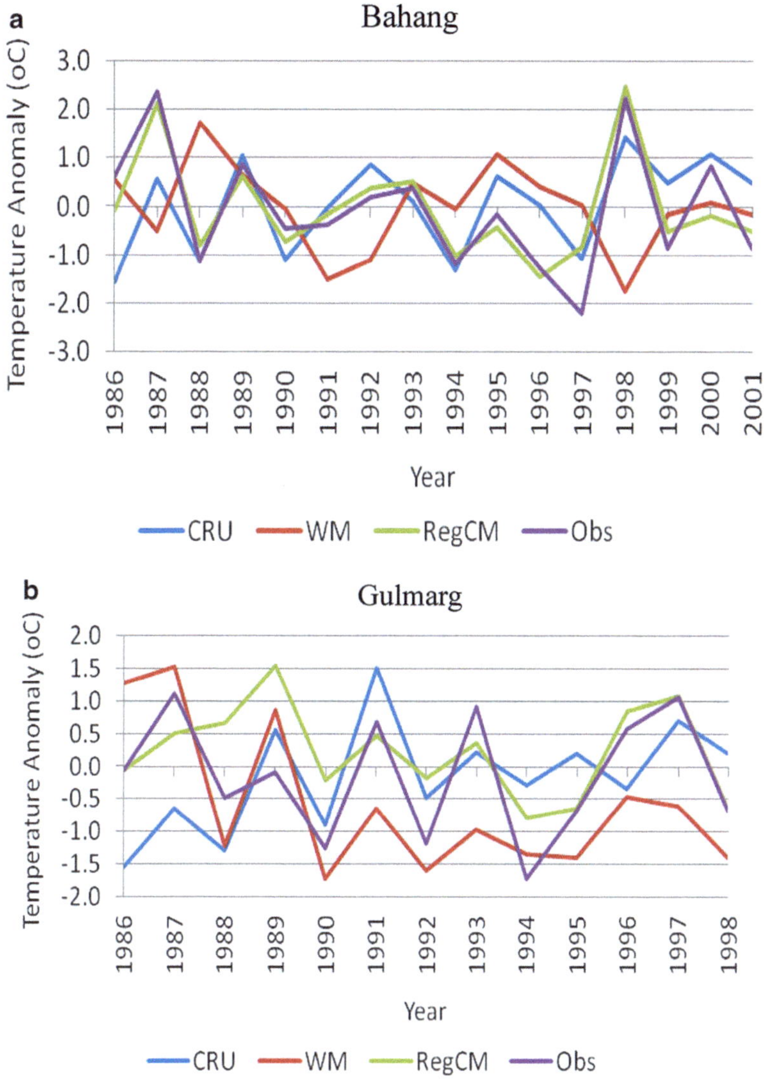

Fig. 7 Same as Fig. 4, but for temperature (°C). (**b**) Gulmarg (**a**) Bahang

As discussed in preceding section, in winter most of the Himalayan region receives precipitation in the form of snow. Spatial variability of snow not only increases, but also do overall snow amounts over the region which can be attributed to the inherent nonlinear nature of snow forming processes—where temperature plays a major role in defining precipitation type. As the temperature threshold for snow formation reached, say at the high elevation of a subgrid peak, snow starts accumulating. Because snow has a higher albedo than a bare soil or vegetation, the overall surface albedo increases and this causes a decrease in absorption of solar radiation at the surface. This in turn inhibits the solar warming of the surface and thus tends to cool the region and increase the lifetime of the snowpack. These feedback processes can be seen in the overall winter cooling of the Himalayan region in simulation.

Summary and Conclusions

The theme of the present paper is mainly focused on the WH regional climate. And impact of topographic heterogeneity and landuse variability is comprehensively illustrated. Present context of the paper represents impact of topographic heterogeneity and landuse variability over two importnant geographical subregions—Bahang and Gulmarg. The main focus of the present study is to demonstrate how well the regional climate model does over the WH region. Though there are such important changes occuring all across Hindukush-Himalayas (HKH) region (Kulkarni et al., 2013). The HKH region is characterized by topographic heterogeneity and landuse variability from west to east. In addition, seasonal patterns over the western, central and eastern Himalayan regions differ, thus having larger spatial and as well temporal variation in precipitation and temperature patterns. Based on model simulation, interpretation with corresponding available observations similar studies are needed to demonstrate comprehensive role of topography and landuse and their influence on the regional climate. Such studies will provide an added tool for socio-economic structuring over the HKH region. Therefore it is important to have (1) a more realistic representation of landuse change and (2) its impact on surface variables driven by the complex terrain features over mountaineous region. Above results over the WH region indicate importance of topography and landuse information to estimate precipitation and temperature variability over the complex terrain, which is generally poor in coarse resolution models for such region (Dimri and Niyogi, 2012). Dimri and Niyogi have comprehensively demonstrated the robustness of the regional climate model over the WH region. Therefore, representation of topography and landuse across the HKH will show improved impact on seasonal evolution of precipitation and temperature patterns. Though it needs higher level of computational machine and manpower to understand the dynamics and physics behind such processes, nonetheless it is important that we look into these aspects at regional level viz., western, central and eastern Himlayas. And indeed, Hahmann and Dickinson (2001) found the effect of subgridscale landuse heterogeneity to be important primarily in tropical regions.

Further, sensitivity to the initialization of surface variables, soil variables, topography and landuse etc., is important to understand as it is highly variable all along HKH region. In addition to this, natural variability in respective seasons is not same.

References

Atkinson, B.W. (1981). Mesoscale atmospheric circulations. Academic Press.

Avissar, R. (1993). Observation of leaf stomatal conductance at the canopy scale: An atmospheric modeling prospective. *Bound. Layer Meteor.*, **64:** 127–148.

Avissar, R. and Peele, R.A. (1989). A parameterization of heterogeneous land surface for atmospheric numerical models and its impact on regional meteorology. *Mon. Wea. Rev.*, **117:** 2113–2136.

Avissar, R. and Verstraete, M.M. (1990). The representation of continental surface processes in atmospheric models. *Rev. Geophys.*, **28:** 35–52.

Dickinson, R.E., Henderson-Sellers, A. and Kennedy, P.J. (1993). Biosphere-Atmosphere transfer Scheme (BATS) version 1e as copied to the NCAR Community Climate Model. NCAR Tech. Note NCAR/TN-387 + STR.

Dickinson, R.E. (1995). Land Atmosphere interaction. *Rev. Geophys.*, **33(Suppl.):** 917–922.

Dimri, A.P. and Mohanty, U.C. (2007). Location-specific prediction of maximum and minimum temperature over the western Himalayas. *Meteorological Applications,* **14(1):** 79–93.

Dimri, A.P. (2009). Impact of Subgrid Scale Scheme on Topography and Landuse for better regional scale simulation of meteorological variables over Western Himalayas. *Climate Dynamics*, **32:** 565–574.

Dimri, A.P. and Ganju, A. (2007). Wintertime Seasonal scale Simulation over Western Himalayas using RegCM3. *PAGEOPH*, **164(8–9):** 1733–1746.

Dimri, A.P. and Niyogi, D. (2012). Regional climate model application at subgrid scale on Indian winter monsoon over the western Himalayas. *Int. J. of Climato.* doi. 10.1002/joc3584

Dimri, A.P., Mohanty, U.C., Madan, O.P. and Ravi, N. (2002). Statistical model-based forecast of minimum and maximum temperatures at Manali. *Current Science*, **82(8):** 997–1003.

Giorgi, F. and Bi, X. (2000). A study of internal variability of a regional climate model. *J. Geophys. Res.*, **105:** 29503–29521.

Giorgi, F. and Shields, C. (1999). Tests of precipitation parameterizations available in the latest version of the NCAR regional climate model (RegCM) over the continental United States. *J. Geophys. Res.*, **104:** 6353–6375.

Giorgi, F., Marinucci, M.R. and Bates, G.T. (1993a). Development of a second generation regional climate model (RegCM2), PartI: Boundary layer and radiative transfer processes. *Mon. Wea. Rev.*, **121:** 2794–2813.

Giorgi, F., Marinucci, M.R., De Canio, G. and Bates, G.T. (1993b). Development of a second generation regional climate model (RegCM2), Part II: Convective processes and assimilation of lateral boundary conditions. *Mon. Wea. Rev.*, **121:** 2814–2832.

Grell, G.A. (1993). Prognostic evaluation of assumptions used by cumulus parameterization *Mon. Wea. Rev.*, **121:** 764–787.

Hahmann, A.N. and Dickinson, R.E. (2001). A fine mesh land approach for general circulatio models and its impact on regional climate. *J. Climate*, **14:** 1634–1646.

Hirakuchi, H. and Giorgi, F. (1995). Multilayer present day and $2 \times CO_2$ simulations of monsoon climate over eastern Asia and Japan with a regional climate model nested in a general circulation model. *J. Geophys. Res.*, **100:** 21105–21125.

Holtslag, A.A.M., de Bruijn, E.I.F. and Pan, H.L. (1999). A high resolution air mass transformation model for short range weather forecasting. *Mon. Wea. Rev.*, **118:** 1561–1575.

Jenkins, G. and Barron, E. (1997). Global climate model and coupled regional climate model simulations over eastern United States: GENESIS and RegCM2 simulations. *Global and Planet Change*, **15(3):** 32.

Jones, R.G., Murphy, J.M. and Noguer, M. (1995). Simulation of climate change over Europe using a nested regional climate model, I. Assessment of control climate, including sensitivity to location of lateral boundaries. *Q.J.R. Meteorol. Soc.*, **121:** 1413–1449.

Kalsi, S.R. (1980). On some aspects of interaction between middle latitude westerlies and monsoon circulation. *Mausam*, **38:** 305–308.

Kalsi, S.R. and Haldar, S.R. (1992). Satellite observations of interaction between tropics and mid latitude. *Mausam*, **43:** 59–64.

Kiel, J.T., Hack, J.J., Bonn, G.B., Boville, B.A., Briegleb, B.P., Williamson, D.L. and Rasch, P.J. (1996). Description of the NCAR Community Climate Model (CCM3). NCAR Tech. Note NCAR/TN-420 + STR.

Kulkarni, A., Patwardhan, S., Kumar, K., Ashok, K. and Krishnan, R. (2013). Projected climate change in the Hindukush-Himalayan region by using the High resolution Regional Climate Model PRECIS. *Mount. Res. Dev.*, **33(2):** 142–151.

Loveland, T.R., Merchant, J.W., Ohlen, D.O. and Brown, J.F. (1991). Development of a land cover characteristics database for the conterminous United States, PhotogrammPhotogram. *Eng. Remote Sens.*, **57:** 1453–1463.

Luthi, D., Cress, A., Frei, C. and Schar, C. (1996). Interannual variability and regional climate simulations. *Theor. Appl. Climatol.*, **53:** 185–196.

Marinucci, M.R. and Giorgi, F. (1992). A 2 × CO_2 climate change scenario over Europe generated using a limited area model nested in a general circulation model, 1: Present-day seasonal climate simulation. *J. Geophys. Res.*, **97:** 9989–10009.

Mohanty, U.C. and Dimri, A.P. (2004). Location-specific prediction of the probability of occurrence and quantity of precipitation over the Western Himalayas. *Weather and Forecasting*, **19 (3):** 520–533.

Mohanty, U.C., Madan, O.P., Rao, P.L.S. and Raju, P.V.S. (1998). Meteorological fields associated with western disturbances in relation to glacier basins of western Himalayas during winter season. Center for Atmospheric Sciences, Indian Institute of Technology, New Delhi, Technical report.

New, M.G., Hulme, M. and Jones, P.D. (2000). Representing twentieth century space time climate variability, Part II: Development of a 1901–96 monthly grids of terrestrial surface climate. *J. Climate*, **13:** 2217–2238.

Pielke, R. and Avissar, R. (1990). Influence of landscape structure on local and regional climate. *Landscape Ecol.*, **4:** 133–155.

Pisharoty, P. and Desai, B.N. (1956). Western disturbances and Indian weather. *Indian J. Meteorol. Geophys.*, **7:** 333–338.

Quinn, P.F., Beven, K., Chevallier, P. and Planchon, O. (1991). The prediction of hill slope flow path for distributed hydrological modeling using digital terrain models. *Hydrol. Processes*, **5:** 59–79.

Rao, Y.P. and Srinivasan, V. (1969). Forecasting manual, Part II: Discussion of typical synoptic weather situation: winter western disturbances and their associated feature. FMU report no III-1.1, issued by India Meteorological Department, India.

Rodriguez Iturbe, I., Vogel, G.K., Rigon, R., Entekhabi, D., Castelli, F. and Rinaldo, A. (1995). On the spatial organization of the soil moisture field. *Geophys. Res. Lett.*, **22:** 2757–2760.

Singh, M.S. (1979). Westerly upper air troughs and development of western disturbances over India. *Mausam*, **30**(4): 405–414.

Walsh, K. and McGregor, J. (1997). An assessment of simulations of climate variability over Australia with a limited area model. *Int. J. Climtol.*, **17:** 201–223.

Willmott, C.J. and Matuura, K. (2001). Terrestrial air temperature and precipitation monthly and annual time series (1950–1999) (version 1.02). Centre for Climate Research, University of Delware, Newark, N.J. USA. http://climate.geog.udel.edu/~climate/hjtml-pages/README.ghcn_ts2.html.

Assessment of Trace Element Distribution in Red-bloom (*E. shafiqii*) and Water of Dal Lake, Kashmir Valley, by Total Reflection X-ray Fluorescence Spectrometry

Shafiq-ur-Rehman, Shaheen, Sangita Dhara, N.L. Misra, and Alok Srivastava

Introduction

Dal Lake is located in Srinagar in the Jammu and Kashmir State of India. It is one of the foremost tourist attractions in the Himalayan valley as well as a lifeline for the local population. It provides fish, aquatic vegetables and drinking water. It has been plagued for the last 20 years by a thin red film during the period of June-August, when the tourist influx is at its peak. This phenomenon is attributed to a rare Euglena species *Euglena shafiqii* (Shafiq-ur-Rehman, 1998), which is now recognized and listed by the International Water Environment Renovation Research Team, Japan and the Society of Protozoologists. The lake frequently receives large quantities of wastes from the surrounding human settlements, hotels, and runoff from agricultural and Dachigam sanctuary catchment lands/area. All these activities make lake pollution prone. Our earlier observations have shown presence of some heavy metals in significant amount in the red-bloomed waters of the lake (Shafiq-ur-Rehman, 2009). The raised concentrations of heavy metals in lakes have also been reported from different parts of the world (Yalcin and Sevinc, 2001; Szymannowska et al., 1999; Elmaci, 2007). There are many spectrophotochemical techniques based on a range of physical principles encompassing the spectral range from Infrared to the X-Ray region (Wobrauschek, 1998). In the case when X-rays are used for elemental analysis their energy can range from

Shafiq-ur-Rehman (✉) • Shaheen
Division of Environmental Sciences, Sher-e-Kashmir University of Agricultural Sciences and Technology of Kashmir, Shalimar Campus, Srinagar 190025, Jammu and Kashmir, India
e-mail: shafiqii@gmail.com

S. Dhara • N.L. Misra
Fuel Chemistry Division, Bhabha Atomic Research Centre, Mumbai 400085, India

A. Srivastava
Department of Chemistry, Panjab University, Chandigarh 160014, India

N.J. Raju et al. (eds.), *Management of Water, Energy and Bio-resources in the Era of Climate Change: Emerging Issues and Challenges*,
DOI 10.1007/978-3-319-05969-3_28

110 eV to about 130 keV (Be to Bk). Total Reflection X-ray Fluorescence (TXRF) is an upcoming trace elemental analysis technique having several advanced analytical features (Klockenkaemper and von Bohlen, 2001). Moreover, TXRF holds several analytical advantages of non-destruction and versatility of sample, capability to analyze multi-elements, and is less time consuming. It has comparatively high detection limits due to high background reduction produced by the absorption and scattering of X-ray beam by the sample and matrix (Misra and Mudher, 2002). In the present study TXRF was applied to determine the concentration of trace elements present in Dal Lake water besides the red-bloom *Euglena shafiqii* which appears in the lake in certain months of the year.

Experimental

The Dal Lake is situated at 34° 04′-34° 11′ N, 74° 48′-74° 53′ E in the north-east township of Srinagar in the heart of Kashmir valley in western Himalaya (1583 m above the sea level). The lake has a total area of about 20 km^2 of which approximately 12 km^2 is the total open water spread area. The lake is shallow in nature with a maximum depth of 3 m. It has a little water-inflow through Telbal nullah in the north and water-outflow channel at the Dal Gate in the south that discharges water into the Jhelum River. An additional outflow channel is situated at the western side of the lake that discharges water into the Nagin Lake. A huge south-west area of the lake is occupied by human settlements, floating gardens and house boats. A total of 277 km^2 area around the Dal Lake falls under the catchments area. The water samples were collected from ten different sites of the Dal lake (Fig. 1): Naseem Bagh (I), Shalimar (2), Nishat Bagh (3), Nishat drain (4), Brein (5), Gagribal (6), Gagribal drain (7), Astana (8), Centaur (9) and Nehru Park (10).

The principle of TXRF Spectrometry is illustrated in Fig. 2. A total reflection effect of X-ray is applied to reduce the background intensity and to double the fluorescence signal. TXRF is an energy dispersive technique wherein excitation of the sample at grazing incidence angle (less than 0.01 degree) is employed. Following the act, the material on the reflector sample disc is excited by both the primary incidence and reflected beam, doubling the fluorescence intensity. Finally, the short distance from sample to detector, a few mm only, allows full energy/signal recognition for detection of fluorescence radiation from the sample thus giving efficient analysis. For trace element determinations an ITAL Structures TXRF spectrometer TX-2000 was used. Mo K_α radiation obtained from a Mo target X-ray tube and monochromatized beam using a W-C multilayer was used for sample excitation. Aliquots of 10-μL of the water samples were mixed with internal standard "Ga" and deposited on flat polished quartz sample supports and presented for TXRF analysis. The bloom samples were first digested in nitric acid and analysed in similar way as the water samples. The elements like K, Ca, Fe, Sr, Zn and Cu were detected in the Dal Lake water samples whereas elements like K, Ca, P, S, Cr, Fe, Mn, Sr, Ti and V were detected in the red-bloom *Euglena shafiqii*.

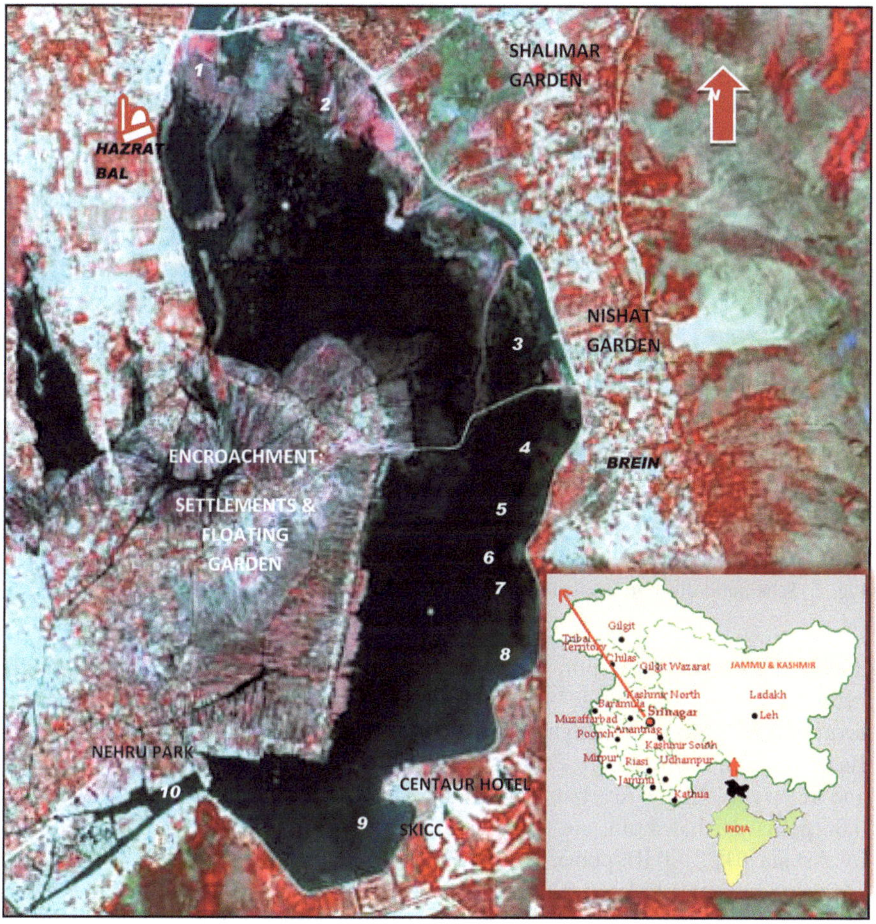

Fig. 1 IRS1D-PAN sharpened LISS-III image of Dal Lake (Srinagar, India) and its surroundings (http://www.isro.org/Imagegallery/satelliteimages.aspx#4), showing sample sites (1. Naseem Bagh, 2. Shalimar, 3. Nishat Bagh (garden), 4. Nishat Drain, 5. Brein, 6. Gagribal, 7. Gagribal Drain, 8. Astana, 9. Centaur and 10. Nehru Park)

Results

The results are presented in Figs. 3 and 4 for the lake waters and the red-bloom, respectively. Table 1 represents trace elements distribution trend in the Dal Lake waters and the red-bloom of *E. shafiqii*. The concentration of K was highest in the lake water at Astana (31 $\pm$ 6 ppm) followed by Gagribal (23 $\pm$ 6 ppm), while at other regions it ranged between 8 and 17 ppm. The level of Ca was high in lake water at Nishat Bagh (120 $\pm$ 7 ppm), while its level in the rest of the lake water was observed to be in the range of 42 to 71 ppm. The water samples from Nishat Bagh (2.4 $\pm$ 0.4 ppm), Gagribal drain (2 $\pm$ 1 ppm), and Centaur (2 $\pm$ 1 ppm) showed nearly

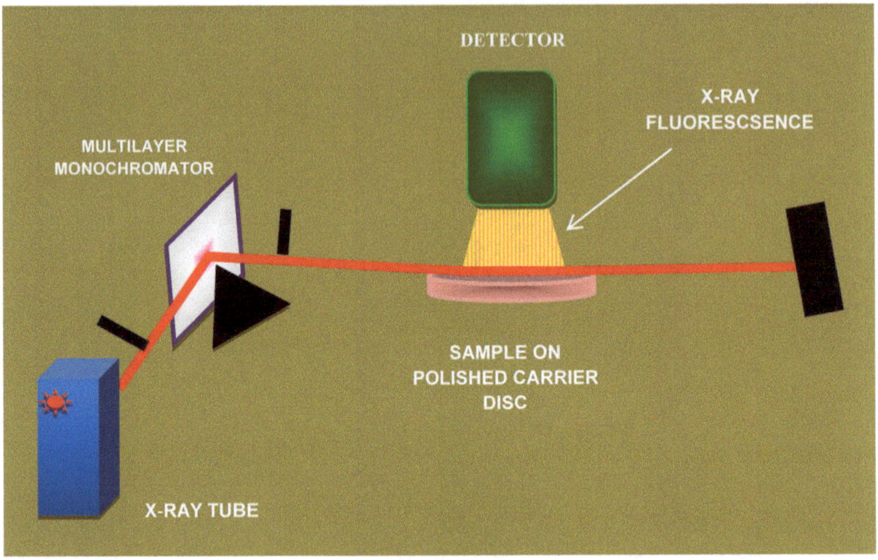

Fig. 2 Schematic illustration displays principle of TXRF spectrometry

double the Fe concen-tration of that obtained from remaining sites which ranged between 1.05 ± 0.02 and 1.7 ± 0.3 ppm excluding Centaur which had minimum level of 0.55 ± 0.02 ppm. High concentrations of Sr were found in Gagribal, Nishat Bagh, Astana, Brein, Nehru Park, Naseem Bagh and Centaur (0.082 to 0.113 ppm) and relatively lower level in Gagribal drain, Shalimar and Nishat drain (0.03-0.06 ppm). Nishat drain site contained maximum Zn (0.27 ± 0.07 ppm) followed by Astana (0.22 ± 0.01 ppm), Nishat Bagh (0.218 ± 0.006 ppm) and lower levels ranging between 0.124 and 0.171 ppm in remaining sites. Nishat drain and Naseem Bagh were the only sites where Cu having concentration of 0.107 ± 0.001 and 0.06 ± 0.03 ppm, respectively was found. The red-bloom of *Euglena shafiqii* was found to contain significant number of trace elements. If one looks at the trend one finds that in the bloom the concentration of Ca (32.1 ± 0.3 ppm) was maximum followed by Fe (23.2 ± 0.3 ppm), K (17.0 ± 0.3 ppm), S (1.6 ± 0.2 ppm), Ti (1.04 ± 0.03 ppm), Mn (0.75 ± 0.01 ppm), P (0.62 ± 0.08 ppm), Sr (0.11± 0.02), Cr and V (both 0.072 ± 0.001 ppm). It was further observed that the bioaccumulation factors in bloom for Fe, Ca and K were 10.9, 0.5 and 0.7, respectively.

Discussion

Trace elements are present in the environment in varying concentrations and forms. They are essentially required by all biological species to sustain life and growth. They are released following natural, mining, industrial, agricultural and many other

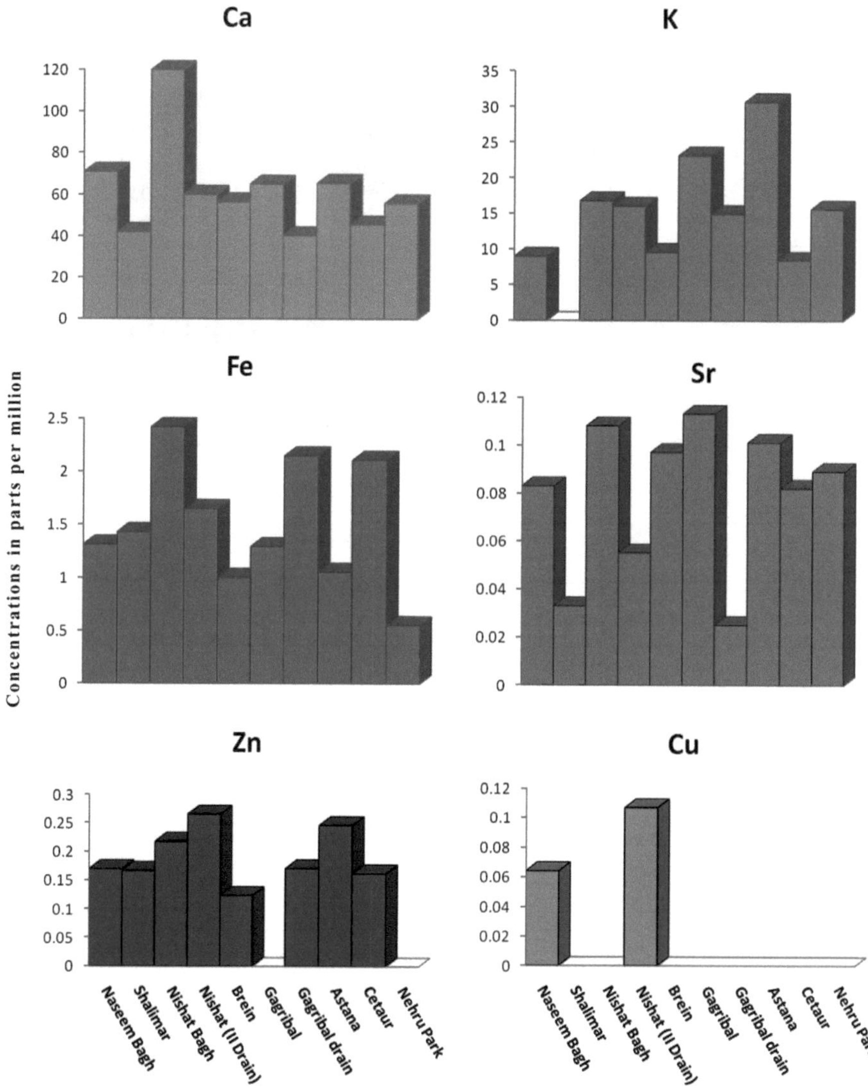

Fig. 3 Concentrations (in ppm) of K, Ca, Fe, Sr, Zn and Cu in different areas of the Dal Lake waters

anthropogenic activities into the environment namely, air, water and soil, to undesirable levels that cause pollution and toxicity. The aquatic ecosystem is highly vulnerable to these events and sometimes gets contaminated to hazardous level. The pollution of freshwaters with trace elements has now become the major problem throughout the world. High amounts of heavy metals in lakes have been reported from different parts of the world (Yalcin and Sevinc, 2001; Szymannowska et al., 1999; Elmaci, 2007). Earlier we have shown that Kashmir

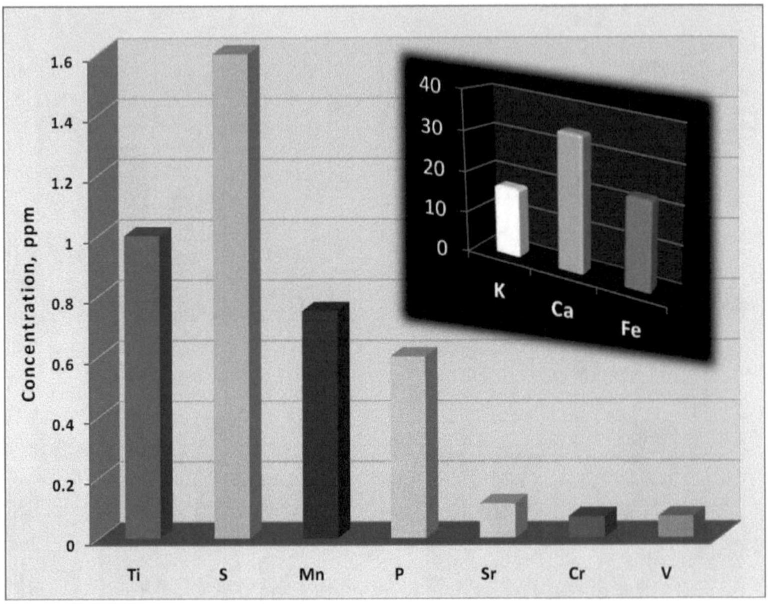

Fig. 4 Concentrations of different minor and trace elements found in the red-bloom of the new species of the *Euglena shafiqii* in the Dal Lake

Table 1 Distribution of trace elements in different areas of the Dal Lake waters and *Euglena shafiqii*

Classification	*Site/Sample*	*Elements*
Lake Waters	Naseem Bagh	Ca > K > Fe > Zn > Sr > Cu
	Shalimar	Ca > Fe > Zn > Sr
	Nishat Bagh	Ca > K > Fe > Zn > Sr
	Nishat drain	Ca > K > Fe > Zn > Cu > Sr
	Brein	Ca > K > Fe > Zn > Sr
	Gagribal	Ca > K > Fe > Sr
	Gagribal drain	Ca > K > Fe > Zn > Sr
	Astana	Ca > K > Fe > Zn > Sr
	Centaur	Ca > K > Fe > Zn > Sr
	Nehru Park	Ca > K > Fe > Sr
Red Bloom	*Euglena shafiqii*	Ca > Fe > K > S > Ti > Mn > P > Sr > Cr = V

valley's most famous fresh water lake, the Dal Lake, experienced an outbreak of the red-bloom of a new species of *Euglena shafiqii*, Shafiq-ur-Rehman, owing to the enrichment of nutrients and pollution of certain trace elements in the lake water (Shafiq-ur-Rehman, 1998). In the earlier study it was observed that the concentration of Fe was significantly high and was found to be equally distributed throughout the lake waters. However, the present study shows that the concentration of Fe is no more uniformly distributed in lake water. It was further observed that in addition to

Fe the lake water of Dal Lake has significant level of Ca, Sr, K and Zn besides Cu. The concentrations of K, Ca, and Cu along with Mg and Na were also found high in the water of lake Texoma in South Korea (An and Kampbell, 2003). Increased Fe and low Zn concentrations were observed in Lake Sapance in Turkey (Yalcin and Sevinc, 2001). Lake Superior, however, was reported to have the lowest concentrations of Fe and Zn (Field and Sherrell, 2003). The trend of trace elements distribution for the most part of the Dal Lake waters was $Ca > K > Fe > Zn > Sr$ (Table 1).

The present study shows that the highest concentrations of elements were apparent in Astana (for K), Nishat Bagh (for Ca and Fe), Gagribal (for Sr), Nishat drain (for Zn and Cu) followed by Gagribal (for K), Naseem Bagh (for Ca and Cu), Gagribal drain and Centaur (for Fe), Nishat Bagh and Astana (for Sr and Zn). It can be stated that Dal Lake waters contain undesirably high levels of some elements, which are potentially toxic to aquatic organisms and might also be the trigger factor in the eutrophication process. The trace element distribution trend in the bloom *E. shafiqii* was found to be as follows: $Ca > Fe > K > S > Ti > Mn > P > Sr > Cr = V$ (Table 1). This showed that *E. shafiqii* could concentrate additional trace elements by binding them which were not detected in the lake waters. Therefore, this specie has the potential to serve as a bio-indicator for detecting hazardous elements present in trace level in the Lake environment.

Conclusion

In conclusion it can be stated that a new emerging analytical technique based on TXRF could be successfully applied in the study of a wide range of elements present at trace level to examine the damages caused by them to a lake ecosystem. This study was also successful in demonstrating the importance of Fe in the appearance of the red bloom in Dal Lake besides postulating that the new species *Euglena shafiqii*, Shafiq-ur-Rehman, showing biological circadian rhythm of its colour change can serve as bio-indicator of hazardous elements present in the lake ecosystem at trace level.

Acknowledgement The authors are thankful to Alexander-von-Humboldt Foundation, Bonn, Germany, for awarding Alok Srivastava a Summer Fellowship in Aachen University of Applied Sciences, Juelich, Germany, where a part of the data was evaluated and the initial manuscript was written.

References

Elmaci, A., Teksoy, A., Topac, F.O., Ozengin, N., Kurtoglu, S. and Baskaya, H.S. (2007). Assessment of heavy metals in lake Uluabat, Turkey. *African Journal of Biotechnology*, **6:** 2236–2244.

Field, M.P. and Sherrell, R.M. (2003). Direct determination of ultra-trace levels of metals in fresh water using desolvating micronebulization and HR-ICP-MS: Application to Lake Superior waters. *Journal of Analytical Atomic Spectrometry*, **18:** 254–259.

Klockenkaemper, R. and von Bohlen, A. (2001). Total-reflection x-ray fluorescence moving towards nanoanalysis: A survey. *Spectrochimca Acta Part B*, **56:** 2005–2018.

Misra, N.L. and Mudher, K.D.S. (2002). Total reflection x-ray fluorescence: A technique for trace element analysis in materials. *Progress in Crystal Growth and Characterization of Materials*, **45:** 65–74.

Shafiq-ur-Rehman (1998). A red-bloom of *Euglena shafiqii*, a new species, in Dal Lake, Srinagar, Kashmir. *Water, Air and Soil Pollution*, **108:** 69–82.

Shafiq-ur-Rehman (2009). Pollution of the Red-Bloomed Himalayan Dal Lake of Kashmir. *In*: Srivastava, A., Roy, I. (eds.) Bio-, nano, geo-sciences: The future challenge. Ana Book Pvt. Ltd., New Delhi.

Szymannowska, A., Samecka-Symerman, A. and Kampers, A.J. (1999). Heavy metals in three lakes in west Poland. *Journal of Ecotoxicology and Environmental Safety*, **43:** 21–29.

Wobrauschek, P. (1998). Total reflection X-ray fluorescence spectrometric determination of trace elements in the femtogram region: A survey. *Journal of Analytical Atomic Spectrometry*, **13:** 333–337.

Yalcin, N. and Sevinc, V. (2001). Heavy metal contents of lake Sapanca. *Turkish Journal of Chemistry*, **25:** 521–525.

Index

N.J. Raju et al. (eds.), *Management of Water, Energy and Bio-resources in the Era of Climate Change: Emerging Issues and Challenges*,
DOI 10.1007/978-3-319-05969-3

MIX
Papier aus verantwortungsvollen Quellen
Paper from responsible sources
FSC® C105338

If you have any concerns about our products,
you can contact us on
ProductSafety@springernature.com

In case Publisher is established outside the EU,
the EU authorized representative is:
Springer Nature Customer Service Center GmbH
Europaplatz 3, 69115 Heidelberg, Germany

Printed by Libri Plureos GmbH
in Hamburg, Germany